QUARRYING, OPENCAST

AND

ALLUVIAL MINING

QUARRYING

OPENCAST AND ALLUVIAL

MINING

JOHN SINCLAIR

M.Eng., Ph.D., C.Eng., M.I.Min.E., Barrister-at-Law

Professor of Mining Engineering and Quarrying
University College, Cardiff

ELSEVIER PUBLISHING COMPANY LTD

AMSTERDAM – LONDON – NEW YORK

1969

ELSEVIER PUBLISHING COMPANY LIMITED
BARKING, ESSEX, ENGLAND

ELSEVIER PUBLISHING COMPANY
335 JAN VAN GALENSTRAAT, P.O. BOX 211, AMSTERDAM,
THE NETHERLANDS

AMERICAN ELSEVIER PUBLISHING COMPANY INC.
52 VANDERBILT AVENUE, NEW YORK N.Y. 10017

Printed in Great Britain by Galliard Limited, Great Yarmouth, England

CONTENTS

PREFACE

Quarrying and all other branches of surface mining rather than diminishing in importance have become of more and more consequence economically, industrially and particularly with the depletion of high-grade deep-mined mineral reserves. Low-grade minerals require low cost extraction and this in many cases necessitates very expensive mechanized equipment with the cost of individual units running into millions of pounds in the case of large-scale operations with high productivity.

There has been, and there still is, a tendency for the smaller single quarries to be amalgamated into groups with large financial resources and therefore with the ability to purchase these expensive machines so necessary to make operations viable. This in turn requires wider administrative and technical knowledge in executives of these groups and as these often handle a wide range of products from widely differing systems of working, this technical knowledge should embrace the exploitation of many different types of deposits.

There is, at present, a great dearth throughout the world of such qualified executives as is apparent from advertisements of vacancies in the technical press. It would appear that these industries offer an attractive career to the widely qualified and experienced technologist in these fields.

This book deals with methods of working in the surface extractive industries, quarry management and power supply—but does not deal with related ancillary processes except where these affect quarrying operations. It should fulfil the needs of those who intend to take the Associate Membership examination of the Institute of Quarrying and degrees and diplomas in Universities and Technical Colleges.

CHAPTER 1

INTRODUCTION

The industries with which this book deals are in the process of adopting increasing mechanization and even automation. They owe no small debt to the opencast coal mining industry for it was in that industry, particularly in the USA, that a number of mechanized methods were developed and which are now so essential for success.

THE DEVELOPMENT OF MECHANIZATION

In America, coal was first won by the Indians many centuries ago from the eroded outcrops of coal seams and records exist of such primitive working in 1680. Later, wheelbarrows, carts and wagons were used to haul away the dirt from over the seam and so uncover it. Where this overburden increased in thickness, horse and mule drawn ploughs were used and later scrapers, as at Grape Creek at Danville, Illinois, in 1866 and in 1875 a similar pit was opened at Hungry Hollow nearby. In the Pennsylvania anthracite field primitive stripping using wheelbarrows for transport began in 1820.

The first mechanical excavator was a British invention when in 1796 a 4 hp James Watt steam engine was installed in a scow or lighter to operate dredging equipment, and in 1805 Oliver Evans produced a similar dredge in the USA.

The construction of canals and railways early in the nineteenth century created a heavy demand for mechanical aids and in 1835 a power shovel was invented by an American, William S. Otis. This was patented in 1839 and was standard in the construction and extraction industries until 1890 and, with major modification, continued to be manufactured until the 1930s.

Bituminous coal stripping by Hodges and Armil using an Otis steam power shovel at Pittsburgh, Kansas, was attempted in 1877 but unfortunately the 8 to 12 ft thickness of overburden was too much for the machine and the attempt had to be abandoned, but in 1881 Pardee and Conner successfully applied a steam shovel to the stripping of anthracite at Hollywood Collieries, Hazelton, Pennsylvania.

In 1885 a second steam shovel was applied successfully by Wright and Wallace to the stripping of bituminous coal at Mission-field, Danville, Illinois. It consisted of a steam dredge to which wheels were applied and with a 50 ft boom it successfully removed 35 ft of overburden to uncover a 6 ft seam, 400 yd^3 being dealt with in a 10-hour shift.

In 1890 the Butler brothers employed draglines in the same area using $\frac{3}{4}$, $\frac{7}{8}$ and 1 yd^3 self-propelled machines with 80 ft booms which did not swing. The bucket was pulled across the surface of the overburden until it was filled when it was run out to the end of the boom and dumped. Later a 2 yd^3 dragline with a 125 ft boom was added.

In 1911 Holmes and Hartshorn designed a steam-driven self-propelled completely revolving shovel with a $3\frac{1}{2}$ yd^3 capacity bucket and persuaded the Marion Co. to manufacture it as their Model 250 shovel. It handled 20 to 30 ft thickness of gravel and shale to uncover a 7 ft seam also at Mission-field. Its success inspired the Bucyrus Co. to put forward two fully revolving machines, one with a 60 ft boom and a $2\frac{1}{2}$ yd^3 dipper and the other with a 75 ft boom and a $3\frac{1}{2}$ yd^3 dipper. Both were steam driven, had three-point suspension, were provided with screw jacks for frame levelling and ran on rail wheels.

Electrically driven shovels were introduced in 1915 when a Marion Model 271 with a 5 yd^3 dipper on a 90 ft boom was installed at Piney Fork Coal Co. in Ohio to be followed in 1916 by the Bucyrus Model 225B with an 80 ft boom and a 6 yd^3 dipper. It had alternative steam or electric drive. Both these manufacturers adopted an improved control on the Ward-Leonard principle in 1919 and they were successful so that most subsequent models adopted this system of control. Crawler mounting replaced rail track mounting in 1925 and improved manouvrability.

In this period tandem stripping was adopted in Illinois. Liquid oxygen was used to break overburden and the truck replaced rail transport, including semi-trailer units.

Shovels increased in capacity in the 1930s to 32 yd^3, alloy steels and aluminium were used in dipper construction and design improved with independent propulsion tubular dipper members, two-piece booms of welded construction and more even weight distribution. Draglines were also improved and took their place as primary excavators rather than as auxiliary plant; the first large walking dragline working in anthracite at Seranton, Pennsylvania, in 1931. The first 'knee-action' shovel was brought out by Marion with a 35 yd^3 dipper in 1935 at Boonville, Indiana.

In 1956 the era of the super-shovel was ushered in with the 60 yd^3 Marion Model 5760 of the Hanna Coal Co. at Cadiz, Ohio, with a 150 ft lattice type boom and a 'knee-action' crowd to reach to a height of 110 ft. A shovel three times as heavy was brought out by Bucyrus-Erie, Model

3850B, with a 115 yd^3 dipper on a 210 ft boom. Giant draglines followed the same trend.

In 1966 Marion brought out a 270 tons per minute shovel, Model 6360 known as the Captain (Figs. 1 and 2), with a dipper capacity of 180 yd^3 with a 215 ft boom for the Southwestern Illinois Coal Corporation's Captain Mine at Percy, Illinois, to strip two coal seams. Four motor

Fig. 1.   *Marion type 6360 shovel with 180 yd^3 dipper and 215 ft boom.*

generators convert the 14,000 volt ac power to dc to supply the twenty main drive motors capable of an output of 30,000 hp (Fig. 2). The shovel can propel itself at $\frac{1}{4}$ mph on four pairs of crawlers each 45 ft long and 16 ft high, each shoe is 10 ft across and weighs $3\frac{1}{2}$ tons.

In 1966 Bucyrus-Erie received an order to build a walking dragline Model 4250 with a 220 yd^3 bucket for the Ohio Power Co. for use in open-cast coal. The cost of the machine was 20 million dollars. It develops 48,500 hp and weighs 30 million lb (13,348 tons).

Wheel excavators are of three types. The Kolbe was used in 1944 at the Cuba Mine in Illinois to dig a 20 ft upper layer of overburden. It can move 2 million yd^3 of material per month.

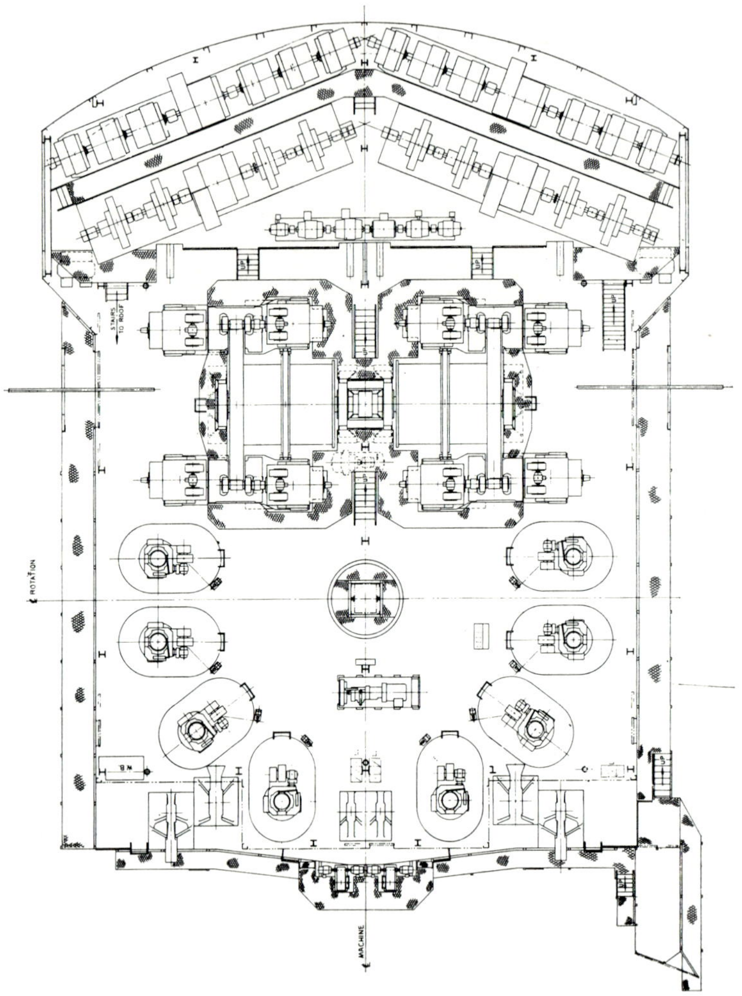

Fig. 2.	*Layout of Marion type 6360 shovel 30,000 hp power deck.*

In the brown coal lignite deposits of West Germany in the Lower Rhine district bucket chain dredges were used in the soft overburden and relatively soft coal. The dredges were mounted on rails or caterpillars and weigh up to 1400 tons. They can cut to a depth of 40 m. The buckets have capacities up to 2·24 m^3 and outputs of 50,000 m^3 per day have been attained.

Fig. 3.   Bucket wheel excavator type AR 220 in sulphur deposit.

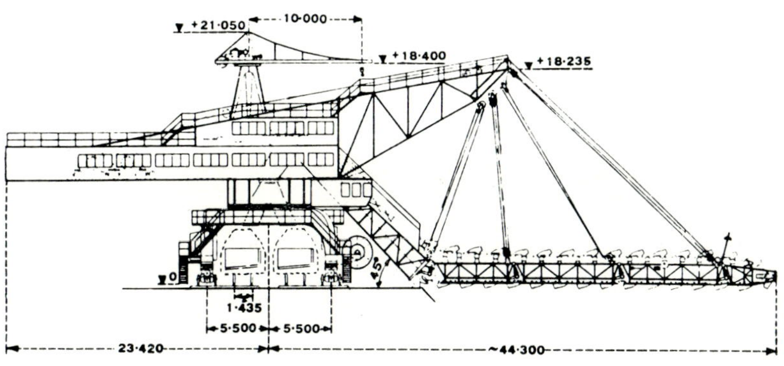

Fig. 4.   The RO 400 multi-bucket chain excavator (Walter).

In the bucket wheel dredge or excavator (Fig. 3) the buckets do the digging only, the material being transported by a series of belt conveyors and the digging action is superior to that of a bucket and chain dredge (Fig. 4). The bucket wheel carries from 6 to 12 buckets.

Crowd action is provided on some bucket wheel excavators. The usual working height is 50 m and the cutting depth 25 m. A typical cell-less bucket wheel is $17\frac{1}{2}$ m diameter with 10 buckets of 3·8 m each. The machine weighs 7400 tons and is carried by three crawler groups each of two twin caterpillar tracks. The power installed is 15,850 kW including 1760 kW for the bucket wheel drive of 1760 hp.

In the United States some 150 million tons of coal are won by opencast methods (151,858,979 short tons, in 1964). In the United Kingdom the maximum annual opencast coal production has reached 11 million tons but production since 1957 has been limited in the interests of deep coal mining to prevent redundancy in the deep coal mining industry. The profit per long ton is much higher from opencast coal, 14s. 3d. per ton in 1965 against a loss of 11s. per ton for deep-mined coal both before charging interest.

In conjunction with the increased size of excavators, trucks both on and off the highway have also proportionately increased in size. In the 1930s the maximum size was some 40 tons but the size steadily increased to the 100 tons semi-trailer of the Hanna Coal Co. of Ohio in 1963 with a 700 hp engine and 21:00:49 tyres. By 1965 the capacity had increased to 240 tons at the Captain Mine, Percy, Illinois. The opencast coal industries can claim:

(1) The development of the world's largest power shovels and draglines.
(2) The world's largest transport trucks.
(3) The world's largest and fastest drills.
(4) The cutting of the cost of explosives by the development of the ammonium nitrate-fuel-oil (AN–FO).
(5) The development of a reclamation programme for stripped lands with the creation of recreation and hunting areas with lakes.

SURFACE MINERAL PRODUCTION

In the USSR much of the mineral production is from opencast mining. Some 22% of coal production is from opencast workings, and 4500 million tons are projected annually for the future. Fifty-five per cent of iron-ore production has this source, as has 50% of non-ferrous metal ores, 40% of manganese ore and almost 100% of non-metallic minerals. Within the next decade it is planned to increase the opencast output up to three-quarters of the total volume of the output of useful minerals worked in the USSR.

While this book deals with the surface extractive industries throughout the world, it is interesting to note the output of British quarries and opencast sites in 1961 and 1964 given in the Chief Inspector of Mines and Quarries Annual Report for 1965. A notable increase in annual output is shown particularly in gravel and sand, and limestone (Table I).

The Ministry of Power's Statistical Digest for 1966 shows that over 280 million tons of surface minerals were produced in that year including 107 million tons of sand and gravel, 65 million tons of limestone, 26 million tons of igneous rock, 18 million tons of chalk and 7 million tons of sandstone; manpower employed was 50,000.

The majority of iron-ore production and a high proportion of copper is produced by opencast methods throughout the world while limestone, chalk, granite and whinstone (diorite) are invariably won by quarrying.

Table I

Mineral	1961		1964	
	Output tons	*Average No. of wage-earners*	*Output tons*	*Numbers of wage-earners*
Gravel and sand including silica sand	80,546,799	10,442	100,994,589	10,364
Limestone	40,669,972	10,042	55,552,466	10,928
Clay shale	30,082,094	3,318	32,236,516	3,180
Chalk, chert and flint	16,767,599	1,450	17,534,378	1,405
Ironstone including gossan	13,774,843	2,911	13,835,286	2,218
Igneous rock including feldspar	15,463,934	5,959	19,315,236	5,535
Sandstone including silica stone and ganiser	4,011,397	2,358	5,539,560	2,414
China clay, china stone, potters, ball and mica clays	2,002,162	2,811	2,351,452	3,454
Fireclay, moulding and pig bed sand	1,466,489	562	1,589,918	484
Slate	82,056	1,762	106,325	1,465
Other minerals	347,459	217	469,165	205
Opencast coal	8,528,017	5,774	6,809,541	4,345
Totals	213,382,811	48,606	256,316,432	45,997

Sand, gravel and crushed rock, so important in these days of construction in concrete, including roads, and for roadstone materials, is also won by surface extraction, including dredging, grab dredgers, draglines and gravel pumps.

The past decennary has seen a remarkable expansion in iron ore production. The most important sources of iron ore are the oxides haematite, magnetite and limonite, the first two constituting the major source of world iron-ore supplies with theoretical iron content of 70% and 72% respectively and commercial ores, particularly if beneficiated, contain 60% or more.

Siderite, carbonate of iron, is the main iron mineral in the deposits which have formed the basis of both the United Kingdom and Western European industries. A replacement product of limestone by solutions descending from the Trias above give the haematite ores of Cumberland, North Lancashire and South Wales. This ore is ferric oxide, Fe_2O_3, and occurs as kidney-shaped concretions, the masses of ore being formed in very irregular shapes related to faults which occurred in Post-Triassic times. The more important sources in the UK are, however, sedimentary deposits of two types: first the clay and black-band ironstones of the coal measures, concretionary in form and composed of siderite upon which the iron-smelting practices of Great Britain were founded but which in the first half of this century ceased to be important; and secondly, but much more important, the Jurassic ironstones exhibiting oolitic structure analogous to that of the limestones of the same system.

Iron sulphide is indirectly an ore of iron since after the sulphur has been roasted off for the manufacture of sulphuric acid the residual oxide can be charged into blast furnaces.

These traditional ores were what is now known as 'direct-shipping' ones being despatched as mined, after crushing and screening to be fed direct to blast furnaces. Blast furnace technology is affecting the type of ore acceptable in the furnace feed sense. Ideally material of maximum iron content, high porosity and uniform size is required and these factors are interdependant. Some high grade direct-shipping ores are so dense that they must be crushed to about $1\frac{1}{4}$ in mesh in order to ensure complete reduction in the furnace and slagging constituents have to be added to the furnace burden. At the other extreme, the low-grade ironstones are acceptable because they are porous and easily reducible and are also self-fluxing.

The need to reduce production costs, or at least minimize their rise, while at the same time avoiding capital investment in new furnaces, called for methods of increasing the capacity of existing furnaces and this indicated a higher grade of furnace feed. The increased efficiency of furnace utilization produced economies that counterbalanced the cost of beneficiating the ore. This led, in particular, to the introduction of pellets, which are made from concentrates and have features of uniform size, uniform high grade and porosity. The latest development is the 'pre-reduction' of pellets—the removal of some oxygen from the iron oxide to produce a pellet containing some 80% iron. Pellets in one form or another constitute the dominant feature of present iron-ore trade. At the beginning of 1965 world capacity for the production of pellets exceeded 40 million tons and should reach 95 millions by 1970.

It has been estimated that iron-ore production will accelerate faster than steel production because of the rapid adoption of the basic oxygen process which takes a furnace charge of 70 to 75% of 'hot metal' (*i.e.* newly smelted iron) and only 25 to 50% of scrap compared with a 50:50 mixture in the open hearth process. British and European iron ore producers are suffering keen competition from high-grade imports carried at low freight rates in large ore-carriers specially designed for this trade. In 1965 the UK consumed a million tons of pellets.

The production of iron ore (Fig. 5) is increasingly of oxides from banded formations. Typically a modern mine has reserves of this type in a remote, generally uncongenial environment which involves the building of port facilities and railways, with townships for employees and staff as well as production and mineral preparation plant and machinery. Such new

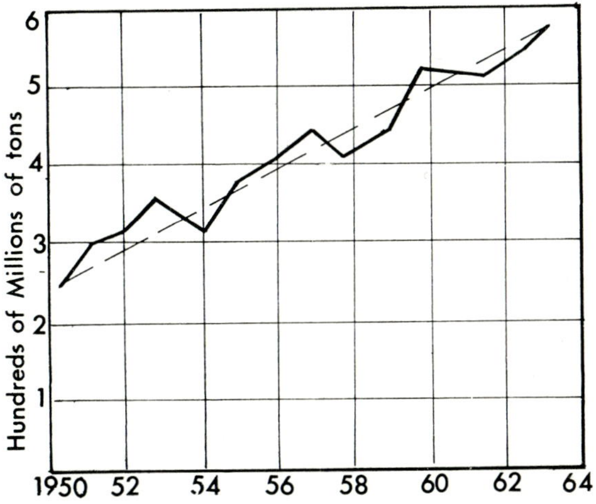

Fig. 5. *World consumption of iron ore.*

mines are estimated to cost from £10 to £13 per ton of annual production capacity in capital investment, the majority of which is required for the ancillary services and only a minority by the mine itself. It is doubtful if the proposition will be viable unless a production of some 2 million tons per annum can be achieved for a period of at least 25 years. Of the nine major installations which have come into production since 1950, the smallest had reserves of 100 million tons, six had potential lives of 25 to 30 years and the remaining three had lives substantially longer.

The pattern of these major opencast ironstone mines is similar in most respects so that they are also similar in layout, appearance and equipment. A typical benched opencast mine would be equipped with blast-hole drills of one type or another, mechanical shovels with bucket capacities of some $2\frac{1}{2}$ yd^3 or more, loading the broken ore into rear dump trucks of 27 tons or larger capacity to transport it to the surge bin at the head of the crusher section, and bulldozers to clean up the benches. The mineral processing plant varies from a simple crushing and screening installation for a direct-shipping ore to simple forms of beneficiation and to the preparation of concentrates. The viability of the installation depends upon the ore-body and the optimum is not necessarily, or even preferably, ore containing large reserves of high-grade direct-shipping ore. The ideal appears to be one with reserves of this high-grade type for a period of some 10 years together with abundant reserves of banded iron formation averaging some 35% of iron of a mineralogical constitution rendering it amenable to concentration and pelletization.

The world production of iron ore in 1964 amounted to 560 million tons to which 56 countries contributed. The tonnages produced by the six largest were:

Soviet Russia	142·7 million tons
USA	81·3 ,, ,,
China	49·2 ,, ,,
Canada	31·5 ,, ,,
Sweden	26·1 ,, ,,
Venezuela	13·7 ,, ,,

In the largest producer, the USSR, the total reserves are estimated at 70,000 million tons mostly of low-grade type.

In 1965 the USSR exported $26\frac{1}{2}$ million short tons of iron ore, most of which went to Communist countries in Eastern Europe. Russia is anxious to export iron-ore pellets from Black Sea ports.

The European Coal and Steel Community has indicated that new iron-stone mines coming into production in West Africa, Canada and Latin America will result in a further increase in iron ore supplies.

The USA on balance is a net importer, large quantities coming from Canada.

Australia has increased in importance since 1960, some 15,000 million tons of high-grade ore, mostly of iron formation type, having been discovered or rediscovered in North Western and Western Australia and five major consortia established. Japan is the sole market at present.

In November 1966, the Rio Tinto Zinc Corporation and Kaiser Steel iron ore project at Hamersley, Western Australia, was officially opened. This plant, including the initial pellet plant, was established at a cost of £60 million and is probably unique in size and for speed of construction in a remote and undeveloped area. Apart from contracts with Japanese steel companies for $82\frac{1}{2}$ million tons of iron ore over 16 years, Hamersley exported 800,000 tons in 1967 to the United Kingdom and to Europe. The company will eventually embark on steel production.

Hamersley is only one of the five consortia developing iron ore resources in Western Australia which together form one of the world's largest mining operations.

The Mount Newnam partnership of Amox Colonial Sugar Refinery and Broken Hill Proprietary is the biggest of the consortia planning to ship 100 million tons of Mount Whaleback haematite to Japan.

Bauxite, the most acceptable ore of aluminium, is a residual deposit which occurs in a restricted range of climatic conditions in which bacterial action has removed quartz and silicates and left the alumina in a silica-free condition. The metal, whose importance to commerce in high-strength alloys seems assured to an increasing extent in the future, is then reduced in electric furnaces often supplied from an adjacent hydro-electrical station as at Kitimat in British Columbia.

Guyana is a major producer of bauxite. At first power shovels, then draglines were used to load into dump cars then into track-mounted wagons hauled by tractors. This was superseded by hydraulic stripping and hydraulicking together with sand-pumping. This in turn gave place to sidecasting of overburden by draglines and then bucket wheel excavators and $1\frac{1}{2}$ miles of belt conveyor were used. New methods are still developing. The working of bauxite is an excellent example of changes in methods of excavation as new equipment and systems become available.

In the fertilizer field the opencast working of phosphate is of prime importance. Phosphate stimulates early root growth while potash is a salt necessary for root strength. From a depression in the 1930s a slow but steady growth in agricultural technology occurred. However, until the early 1950s brought the threat of famine to the dark corners of emergent and developing countries the importance of a marked increase in both production and consumption of fertilizer ingredients, the chief being potash and phosphate, was not realized.

The annual production of potash salts was 15 million short tons in 1965 and is expected to increase to 20 million tons by 1970 and 27 million tons by 1975. That of phosphates was 68 million short tons in 1965 and is expected to reach 92 million tons by 1970.

In 1860 Germany was the main supplier of potash from the Stassfurt deposits and founded a cartel for the control of supply and price which was not broken until World War I in 1914. After 1911 some 100 plants were brought into production in the USA but the trade returned to Europe in 1918.

In 1943 the Canadian Saskatchewan potash field was discovered by an oil-drilling crew and after sundry failures the Potash Corporation of America successfully pioneered this field. Potash is generally won by deep mining using the pillar and stall system of working, often employing mechanization including continuous miners delivering to shuttle cars.

Phosphates, however, are generally worked by opencast methods. Phosphate rock has no fixed chemical composition but calcium phosphate is usually one constituent, the principal minerals have the apatite structure $Ca_{10}(X_2)(PO_4)_6$ in which X represents a hydroxide of fluorine or chlorine.

PO_4 can partially be replaced by small quantities of VO_4, AsO_4, SiO_4 and CO_2. The USA has about one-third of the world reserves of phosphates. Where opencast mining is adopted scrapers pushed by caterpillar tractors strip the overburden from the deposit. Power shovels of $2\frac{1}{2}$ to $3\frac{1}{2}$ yd^3 capacity load into 35 ton capacity trucks.

The road ahead

It has been estimated that more of the earth's minerals were consumed in the immediate past half century than during the previous history of man and that twice that quantity will be consumed in the next half century from an increasing demand of industrialized and emergent nations and from the growth of world population.

In the past the accent has been on underground mining. Today, wherever possible, it is on surface mining brought about by the advent of heavier earth-moving equipment such as shovels, draglines and loaders, large trucks and improved drilling equipment bringing in their train increasingly large production capacities with decreased costs enabling lower grade material to be won economically.

The Operation Ploughshare experiments in the USA are designed to develop a technology of using controlled nuclear explosives once the radioactivity hazard has been overcome. The laser ray may be used to bore holes for explosives.

Bacterial leaching of the metal content of ores has as yet only been applied to copper but its extension to other metals is expected in the near future.

THE RECOVERY OF MINERALS FROM THE SEA

Many minerals are already being won from the bed of the sea, mainly near shore, alluvials forming placer deposits of drowned beaches and other continental shelf areas. These include the exploitation of diamond-bearing gravels off the South-West coast of Africa, that in which De Beers has a substantial holding is a notable example. The presence of substantial reserves of gold-bearing sands has been revealed by prospecting off Nome and Juneau in Alaska and these are expected to be exploited in the near future.

Mineral ores at depths to 100 ft in drowned beaches and alluvial deposits are being won by dredging off the coast of Malaysia, Thailand and Indonesia and similar reserves may exist off the coast of Cornwall. The Japanese are winning magnetite-rich sands and in the last few years have dredged some 7 million tons of relatively high-grade iron ore from a depth of 90 ft in Tokyo Bay some 3 miles off shore to help supply the heavy demand for her steel industries which have made great strides since the war with heavy iron ore imports particularly from Australia. Similar iron-rich sands and chromium-bearing sands also exist off the Alaskan coast and titanium-bearing sands await exploitation off the coasts of Florida, India, Ceylon, Japan and Australia.

In the continental shelves of Brazil and India monazite sands bearing thorium and rare earths are also believed to exist.

All these reserves are in relatively shallow water and are, of course, the result of land erosion, concentrated by wave or current action.

The outer edge of the continental shelf generally lies at a depth of some 600 ft and here reserves of calcium phosphate occur off the coast of Southern California, Australia and India. At this depth in these areas drilling for natural gas and oil is proceeding including sites in the North Sea.

Outside the limits of the continental shelves and in waters as deep as 15,000 ft mineral deposits are also attracting attention though the apparatus for recovery will probably be expensive and need special design particularly to be able to position the surface craft in spite of currents, wind and other factors which will tend to deflect it. The suction dredge seems to present the best possibilities at present.

The potential reserves in the ocean depths are immense. The dissolved salts amount to 150 million tons per cubic mile including appreciable amounts of lead, tin, copper and most other known elements. The gold content alone has been valued at between £8000 to £8 million per cubic mile depending on the locality. Important nodular reserves in the ocean deeps are those of manganese and phosphorus which extend over large areas of the Atlantic and Pacific Oceans. The manganese potato-shaped concretions lying at depths between 3000 ft and 19,000 ft contain 25 to 35% of manganese, 18 to 20% of iron and approximately 1% each of cobalt, copper and nickel formed by ionic precipitation of colloidal manganese and iron oxide on any hard surface, other metals are scavenged by the iron and manganese during transport from their point of origin. It has even been suggested that these deposits are being formed at a faster rate than world consumption of these metals.

The phosphorite nodules are formed in a similar manner and contain up to 30% of tri-calcium phosphate at depths between 400 and 8000 ft off the coasts of the USA, Mexico, Argentine, Peru, Chile, Spain, South Africa, Japan and in the Indian Ocean.

The Oceanographic Commission of Unesco give the following reserves of oceanic minerals:

Material	*Estimated tonnage*	*Elements of interest*
Manganese nodules	10^{12}	Mn, Cu, Co, Ni, Mo, V, Zn, Z
Phosphorite nodules	10^{10}	P, Zr
Red clay	10^{15}	Cu, Al, Co, Ni
Magnetic spherules		Ni, Fe

REFERENCES

'Technical Progress in Opencut Mining', Fifth International Mining Congress, 1967, Moscow, *Colliery Guardian*, August 25th 1967, p. 223.
'Potentialities of Ocean Mining', *Mining Journal*, March 24th 1967, p. 212.
'An Analysis of Mining Activity', *Mining Magazine*, September 1967, p. 160.
'Geobotany—A Valuable Aid in Prospecting and Mining', J. Sholto Douglas, *Mining Magazine*, February, 1968, p. 93.
The Mineral Resources of the Sea, J. L. Mero. Elsevier Publishing Co. Ltd, 1965.
Submarine Geology, F. P. Shepard. Harper and Row, 1963.
Submarine Geology and Geophysics, W. F. Whittard and R. Bradshaw. Butterworth, 1965.
Surface Mining. American Institute of Mining, Metallurgical and Petroleum Engineers Inc., New York, 1968.

CHAPTER 2

PROSPECTING

Heavy capital investment in plant and machinery is a characteristic of the modern quarrying and surface mining industries in order to obtain efficiency and low cost of production. Increased financial reserves and capacity explains also the take-over of the smaller quarries in the United Kingdom into groups with centralized control, the group having access to capital and plant to enable resources to be made available so that optimum results may be obtained consistent, of course, with the size of the reserves of mineral available in each case.

In order to make the load factor of expensive equipment as high as possible so that it operates for the maximum possible number of hours per day, planning is essential and it is necessary to know as precisely as possible the nature and extent of the different strata and minerals which will be encountered. This entails a programme of exploration and prospecting, and the thoroughness and skill with which this is carried out will affect the assessment of the quarry or opencast site's potential and the planning of its most profitable and efficient working. Such a programme will depend on the particular conditions and circumstances of each case but will probably include surface mapping, aerial photography, geophysical prospecting, photogrammetric methods of measuring volumes of overburden and excavations, and drilling.

Much of the solid geology of the site under examination will be covered by rainwash vegetation and alluvium so the best possible use must be made of any existing exposures to enable a geological map and section of the region to be produced. These will include rocks on hills or exposed by streams, crags, colour of soils and soil thrown out by burrowing animals and civil engineering cuttings for railways, roads, sewers and other works.

Trees often have a preference for certain strata and aerial geological reconnaissance has made use of this selection of certain formations by distinctive plants or trees, while the difference in the colouring of grasses and other vegetation on different strata has been useful in determining the junction between dissimilar rocks.

The ores of metals or the natural elements are often more resistant to weathering and down the ages prospectors have traced 'float' back to its source, or 'panned' river gravels in frying pans or other pans for native metals or heavy sulphides.

Trenching to a depth of 1 to 7 ft may have to be resorted to reach 'float' under silt or vegetation or to uncover the bedrock, a particular horizon or the junction of beds. Trenching at definite intervals of 50 to 500 ft at right angles to the strikes of the beds, or of a mineral vein is often adopted in mineral prospecting.

Pipes of one or two inches diameter with a cutting edge at one end may be used for probing to determine harder or softer beds to test the thickness of silt or soft strata to bedrock and to obtain samples of the soil and soft strata until hard strata is reached. Pointed steel rods may also be used for these purposes.

Hand-operated drills may also be used to penetrate to a relatively shallow depth. Where a considerable depth of alluvium is encountered too deep for trenching, test-pitting to a depth of 100 ft may be adopted. A hand windlass is used to raise the material excavated which is loaded into a bucket with a hemp or $\frac{1}{4}$ in diameter wire rope. Short-handled picks and shovels are used as the pits are only $2\frac{1}{2}$ to 3 ft in diameter. Drilling by power-driven rigs is used for definite evidence of the presence of minerals but as the method is comparatively costly it is only used when geological or prospect indications of a probable viable field are present.

The old-time prospector, greatly experienced and tough, with very primitive equipment must be credited with the discovery of mineral fields of great richness from which gold, diamonds and copper of great value have been won, often by a happy chance, but these more or less easy plums can in future be expected to be few and far between. The Eureka diamond which was picked up by children on a Boer farm and at first not recognized as a diamond, was the forerunner of the South African diamond industry and has now a century later, been returned to South Africa by De Beers. Originally 21 carats in the rough, it is now $10\frac{3}{4}$ carats.

New fields are now the result of carefully planned prospects in which sophisticated and expensive scientific equipment is used by highly expert specialists when the useful mineral is covered by a considerable thickness of overburden which may give no indication at the surface of the type of geological structure at depth with which the occurrence of the mineral is usually associated.

GEOPHYSICAL PROSPECTING

In conjunction with geological co-operation which indicates the limited areas in which the type of formation required is likely to be found, these scientific methods of prospecting, some of which are carried out from the air together with aerial photography of the area under examination, are known as methods of geophysical prospecting. This may also indicate the method of geophysical prospecting most likely to be successful although quite often two systems are used to confirm and supplement each other. In remote parts of the world such prospecting may require the setting up of base camps and in any case, geophysical prospecting is not cheap although the use of aeroplanes and helicopters speeds up the operation and reduces the expense.

The differentiation, usually abrupt, of some physical property as between rocks known as 'anomalies' is the basis of systems of geophysical prospecting, but since these local variations, naturally produced or artificially stimulated, are generally small they require delicate and expensive measuring apparatus and the results obtained need specialist interpretation.

Geophysical methods are available to solve the following types of problems in the quarrying, opencast and alluvial mining industries:

1. Airborne magnetic, electromagnetic and radiometric surveys for base and precious metals, iron ore and radioactive minerals.

2. Marine geophysical surveys for the location of off-shore minerals for the sand and ballast industries and mining companies.

3. Ground geophysical surveys for the location of ore-bodies, faults, sheer zones, overburden thickness using gravimetric, magnetic, seismic, resistivity, induced polarization, self-potential, electromagnetic and radiometric methods.

Other problems which may be solved by such methods are:

Assessment of gravel deposits
Thickness of clay resting on limestone
Location of solution channels in limestone
Thickness of overburden
Location of faults
Position and width of volcanic dykes.

By themselves, however, geophysical methods are not usually the complete answer and their importance is as a guide to enable drilling programmes to be planned to obtain the maximum amount of 'hard' information in the most economic manner to enable the future overall development planning to proceed with expedition and the maximum possible information.

It is not proposed to describe the different methods of drilling available and the layout of drilling programmes. The factors involved are dealt with by the author in *Geological Aspects of Mining*.

The five main methods of geophysical prospecting employed in quarrying and in mining are:

(1) Gravitational—depending upon difference in density of rocks below the surface
(2) Magnetic—depending on magnetic susceptibility of different rocks.
(3) Seismic—depending on the velocity and the reflection or refraction of elastic wave propagation in different rocks and their junctions.
(4) The electrical conductivity or the self-potential of different beds.
(5) Radiometric—in which the radioactive mineral content of the rocks is measured by a very sensitive scintillation counter.

Gravitational methods

Minute changes in the force of gravity are used to interpret the distribution of strata underground such as those produced by the denser rocks in the core of an anticline compared with the rocks on its flanks, rock faults, haematite and brown coal beds. The instruments used include pendulums and crystal clocks for the measurement of the absolute gravity at stations; gravimeters which measure the change in the gravimetric attraction by the extension of spring and torsion balances. These were first invented by Baron Roland Von Eötvös in 1915 and the basis is a torsion wire suspending two weights on a light beam, one weight above and the other below the point of suspension (Fig. 6). Although both weights are attracted by the strata, the lower weight m_1 being nearer is, by the law of inverse squares, more strongly attracted and the suspended system will turn until the torsion in the wire T balances the differences in the couples due to the two weights m_1 and m_2. The balance must be orientated in several directions in order to determine the direction and magnitude of the force to be calculated. It is obvious that the changes in gravity measured will be minute, amounting to one thousandth of a gal (Galileo), the gal being the field strength acting on a mass of 1 gram with a force of 1 dyne, thus 1 gal is 1/981 of the earth's field and a milli-gal is approximately one-millionth of the earth's field.

Magnetic methods

This method, worked out by De Castro in the seventeenth century, apart from being the oldest is also the simplest and cheapest to apply in the field. Portable magnetometers or variometers are used to measure local variations of the vertical and horizontal components of the earth's magnetic field, that is the changes in the magnetic inclination and declination respectively. The method thus depends upon the differentiation of the magnetic susceptibilities of the different rocks. The distortion of the earth's field (the anomaly) will only be measurable if the rock structure concerned has a large susceptibility such as a ferruginous mass or a dyke of basic igneous rock such as basalt containing a considerable amount of iron.

The chief corrections to be applied, which are essential when rocks of relatively low susceptibility are concerned with readings approaching the limit of the variometer, are those of the diurnal variation of the earth's field and for temperature changes.

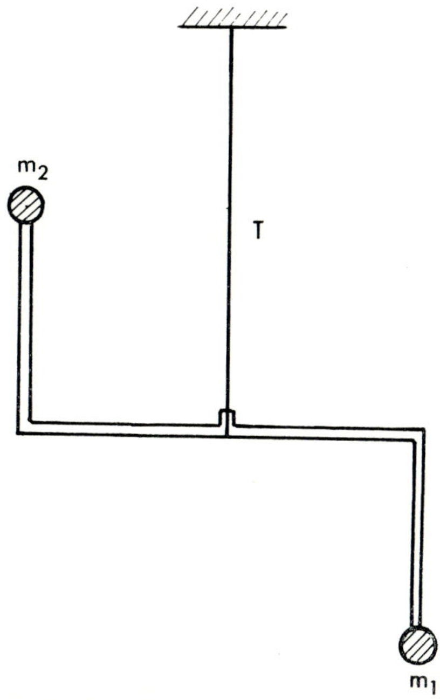

Fig. 6.   *Essential elements of Eötvös torsion balance.*

Two methods may be employed. In the first the swing of the magnetic needle is in the direction of the earth's magnetic field when the susceptible structure is at one side of the magnetic profile. In the second the needle swings at right angles to the earth's field so that only the vertical component of this field is effective. The maximum deflection then occurs directly over the susceptible structure.

Modern instruments are generally of the magnetic balance type in which the initial deflection of the needle due to the earth's field is counterbalanced either magnetically or by means of a weight. A magnet is supported so that its axis lies in a horizontal plane and perpendicular to the magnetic meridian thus eliminating the effect of the earth's field's horizontal component so that only the vertical component tends to tilt it. This tendency is neutralized by a small counterbalance weight, the position of which gives the reading taken by means of an optical system. The swing of the needle is damped by means of pure copper dampers.

When an airborne magnetic survey is undertaken for minerals the *modus operandi* is in some ways different from when oil is being sought. The aircraft flies very much lower, about 500 ft above the ground instead of about 2000 ft, and the lines of the flight are usually more closely spaced. It is necessary to fly much lower and closer because an ore-body, however big it may seem on the ground, appears very small from the air.

The problem of interpretation is no longer the depth of the magnetic rocks since it is already known that they are at, or very near, the surface, but what is the pattern of the structure of the rocks, and how is this to be related to the bodies, copper, lead or gold ore, which are being sought.

If the ore-body sought contains magnetite or pyrrhotite and is magnetic, interpretation is relatively easy since the size of the magnetic anomaly can be calculated and indicate what shape of anomaly might by expected over the ore-body. The shape depends on the geographical position on the earth's surface. It might be expected that over a body of magnetic iron ore the magnetic field strength would always be high but near the equator it is in fact low over the ore-body and high over the edges of the body. The anomaly due to such an ore-body stands out very clearly if none of the surrounding rocks are strongly magnetic. If the rocks in which the ore-body occurs are also magnetic it is necessary to study each anomaly in turn and to decide from the size and shape whether it could be an ore-body. Unfortunately not many ore-bodies are magnetic. Some bodies of iron ore are, but by no means all. A few bodies of copper and of gold are magnetic but most are not. One of the other methods must then be tried.

On the other hand the magnetic method may be used in a completely different way which will lead to the discovery of the ore-body. Ore-bodies do not occur in a haphazard way in the rocks, but are related in a very complex manner to the type and the structure of the rock. If it is possible to find what the geological pattern of the area is and what is the relation of ore-bodies to this geological pattern it is possible to predict in which parts of the area new bodies are most likely to be found.

Magnetic surveys provide information which may be used for the discovery of road metal, building stone, sand, gravel and water.

Seismic methods

The principles used in these methods are the same as those applied in permanent recorders at seismological stations all over the world for recording earthquake shocks. Energy is put into the ground by explosive charges or by striking the ground with a hammer or a machine. The energy travels through the ground and is recorded by a number of detectors and the travel times of the arrivals of the energy at the detectors are measured. The observations obtained enable time-distance graphs to be plotted of the paths of the compression waves which have been reflected or refracted at the surfaces of discontinuity between strata before arriving at the recorders. The arrangement of the energy source and the detectors varies with the type of information required. The most common is one in which usually twelve detectors are equally spaced in a straight line and the energy sources are along the same line. The elastic waves travel at different speeds in different formations. In weathered surface deposits and soils the velocity may be 2000 to 3000 ft/sec, in unconsolidated sands and clays 4000 to

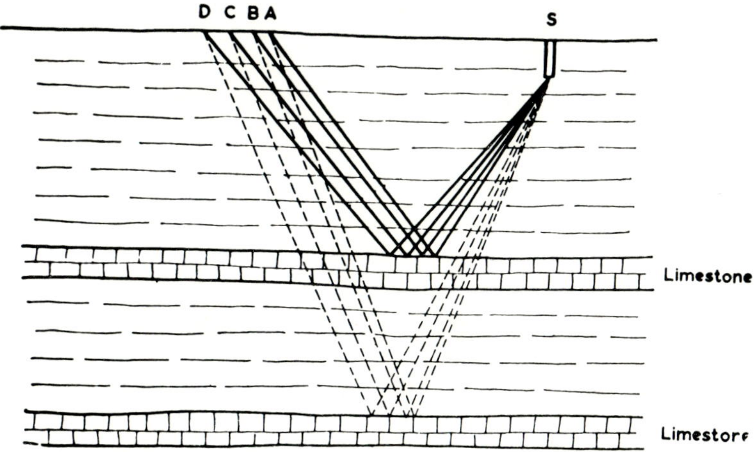

Fig. 7.   *Reflection method of seismic prospecting for deep-seated deposits.*

6000 ft/sec, in shales 8000 to 10,000 ft/sec, in normal sandstones 10,000 to 14,000 ft/sec, and in crystalline limestones and igneous rocks 18,000 to 22,000 ft/sec.

Explosive is generally fired in a borehole from 10 to some hundreds of feet deep in order to penetrate the zone of weathered strata in which the waves would travel at slow speed. The recorders which pick up the reflected or refracted waves are also generally buried to shield them from wind and other sources of interference.

The mechanical movement imparted by the waves to the recording seismographs is converted to an electric impulse which is amplified, particularly the weak wave from deep-seated structures where the reflection method is used. An oscillograph is fitted to record the instant of firing the explosive charge which breaks a wire wound round the charge. The electrical impulses from the amplifier cause deflection of the light beam of the oscillograph which then passes through a condensing lens and is photographed on a moving roll of sensitized film. At the same time a metronome deflects light from a second beam and records time intervals simultaneously on the film.

The application of the reflection method is shown in Fig. 7. Two limestone bands give indications on four recorders A, B, C and D of the echoes produced by a charge fired at S, the echoes arriving in two separate groups, that from the upper band first.

Seismic refraction surveys may be used to enable the rock-head to be contoured with an accuracy of 6% in an area covered by drift. The thickness of the unconsolidated deposits is calculated by the time intercept method and the critical distance at which the direct wave arrives at the same instant as the refracted ray through the harder strata where the speed is higher. The thickness of drift to the rock-head H ft below the shot point is measured by:

$$H = \frac{t_1}{2} \ \frac{V_1 \, V_2}{\sqrt{(V_2{}^2 - V_1{}^2)}}$$

in which V_1 = velocity of transmission of elastic waves in the upper, unconsolidated deposits in ft/sec.

V_2 = velocity of transmission of elastic waves in the underlying rock formation in ft/sec.

t_1 = intercept on time axis of the V_2 branch of the time/distance curve in secs.

In a typical example,

Let V_1 = velocity in weathered layer of surface deposits = 2500 ft/sec.

V_2 = velocity in glacial deposits = 4500 ft/sec.

V_3 = velocity in rock-head = 9000 ft/sec.

t_1 = time intercept (V_2 branch) = 0·002 sec.

t_2 = time intercept (V_3 branch) = 0·015 sec.

Depth of shot beneath ground = 4 ft.

The thickness of the weathered deposits and the depth to the rock-head may be calculated as follows:

Thickness of the weathered layer

$$H_1 = \frac{t_1}{2} \frac{V_1 V_2}{\sqrt{(V_2{}^2 - V_1{}^2)}}$$

$$= \frac{0 \cdot 002}{2} \frac{2500 \times 4500}{\sqrt{(4500)^2 - (2500)^2}}$$

$$= 0 \cdot 001 \frac{11{,}250{,}000}{\sqrt{14{,}000{,}000}}$$

$$= \frac{11{,}250}{3742}$$

$$= 3 \text{ ft.}$$

Thickness of glacial deposit

$$H_2 = \frac{t_2}{2} \frac{V_2 V_3}{\sqrt{(V_3{}^2 - V_2{}^2)}}$$

$$= \frac{0 \cdot 015}{2} \frac{4500 \times 9000}{\sqrt{(9000)^2 - (4500)^2}}$$

$$= 39 \text{ ft.}$$

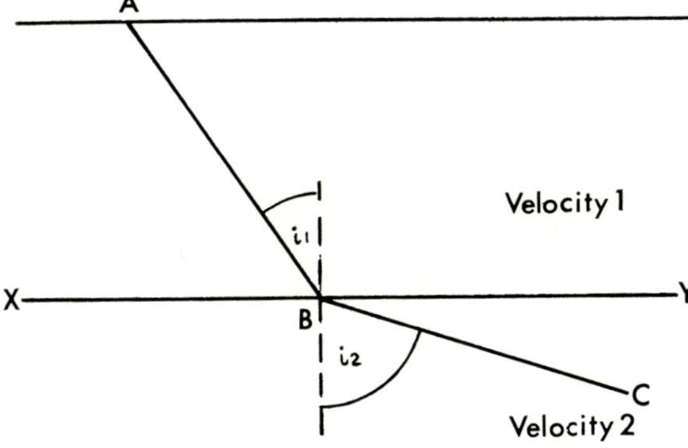

Fig. 8.

Correction for depth of shot below surface depth to rock-head

$$= 3 + 39 + 4 = 46 \text{ ft.}$$

The formulae used in the interpretation of the above problem depends on the application of Snell's law in optics. In Fig. 8 velocity V_1 is that of the ray ABC in its journey from A to B and V_2 during its journey from

B to C and $sin\ i_1/sin\ i_2 = V_1/V_2$, XY being the surface of discontinuity between the two beds.

When the ray BC is refracted and travels along the boundary XY then $i_2 = 90°$ and $sin\ i_1 = V_1/V_2$ and i_1 is the critical angle and AB is the critical ray. Defraction along surface XY can only occur if V_2 is greater than V_1 (Fig. 9).

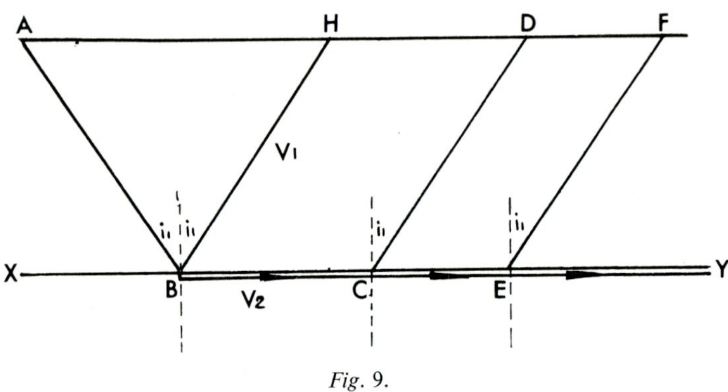

Fig. 9.

The energy from the refracted ray reaches the ground surface from all points on the boundary surface XY along critical ray paths, as at BH, CD and EF in Fig. 9. It should be noted that no refracted energy can be received at ground surface between A and H and AH is defined as the critical distance.

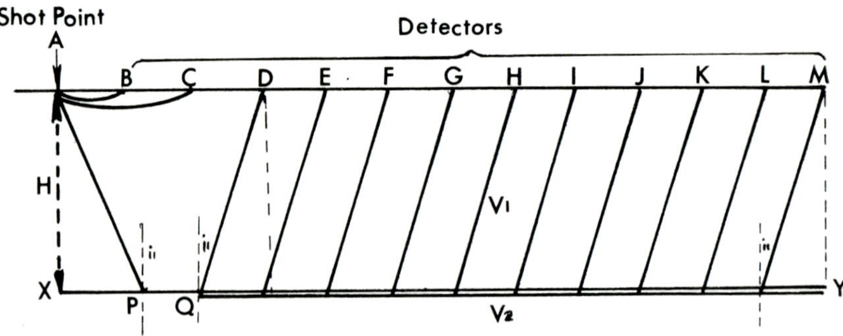

Fig. 10. Spacing of detectors.

The spacing of detectors depends upon the depth of strata to be investigated and the amount of detailed information required. In the layout shown in Fig. 10, A is the shot point and twelve detectors are shown and

XY is again the surface of discontinuity at depth H ft. The ray paths are indicated and in Fig. 11 a graph of distance from shot point A to the detectors is plotted against the time of the first arrival of the energy at the detectors. No refracted energy will arrive at detectors B and C (Fig. 10), the energy arriving at these has travelled just below the surface along the curved paths shown at a velocity V_1. Energy which has travelled along the path APQD arrives at detector D before the energy which has travelled

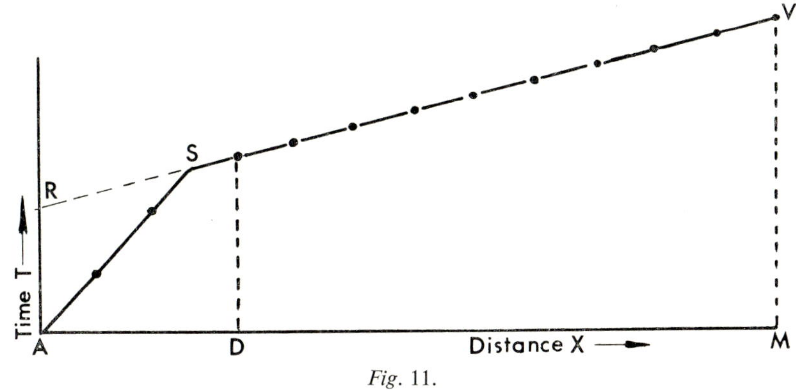

Fig. 11.

directly from A to D. If the boundary surface XY is parallel to the ground surface AM (Fig. 10), then the time of arrival, *t*, of the refracted ray at distance X ft from A

$$t = \frac{X}{V_2} + \frac{2H\sqrt{(V_2{}^2 - V_1{}^2)}}{V_1 V_2}$$

This represents the line RSV of Fig. 11. The intercept AR on the time axis is given by

$$\frac{2H\sqrt{(V_2{}^2 - V_1{}^2)}}{V_1 V_2}$$

V_2 may be determined from the slope of RSV and V_1 from the slope of AS so that the depth H can be found from AR. The critical distance X_C is given by $X_C = 2H (V_2 + V_1)/(V_2 - V_1)$ and may be obtained from the graph (Fig. 11), as it is the distance at which the discontinuity S occurs thus again giving H. This method of analysing the results may be adopted for any number of horizontal layers provided velocity increases with depth as in the example of the unconsolidated weathered surface deposits and glacial deposits over a rock-head. In actual recording the usual procedure with a simple layout as in Fig. 10 is to fire two shots at different distances from B, two shots at different distances from M and one shot from the point midway between B and M. Information can then be obtained of the lateral changes of velocity within the different layers and to determine their inclinations. The accuracy of depth determined by the seismic refraction systems approaches $\pm\ 10\%$.

Resistivity surveys

These methods determine the apparent resistance in ohms/cm^3 of the ground through which an electrical current is passed.

Resistivity field measurements are divisible into two groups to determine: (a) changes in resistivity with increase of depth using an expanding electrode layout and (b) lateral changes in resistivity using constant separation traverses.

Fig. 12.   *Successive positions of electrodes in traversing system.*

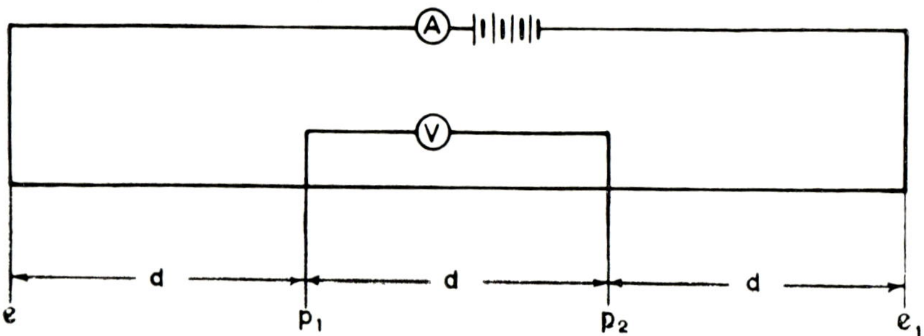

Fig. 13.   *Wenner arrangement of current and potential electrodes.*

(a) In this group may be found the Wenner and the Schlumberger configurations (Figs. 12, 13 and 14). Four electrodes are placed initially on the ground in line at 2 or 3 ft intervals. Current is passed between two of the electrodes e and e_1 (Figs. 12 and 13) and e_1 and e_2 (Fig. 14) and the resulting potential difference between the other two electrodes p_1 and p_2, is measured. The apparent resistivity is a simple function of the electrode spacing, the current and the potential difference.

In both systems, successive measurements of the current and potential difference are made at increasing electrode spacing, for example, multiples of 2 ft to 24 ft. The variation of apparent resistivity with electrode spacing is then plotted as a graph, which forms the basis of the ultimate interpretation.

In the Wenner configuration (Fig. 13) and its later modifications the distances apart of the current and potential electrodes are equal as indicated at d in Fig. 13. Deeper penetration is obtained by increasing the spacing of the current electrodes e and e_1.

In modifications of the Wenner system, in particular the Gish-Rooney, low-frequency alternating current is used instead of dc by using a hand-operated commutator system, to eliminate self-potential.

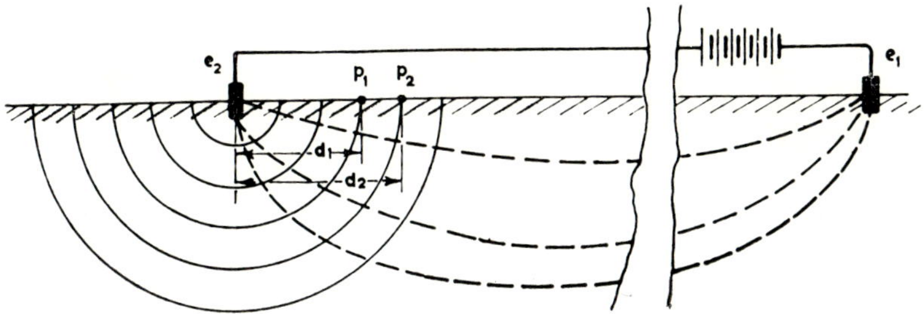

Fig. 14.   *Schlumberger method of ground resistivity prospecting.*

In the Schlumberger system of ground resistivity prospecting (Fig. 14) the electrodes e_1 and e_2 through which the current from an accumulator of some $220V$ flows, are placed far apart and the potential drop measured by two non-polarizing electrodes p_1 and p_2 near electrode e_2. The resistance of deeper rocks is measured when p_1 and p_2, which are kept a constant distance apart, recede from e_2. If d_1 and d_2 are the distances of p_1 and p_2 in cm from e_2, then the resistivity of the strata in ohms per cm^3 = Res and

$$\text{Res} = \frac{2\pi d_1 \, d_2 \, V}{(d_2 - d_1)I} = \frac{2\pi d_1 \, d_2 \, R}{(d_2 - d_1)}$$

the voltage V being measured with a potentiometer and the current, I, by a milliammeter and the resistance, Res, by means of an adaptation of the megger testing set.

In the induction or electromagnetic methods ac is used with a uniform frequency, generally about 500 c/s, and two current electrodes are grounded as in the Wenner and related systems. Secondary currents are induced in the strata by the primary current passing through it and the magnetic field at the surface which results is distorted in phase and direction in the presence of anomalies. Two systems predominate. One employs a coil related in size to the depth to be penetrated and the induced magnetic field is measured by either a single or a two-coil system.

In the Turam system, the current electrodes are connected by a long insulated cable passing over the ground to be investigated and short traverses at right angles to its length are made with a pair of coils a fixed distance of some 60 ft apart to measure changes in phase and amplitude between induced voltages in each coil. Such phase differences are contoured by iso-phase-difference contours and when these approach each other, indicating a steep gradient in the rate of phase-difference change, minerals may be expected. A detailed account is given in Appendix I.

(b) In the constant separation traversing system (Fig. 12) the four electrodes are spaced equally along the survey line at a constant spacing of say 10 ft for a shallow investigation and more for a deeper test (Fig. 14). After the first measurement, in order to determine lateral changes of strata, the set of four electrodes are moved progressively along the survey line (Fig. 12) and successive measurements taken. This procedure can be repeated indefinitely and enables sub-surface boundaries such as the edge of a gravel deposit, the location of a fault or the posistion of a swallowhole or vugh or fissure in limestone to be determined. In order to eliminate self-potentials, reversing switches are provided on both current and potential circuits.

The interpretation of resistivity field measurements has been expedited by the use of electronic computers which enable 'Master Curves' to be calculated. These show the variation of apparent resistivity with electrode separation experienced under different geological conditions and the field results are compared with these.

Electrical potential methods

These may be divided into two: (a) self-potential system and (b) surface potential method.

(a) Oxidation of sulphide ores may give a potential difference of 20 millivolts per 100 ft in the electric field produced by the oxidation of sulphide ores. In investigating by the self-potential method the area is traversed by parallel lines of observation points and then points of equal potential are joined to give isopotential contours.

Two porous pot non-polarizing electrodes *ee* (Fig. 15) are connected by insulated leads to a sensitive potentiometer reading to one millivolt. When the traversing method of prospecting is adopted, the electrodes are a short distance apart (Fig. 15) and the survey proceeds in one direction, the rear electrode occupying the position previously occupied by the leading electrode in the previous reading. This method indicates the mineral sought directly beneath the greatest anomaly. It has been extensively and successfully used to prospect mineral ores such as pyrite, chalcopyrite, galena and insulating materials like cinnabar and stibnite.

In the fixed electrode method, on the other hand, one electrode remains fixed while the other is grounded at progressively greater distances from it.

(b) The surface potential method employs earthed electrodes connected to a source of electricity and the current distribution on the surface is mapped by means of equipotential lines.

Electromagnetic geophysical prospecting is dealt with in Appendix I.

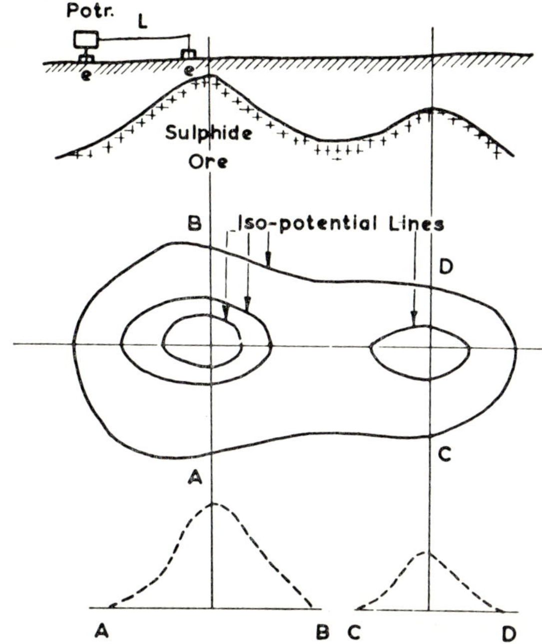

Fig. 15. Electrical self-potential system.

Radiometric surveys

Radiometric surveys of soils and particularly of phosphates are being increasingly adopted in these days of a growing world population with limited agricultural development, and in many surveys apparatus is airborne, a very sensitive scintillation counter especially designed for airborne radiometric surveys being used. Where point sources are being investigated the ratio of significant anomalies to background noise is very low at integrating times, necessary when surveying by aeroplane, In such circumstances an extremely sensitive airborne radiation detector is necessary such as the ARD Type Harwell 1531A. This consists of three thalium activated sodium iodide crystals 4·65 in (118 mm) in diameter and 2 in

(50·8 mm) thick which feed a common ratemeter. The calibration against radium is 180 c/s per r/hr. An advantage of the large phosphor area is that of the surveying speeds commonly used, a shorter time constant gives satisfactory results or, conversely, surveying speeds can be increased using usual time constants, thus making for more economic surveying. In the case of marginal anomalies, the increase in actual sensitivity is critical. The detector unit is stabilized against input voltage variation from 11 to 14 volts. Variable EHT for the photomultipliers from 1 kV to 2 kV is available so that the variable gains of the three tubes can be marked.

The three cylinders of the detector unit can be grouped or mounted side by side but being joined by coaxial cable they must be mounted within 5 ft of the ratemeter with nominal full-scale values of 1000, 2000 and 4000 c/s. The operating supply required for the ratemeter is 12 volts with a negative earth, the current required being 0·12 amp.

A Texas dual channel rectilinear recorder is used, the second channel being for recording the radio altimeter. A similar equipment, the type 1531B, has the output from the ratemeter of 1 mA dc which can be used direct to a recorder without further amplification.

AIRBORNE PROSPECTING FOR PHOSPHATES

Radioactive phosphates can be detected by sensitive airborne radiation detectors. Ground investigation can be concentrated in areas where near surface phosphates have been located by rapid and systematic geophysical flying over large areas coupled with photogeology.

Because of their high density and atomic number sodium iodide crystals absorb a high proportion of incident gamma radiation and are used on airborne surveys. The greater the diameter of the crystal the higher the count which enables a shorter time constant to be used, and this, in turn, allows the ground speed of the aircraft to be raised and the survey is cheaper. Since the intensity of radiation decreases with increase of height it is essential that a continuous record of the aircraft's height above ground should be kept and for this a radio altimeter may be used. From the scintillation counter also a continuous record is obtained and related to ground features by 35 mm vertical photography and a system of co-ordinating reference points.

It is important that a proper relationship should exist between the flying height of the aircraft, the spacing of the flight lines, the ground speed and the integrating time of the scintillation counter. The nature of the deposit, aircraft safety and the need to plot the tracking photography will determine flying height and commonly this is between 150 and 500 ft (50 to 150 m).

The strength of the radioactive source measured by grade multiplied by aerial extent, its geometry and the economics of flying will determine the flight line spacing, 600 ft (200 m) is the closest effective line spacing which can be flown in fairly flat and well mapped areas with little topographic detail. The best integrating time constant is that which gives the best signal to background ratios. The Doppler techniques may be adopted in poorly mapped areas to improve navigational accuracy. The Doppler effect in physics is best illustrated by the change in the pitch of the whistle of an express train as it approaches, passes and recedes from an observer due to the addition of the forward speed of the train in the one case and the subtraction of it as it recedes. Radar waves are switched on the Doppler navigation system so that beams are reflected from the ground fore and and aft of the aeroplane and from both sides. The frequency difference is proportional to the ground speed in the case of the fore and aft beams and is used in the case of the sideways beams to turn the aerial systems so that it is aligned along the true track of the aircraft. With the addition of compass reference all the required information is available for a computer to give continuous information of the aeroplane's position with reference to its starting point. The frequency of the reflected signal is compared with that of the transmitted signal.

The level of radioactivity in the atmosphere is varying and it is necessary to maintain strict control over the flying procedures and instrument calibrations from day to day.

Although small amounts of radioactive minerals are present in all igneous and sedimentary rocks, measureable gamma ray activity is principally associated with members of the uranium, thorium and potassium K40 series. These are present in varying quantities in many different types of rocks and the mapping of these variations by airborne surveys has proved very useful in regional surveys in conjunction with photo-geology and airborne magnetometer surveys. Significant quantities of uranium are found in the majority of marine phosphate deposits mainly in the form of carbonate, fluorapatite and crandalite. Equivalent uranium contents vary from less than 10 to 1500 parts per million and the content of marine phosphates increases roughly with increase of phosphate content. Apatite is more resistant to weathering than the carbonates and other minerals with which it is commonly associated. Weathering thus generally leads to the increase of P_2O_5 and may lead to a viable phosphate deposit derived from phosphatic limestone not viable in its unweathered state. It is obvious, therefore, that the success of airborne prospecting for phosphates is partly due to contrast in radioactive content between phosphates and adjacent sediments, partly to the broad target presented by bedded phosphates and finally to control of the survey variables.

Other physical methods of prospecting are geochemical, temperature surveys for the discovery of formations with high thermal conductivity and micro-gas analysis of soils particularly as an indicator of potential oil deposits.

GEOCHEMICAL PROSPECTING

Outcrops at the surface still form the index to useful mineral deposits now being worked but the rate of geological discoveries of useful minerals has seriously declined and with the alarming rate of increase of the world population and the rise in the standard of living and therefore the comfort and other requirements of the industrial nations the gap between the exhaustion of mineral reserves and production demands and output is closing. Although many important finds have been made within recent years, such as those of iron ore, uranium and copper, they do not redress the balance and reserves are being reduced at what many regard as an alarming rate. It is certain, however, that very considerable reserves of mineral wealth lie hidden beneath a thin mantle of soil, clay or barren rock and which have, so far, escaped detection by the older methods of prospecting and exploration. Increasing attention is therefore being devoted to geochemical prospecting to assist in the search for concealed ore-bodies and reserves.

In many of the areas in which minerals occur detectable traces of ore metals are dispersed outwards for considerable distances from the parent ore-body. These dispersions may occur either during the period of mineralization itself, so that a halo of trace metal values is developed in the country rocks around the ore-body, or later during the process of weathering when vestiges of metals from the deposit are incorporated in nearby soils, glacial till, vegetation, stream sediment and other surface material.

Systematic sampling and analysis of surface material on a large scale in geologically favourable areas may, therefore, reveal significant patterns in the distribution of trace metals and so guide exploration to valuable concentrations of ore minerals even though they may be concealed by a deep cover of barren superficial material. Rapid and economical methods of trace analysis which nevertheless must be sensitive are essential, the most useful of which are specially adapted techniques of chemical colorimetry, chromatography, spectography and radiometry.

Geochemistry is, of course, a supplementary aid to geological mineral exploration but when standard methods fail to yield adequate information, geochemistry may supply the missing clue and since every mineralized area has its own peculiarities, or suite of minerals, the practice is to carry out detailed preliminary tests, preferably in the vicinity of known ore, in order to assess the applicability of geochemical methods and to establish a satisfactory technique before embarking on actual exploration surveys. The formation of dispersion patterns, their precise nature and general significance in relation to particular ore-bodies of the different metals is rapidly being established.

The most spectacular successes have occurred where the ore is overlain by a cover of residual soil so that soil analysis has become a standard aid to prospecting in some districts in America where it has helped to locate base-metal deposits in an area of cobalt mineralization.

Geochemical maps based on analysis of stream sediment are prepared, their principal purpose being to provide reconnaissance geochemical data to aid in planning mineral exploration programmes. The maps will show the cold acid-extractable copper and citrate-soluble heavy metals of stream sediment. Subsequent maps based on semi-quantitive spectrographic analyses will supplement this information for the possible presence of other elements of economic importance.

Results of geochemical surveys of volcanic complexes indicate that volcanic components are present in sequences or cycles which range up to 20,000 ft in thickness and display a compositional progression from predominantly andesite-basalt flows in the lower part to predominantly rhyolite-dacite pyroclastics with associated sediments in the upper part.

Values for copper exceeding 2 parts per million and for heavy metals exceeding 10 parts per million are considered anomalous. Some heavy metal anomalies undoubtedly reflect non-significant enrichment of metals by the scavenging action of manganese-iron oxide precipitates in streams.

Attention has been directed to 'pathfinder' elements and their application in exploration problems. Mercury has shown usefulness in this direction and field mercury detectors are in successful use for this purpose.

Copper nickel mineralization investigations by bio-geochemical methods have shown black spruce. Labrador tea trees and sphagnum moss to define sub-outcrops under a swamp.

The temperature of the earth increases with depth, the rate of increase at any place is known as the Geothermic Gradient and is expressed in the increase of depth in feet for a rise of strata temperature of $1°$ F, and in the British Isles this amounts to 70 ft per $1°$ F rise although it is very variable throughout the world, being only $1°$ F for 250 ft increase in the Witwatersrand, $1°$ F in 223 ft in the Hollinger and McIntyre goldfield in Canada and $1°$ F in 166 ft increase of depth in the Kolar goldfield in Mysore, India.

Because of the dependence of geothermal gradients on the heat conductivity of rocks, measurement of the temperature at various depths in boreholes and wells in which thermal equilibrium has been established, has been used sometimes for strata correlation purposes. The temperature gradients are in general low in formations having high heat conductivities and high if the heat conductivity is low.

Temperature measurements in drill holes are made with a thermometer whose sensing element is a temperature-sensitive resistance. A voltage is transmitted to the surface by cable and recorded photographically on film. The thermometer is lowered slowly in the hole so that the instrument has time to attain strata temperature. Readings are taken going down only to eliminate the disturbing effect of the cable.

SUBMARINE DEPOSITS

As five-sixths of the earth's surface is covered by the oceans, submarine deposits are assuming increasing importance as a source of useful minerals in a world with a rapidly escalating population with its increasing demands for commodities of all kinds.

Seismic refraction surveys were used by the National Coal Board to determine the depth to the rock-head under the Firth of Forth and to discover the course of any deep pre-glacial channels in the old Forth river system. The apparatus used was portable AB Electrik Malmetning equipment for seismic refraction shooting with 24 recording channels. A 'sea-snake' consisting of a special arrangement of 24 seismometers and their connecting cables on a tow-line of hemp 1300 ft in length floated on special buoys each weighted to maintain it at a known height above the sea bed. Marconi sea-graph echo-sounding equipment was used to determine the depth of water to the sea bottom. Depth to the rock-head did not exceed 500 ft and an interface appeared to exist at a depth of 570 ft which may be that of the barren red measures above the Coal Measures or boulder clay.

Similar surveys off the Durham, Northumberland and Cumberland coasts and a further survey of the Firth of Forth have been carried out by electric spark or explosion of propane and oxygen mixture in place of the firing of explosives to supply a pulse for seismic recording. Similar surveys have been carried out for the determination of other submarine minerals. In the E.G. and G. Seismic Profiling System used by Hunting Geology & Geophysics Ltd., the unit consists of a Sparkarray or a Boomer sound source transducer, suitable hydrophones, a signal processor and a suitable source of power. It is designed to give continuous seismic profiles of sub-bottom geological structure with an average penetration of 100 m. Discharge of electrical energy pulses into salt water generates bubbles of ionized gas which in turn cause acoustic pulses. The firm also used Hydrosonde Continuous Seismic Profiling equipment consisting of an acoustic sound source and a recording device synchronized to record the echoes received from the sea floor and sub-bottom. The recording device records the echoes received on an accurately controlled time scale, the actual strength of the printed echo being related to the strength of the received echo. The hydrophones which convert the sound in water to usable electrical signals normally lie some 5 to 20 ft from the sound source, thus, except in very shallow water the first signal to be received by the Hydrosonde is the direct wave travelling through the water to the hydrophone direct from the sound source. The second signal will be a reflection from the sea floor and this will be followed by a continuous stream of echoes from reflectors below the sea floor. This first reflected signal from the sea floor will again be reflected towards the water surface and so on.

In an area where the water is 48 ft deep, the first reflection from the sea floor will be received $2t$ sec after transmission of the sound-pulse where $2t$ is given by:

$$2t = \frac{\text{depth of water}}{\text{velocity of sound in water}}\text{sec}$$

The velocity of sound in sea water is 4800 ft/sec.

$$\text{Therefore } 2t = \frac{48}{4800} = \frac{1}{100}\text{ sec}$$

so that the 'two way' time = ·01 sec.

As a general rule it can be safely assumed that superficial deposits of interest to the sand and ballast industry have a velocity of the order of 4800 to 5300 ft/sec so their thicknesses can be calculated. Using the Hydrosonde offshore, sand and gravel is normally penetrated to bedrock and beyond with ease. Certain types of sediment are acoustically absorbent and reduce penetration to some tens of feet. Where gently dipping bedrock is exposed at the sea floor consistent penetration is obtained from 100 to 500 ft enabling structural studies to be carried out to these depths. Deep penetrations give resolutions of 2 to 12 ft and shallower penetrations to about 1 ft.

Prospecting the sea bed

As already indicated submarine minerals are attracting increasing attention. It is necessary however to be able to sample submarine deposits without deterioration or loss of part of the sample. Various devices are available for this purpose. The Hydrop submersible prospecting and sampling unit has been designed for this purpose by Alluvial Mining & Shaft Sinking Co. of Belvedere, Kent. Three vessels have been equipped using a 3 ft diameter unit which cuts and raises a 3 ft diameter sample from the sea bed and ensures that only the material lying within the area of the cutting shoe is raised thus giving a truly representative sample for assaying. A Hydrop air-lift system with intergral buoyancy chambers is used by which it is submerged and maintained in a vertical drilling position. High-pressure water jets disintegrate the material and allow high drilling rates to be achieved. All operations are controlled from the vessel without the necessity for divers and thus a greater number of holes can be bored and more cheaply than by some other methods.

An electric repeating clinometer and a hydrostatic depth gauge with their recording instruments in a control panel is installed on the ship's bridge, giving the operator precise information of underwater conditions. The equipment has been used in 150 ft of water, can deal with particles to 7 in diameter and drill 25 to 30 ft into the sea bed. It has also achieved 28 holes in a 14-hour double shift in 75 ft of water.

The Shipek sediment sampler was designed by Carl J. Shipek of the US Navy Oceanographic Electronics Laboratory, San Diego, California. It was specifically designed for sampling unconsolidated sediment from soft ooze to hard packed coarse sand and is capable of bringing virtually undisturbed unwashed samples to the surface from any depth. The unit consists of two concentric half-cylinders, the inner or sample bucket is rotated by two helically wound high torque external springs. When the the sea-bottom is contacted triggering is automatically accomplished by the inertia of a self-contained weight upon a shear mechanism. At the end of its half-circle track the sample bucket is stopped and held in the closed position by residual spring torque. The sample taken is 64 sq in in area (0·04 sq m) to a depth of 4 in at the centre.

The sample is given maximum protection from washout during the return traverse by the cylindrical configuration of the unit since closure is made at the side rather than at the bottom. Once on deck, the sample bucket may be disengaged by releasing two retaining latches at each end of the upper half-cylinder. The sample is then available for examination or transport to a laboratory.

PHOTOGRAMMETRY

This science is being increasingly used in conjunction with aerial surveys and computers in reducing the time involved in surveying and in the determination of volumes in the surface mining industries such as the volumes of excavations, overburden, stockpiles and dumps. The size of the survey must warrant the use of photogrammetric methods when volumetric assessment can yield practically the same accuracy as assessment by conventional methods but much more cheaply and quickly.

The main problem in making maps from aerial photographs is the transformation of the perspective photograph into the orthogonal projection of a map and this is achieved (a) by the restitution of an apparent stereo model from the overlapping photographs and (b) by the subsequent evaluation of the stereo model by a mechanism which gives the orthogonal projection of the information on the stereo model on to the map sheet. The stereo model as restituted from the overlapping aerial photographs is a geometrically true replica of the terrain photographed. A floating dot is employed to scan the model and a pencil point follows accurately every movement of the dot mark tracing the required map on to a drawing board. Aerial photography for mapping purposes is taken with highly specialized cameras having practically distortion-free lenses. To establish the scale and the height datum for the final map at least three ground reference points have to be provided for each pair of stereoscopic overlapping photographs. These points can take the form of precisely identifiable topographic detail or specific points, clearly marked on the ground prior to photography.

Certain pre-requisites are necessary to establish a geometrically true model:

(1) The aerial photographs must be taken with a camera which guarantees distortion-free photographs or with a camera of known distortion for which allowance can be made.

(2) The base of the photographic emulsion must be stable and not shrink or expand differently in any direction.

(3) The stereo plotting instrument must be designed to enable the correct geometrical restitution of the stereo model and its mechanical deficiencies must be of a lower order than the pointing accuracy which is governed by the restitution of the photographic image.

(4) The necessary ground reference or control points must be identifiable with accuracy and must be surveyed to a comparable degree of accuracy with regard to the scale of the stereo model.

(5) The skill of the photogrammetrist should be such that the potential of the instrument can be fully realized.

(6) The greatest advantage of the method is the facility of plotting contour lines directly on the final map without having to derive them from interpolation or as joins between surveyed points. The contour interval may be selected to suit the particular operation such as in irregular surfaces in opencast mining for overburden determination. The scale of the photography will be selected to guarantee contour accuracy and at the same time cover as large an area as possible. Normally the contour accuracy is within half the contour interval, scale is such that measurements from the photographs properly set up in suitable apparatus meet all normal accuracy requirements for plans up to 40 ft to 1 in scale for graphical presentation and contours may be plotted at intervals down to 1 ft. The precision is largely governed by the scale of the plan, the larger the greater the precision and at 1 in 500 the readibility is of the order of $\pm$ 3 to 6 in.

The instruments themselves are analogue computers and simulators into which stereoscopic aerial photographs may be introduced in a manner that exactly reproduces the relative and absolute conditions of the cameras in space. The prime function of a photogrammetric plotting instrument is to simulate flight conditions of the actual camera at its moments of exposure, using pairs of overlapping photographs, generally 9 in by 9 in. The speed achieved is greatly in excess of ground methods.

Improved methods for the rapid processing of data by electronic computers has not only extended the scope of photogrammetry within the range of projects already committed to its use but has led to its introduction to spheres where previously little or no advantage was to be gained from it, such as data much nearer the end result for statistical design or planning purposes which come direct from photogrammetric instrument/computer combinations, having by-passed completely the preparation of the plan where it was only the means to an end.

As an example of the importance of aerial photography in mineral prospecting, in Saudi Arabia in 1966, although there was no mining activity of note, there was an intensification of mineral exploration and related activities.

Some 95% of Saudi Arabia has now been mapped by aerial photography to scales of 1:50,000 or 1:100,000 and this includes all the Arabian Shield, which offers the greatest potential for mineral discoveries and specific areas for detailed geological survey have been allotted to Japanese, French and United States geological missions, and geochemical surveys are in progress.

The aerial photography in Saudi Arabia is being linked in a geodetic net under a three-year contract of £1·7 million to the Arabian Geophysical and Surveying Company which controls sub-contractors from Japan, Canada, Holland and the United Kingdom.

In the north at Wadi Sawawin near the Gulf of Aqaba, jasper iron deposits estimated at 800 million tons are being reassessed with a view to exploitation.

A copper deposit at Jabal Samran, 60 miles north of Jeddah, was drilled and mapped in 1966 and a pyrite and a chalcopyrite gossan was discovered 7 miles distant which is being drilled and sampled.

Extensive deposits of phosphate rock have been indicated in the extreme north which are believed to be 50 ft thick and related to the nearby Jordanian deposit worked since 1950.

South of Bir Idimah and 140 miles from the coast is a gossan of pyrite 11 miles on strike which was investigated in 1966 by geophysical surveying, geochemical analysis and core drilling. The pyrite content is estimated at 90% with a minor precious metal content, which, though inaccessible at present, is potentially a large future reserve of iron and sulphur.

PROSPECTING CONSORTIA

In many countries such as Britain, Canada and the USA mining companies have expanded and diversified their interests and have become mining finance houses rather than mining companies. Examples include Rio Tinto-Zinc, McIntyre Porcupine, Consolidated Mining & Smelting, Selection Trust and Anglo-American. All such houses have large prospecting and development departments and are ready to take advantage of any find that is made anywhere in the world. But it is significant that a high degree of co-operation as well as rivalry exists. Prospecting is so expensive and the risks so high that many mining houses enter into consortia to prospect in new areas and it is common for four or five mining finance houses to join together to prospect or develop an area. This stems from the risks and costs of prospecting and because the sort of project which is sought normally requires very large amounts of finance. Low-grade disseminated ore deposits must be worked, generally by surface mining, on a very large scale to be profitable.

In addition finds are increasingly in more isolated and often all but inaccessible places, and in the middle of frozen wastes, jungles or barren deserts. Consequently the deposit must be capable of being exploited on a massive scale to be viable and the project becomes very costly in capital expenditure. For example, even in a relatively developed area of Africa the Palabora Mining Co. required £37 million to reach production and of the iron ore projects in Western Australia, Mount Goldsworthy anticipated laying out more than £38 million, Hamersley Iron Prop. Ltd. ultimately £62 million and Mount Newman Iron Ore Co. £45 million. The need for partnership and co-operation is apparent.

REFERENCES

'Location and Evaluation of Sand and Gravel Deposits by Geophysical Methods and Drilling', G. Vann, *Opencast Mining Quarrying and Alluvial Mining*, Institution of Mining and Metallurgy, 1965, p. 3.

'Prospects of Offshore Mineral Deposits on the Eastern Seaboard of Australia', W. Layton, *Mining Magazine*, November 1966, p. 344.

'The Ward Hand Drill', R. W. McCallum, *Mining Magazine*, March 1967, p. 166.

'Determination of Volumes in Opencast Mining by Aerial Surveys', R. F. Rawiel, *Opencast Mining, Quarrying and Alluvial Mining*, Institution of Mining and Metallurgy, 1965, p. 483.

'Volume Assessment by Photogrammetric Methods', H. G. Dawe, *Opencast Mining, Quarrying and Alluvial Mining*, Institution of Mining and Metallurgy, 1965, p. 493.

Geological Aspects of Mining, J. Sinclair. Pitman, London, 1958, pp. 201–244.

'Prospecting Tin Placers in Indonesia', R. Osberger. *Mining Magazine*, August 1967, p. 97.

Principles of Applied Geophysics, D. S. Parasnia. Methuen & Co., London, 1967.

'Photogrammetry and Coalfield Surveying', W. D. Evans. *Proceedings of the South Wales Institute of Engineers*, Volume LXV, No. 2, 1949, p. 29.

'Changing Patterns of Mining Finance', R. H. MacWilliam. *Transactions of the Institution of Mining and Metallurgy*, Vol. 75, 1966, p. 36.

'Quarry Site Investigations and Evaluation', Staff of Mackay & Schnellmann Ltd. *Quarry Managers' Journal*, October 1967, p. 385.

CHAPTER 3

PLANNING AND DEVELOPMENT

An early pioneer in the development of techniques of management was Henri Fayol, a French mining engineer who took over a group of mines facing bankruptcy and guided it to success. Fayol analysed the essentials of successful management as: forecasting, planning, organizing, command, co-ordination and control.

It will be noted that Fayol placed 'forecasting' and 'planning' in the forefront of his essentials for successful management and in these days of large surface mining enterprises employing expensive machinery in mechanized production 'planning' assumes even greater importance.

A problem which is attracting attention, but which is constantly changing in its implications with the invention of increasingly efficient surface mining machinery, is the depth at which surface mining should cease and underground exploitation be adopted.

In the past, technical methods provided for only relatively small portions of outcrop to be uncovered by opencast methods, so that in most cases underground exploitation soon took over, but progress in opencast practice has changed the position. Comprehensive mechanization of operations, the use of particularly powerful appliances, in many cases working continuously twenty-four hours a day to achieve maximum utilization of large expensive capital items, and the application of simple technical design have paved the way for opencast working with large productive capacity and considerably reduced working costs. Modern opencast operations may produce 10 to 40 million tons a year from depths of 200 to 250 yards and in some cases even 500 to 700 yards, while the ratio of overburden to useful mineral reaches 10 to 15 to 1 and even 20 to 1. Consequently in many countries exploitation by open pits has been extended considerably to take the place of underground mining.

The maximum depth of exploitation by open pit is generally determined by the stripping ratio established as the limit.

In open pit production total cost per ton increases continually with depth because of the greater volume of overburden to be removed in order to obtain a ton of ore, whereas in the case of underground working the cost per ton remains nearly constant for a longer period. Where open pit mining is less costly than that of underground mining under comparable conditions

$$C_u + KC_d < C_m$$

or
$$K_L = \frac{C_m - C_u}{C_d}$$

where C_m = cost price of useful mineral from deep mine.

C_u = cost price of useful mineral from open pit.

C_d = cost price of waste from open pit.

K = waste/ore ratio, stripping ratio.

K_L = economic limit of waste/ore or stripping ratio.

This formula is perhaps somewhat of an oversimplification and more detailed study of the variants is necessary. However, it is a useful indication of the depth limit in terms of the stripping ratio.

Before the open pit operations can be planned it is essential that the outlines of the ore-body be established in size and shape, perhaps in part by geophysical prospecting, but certainly by drilling. Planning then steps in to minimize the distances ore and waste must be hauled and to keep the ratio of waste or overburden to ore as small as possible since on these depend, to a large extent, the viability of the operations. In rough country the vertical distance that waste must be lifted before disposal is also of importance.

Where the mineral worked is of relatively low value but heavy, such as aggregates for road construction, the relative position of the proposed quarry, and the market it is expected to command is important even in these days of modern roads and means of transport. Transport is expensive and its cost often exceeds the production costs, and economies in production can be absorbed by a quarry being at a geographical disadvantage to its competitors and in the future the general improvement in the efficiency of quarries will make this factor more critical. The process of choosing an area for investigation will at the same time have enabled the basic factors such as output, type of rock, anticipated production costs and probable market requirements, perhaps from market research, to be settled, but before the tentative proposal can proceed further a geological survey is required to define precise areas and enquiries should be made discreetly of the availability of sites for acquisition or leasing and probable cost or lease terms. Landlords are usually quite receptive to approaches made to survey their property for redevelopment but local residents, with no chance of benefitting financially from quarrying operations, are often antagonistic and this is often one of the reasons why Town and Country Planning applications for new quarries lead to a public enquiry at which the application is opposed by local residents. The area, of course, has only a speculative development value so that the right to purchase or to lease is made conditional on the obtaining of planning permission.

The outright purchase of mineral reserves means the freezing of capital for long periods and the future attitude of Governments in relation to land and mineral values is difficult to anticipate even with the modifications in the Finance Act of 1963 permitting depletion allowances. A lease, on the other hand, effectively reduces the amount of capital involved and is often preferred by the landowner who may wish to keep an estate intact. A long lease period is desirable in view of the heavy expenditure on fixed plant and the build-up of goodwill during the life of the quarry, although landlords are generally willing to renew leases it is difficult, even if possible, to compel a reluctant landlord to renew. It should be remembered that agreed royalty values provide one of the bases of rating valuation.

At this stage decisions can be made on the siting of overburden and topsoil dumps. A plan can be made for the main quarrying operations and the possible plant sites considered.

The geological survey will have defined the dimensions of the deposit and these will indicate the plan of operations for the future running of the quarry, bearing in mind the following criteria:

(1) The removal of minimum overburden.
(2) Maximum length of quarry face.
(3) Avoidance of bad areas with faults or poorer quality stone.
(4) To arrange for the maximum continuity of face operation before a new face has to be opened up.
(5) The reduction of uphill transport of loads to a minimum.
(6) Distance from the potential plant sites to the quarry face should be reduced to a minimum.

There is often little difference or variation in the type of equipment used in different quarries working the same or similar materials. In hard rock quarries, drilling and blasting is generally necessary and the use of shovels or draglines and dump trucks is fairly standard.

Although the planning and choice of the plant for a new quarry undertaking may be carried out by giving to one or more plant manufacturers the job of designing and quoting for the complete installation to meet the production, preparation and marketing programmes, the availability of plant will be limited to a large extent to the products of the manufacturer or manufacturers chosen and these might not be the most suitable for the conditions.

The alternative and generally preferable method is to draw up a detailed specification for the whole plant after selecting the most suitable items of plant. The complete installation is then carried out by one main contractor, although responsibility for the correct functioning of nominated supplier's equipment still rests with the supplier. Close liaison between the various contractors under the supervision of the company's engineer is thus essential for the efficient operation of this system.

The future management should always be brought fully into discussions or investigations at all stages of planning. Their knowledge of quarrying technique will be of great assistance and since they will ultimately be responsible for the operation of the quarry they should be consulted on layout and choice of plant.

For a plant dealing with aggregates the seven main design features to be considered would be:

(1) Output required attained over what period.
(2) Choice of crusher and screens to be installed.
(3) Method of dirt treatment to be installed.
(4) The provision of stock pile facilities for primary and secondary crushed stone.
(5) Storage capacity required for finished products.
(6) The siting of the plant for the most economic operation.
(7) The operating personnel required.

The preliminary surveys will have established the positions of the quarry and potential plant areas and detailed plans of these areas must now be prepared. The basic requirements for the ideal plant site are:

(1) The quarry face must be as close to the plant as blasting operations will permit.
(2) The site should be reasonably flat to minimize preparatory work and at a lower level than the quarry so that quarry to crusher loaded transport is downhill.
(3) Supplies of water and electric power should be within easy reach.
(4) Loading points and storage bins should be close to a public highway.
(5) The requirements of the Town and Country Planning authorities must be complied with and the plant and quarry face should be as remote as possible from houses and other buildings in view of the potential noise and dust nuisance.
(6) The ground on which the plant is to be erected should, if possible, have a good bearing capacity for the heaviest loads, such as the primary crusher and storage hoppers.
(7) Care should be taken that future stone reserves are not unnecessarily sterilized.

With the high capital cost of the complex mechanical and electrical equipment a high degree of training of foremen and operatives is required to ensure optimum results. Only limited personnel with the necessary qualifications may be available and as much economy in labour as possible should be the aim, by mechanization, remote control and even automation. Not more than two men should be required to run a well laid-out aggregate plant, though two further will probably be required for cleaning up and to relieve the plant men at meal times.

In the preliminary surveys, rough estimates of capital and operating costs should be made, founded on operations of other quarries making similar products, and based on probability theory and law of averages. The greater the volume of information that can be obtained of operations at other quarries the more accurate the estimated costs are likely to be.

The market survey should enable the sales potential of the proposed quarry to be assessed with probable average selling prices for each product. From these, the anticipated income and profit can be calculated with maximum and minimum figures to allow for good and bad conditions.

Potential output figures will govern the operational costs and it should be realized that the smaller part of costs is proportional to output since overheads or standing charges are more or less independent of output so much higher costs per ton will result from a drop in output. On the other hand, income is almost proportional to output so that profits can fluctuate widely.

Before the final decision to embark on the new venture is taken its investment potential should be assessed. The minimum percentage return that is acceptable should be decided on financial grounds and it is then usual to assess each section as well as the whole undertaking against this figure. Any section which shows a low return should be re-examined and unless essential to the functioning of the rest of the quarry's operations should be reconsidered, trimmed or not proceeded with initially.

SOIL AND ROCK MECHANICS AND GROUNDWATER CONTROL IN QUARRYING AND OPEN PIT MINING

The planning and design of open pits can gain much from a study of soil and rock mechanics and a reduction in the hazards of flooding by a study of groundwater control.

The extraction of mineral with the minimum excess excavation is the objective of surface mining and this implies that slopes must be as steep as possible consistent with safety.

Water fills pores in the rock below the water-table and there is also a seasonal fluctuation in the height of the water-table. The drained rock does not have the same physical properties, such as strength, as the undrained rock and as the pit develops there is a lowering of the water-table into the pit as soon as it is deeper than the water-table and flow is then into the pit and a pocket is created in the water-table. Seasonal changes will sometimes result in the areas of maximum permeability of the rock, which is variable according to direction, being the lowest and driest in the dry season and wettest and highest in the wet season so that these areas will have maximum spread between dry grained rock conditions and wet flooded rock conditions. Cracks along shear planes in the rock permit standing water to occur in areas of previous low permeability and the water pressure per square inch must be added to the stress in the rock involved. Average pressure could be as high as 44 lb/in^2 at a depth of 100 ft and the total load in pounds would be 44 times the number of square inches exposed.

Pit drainage must, then, be part of open pit design and information on water-tables and porosity of rocks can readily be obtained by standard techniques. There is a dearth of information regarding some other sources of load such as tidal forces, thermal loading, ground vibration from blasting but these are of very secondary importance and are problems of maintenance and control as information becomes available during the operation of the pit.

'Soil' is generally used to describe relatively weak and unconsolidated strata encountered near the surface while the term 'rock' is used for the harder, well-cemented materials. Soils may most conveniently be grouped into residual soils produced by the *in situ* weathering of underlying rocks, wind-blown deposits, waterborne sediments either marine, lacustrine, estuarine, deltaic or fluvial and glacial soils, such as boulder clays and moraines, formed by the action of ice during the Pleistocene age. From a strength point of view, there is a distinction to draw between granular materials (sands and gravels) and cohesive (silts and clays), based particularly on the relative rates at which pore-water pressures are able to dissipate in the respective materials.

The shear strength properties of the soil and the pore-water pressures in the ground are the factors which govern the stability of slopes in soils which may be considered as a two-phase system consisting of a solid phase, the skeleton of soil particles, and a fluid phase, water and air in a partially saturated soil and water alone in a saturated soil. Thus the stress across any plane will have two components, one carried by the soil particles and known as the effective stress σ' and a fluid or pore pressure μ, the sum constituting the total normal stress σ. The shear strength S is then

$$S = c' + (\sigma - \mu) \tan \varphi'$$

where the cohesion intercept c' and the angle of shearing resistance φ' are the strength parameters in terms of effective stress.

For a stability analysis of a soil slope, the ratio to that required to maintain limiting equilibrium is a measure of the factor of safety of the slope, a factor of unity indicating a failure condition. For a homogeneous soil slope the shape of the failure surface may approximate to a circular arc but will depart from it depending on stratification and on variations in shear strength and pore-water pressure.

The 'long-term stability', may be estimated using the effective-stress shear parameters and the pore-water pressures corresponding to the final equilibrium groundwater conditions.

The geology of the site must be fully investigated so that all possible mechanisms of failure may be examined, shear-strength parameters must be assigned to all materials within the possible failure zone and groundwater conditions must be established. Shear-strength parameters can be measured in laboratory tests on undisturbed samples obtained from bore-holes or trial pits. Groundwater conditions can be investigated using bore-holes and field evidence of seepage. Groundwater levels, in simple cases, can be found from measurements of the head of water in a bore-hole but in complex ground conditions the particular stratum or zone under investigation must be isolated from the effects of water in other strata.

The shear-strength of soils is generally measured in a triaxial compression apparatus.

Rocks also consist of grains or particles but these are cemented or glued together by calcaneous, ferruginous, argillaceous or siliceous material. During their histories rocks have been subjected to changing conditions and loads with resulting continual adjustment. There is a movement towards a new state of equilibrium when a new face is created in a rock and this occurs in two phases—an immediate elastic rebound and time-dependent further relaxation. A strain gradient is established from the outside in to a region where the original strain condition is approached.

The shape of the ore-body determines the shape of the open pit but also depends on the characteristics of the rocks and the kind of changes imposed on the existing force field and strain pattern by extraction operations. Good open pit design therefore depends on a knowledge of the force field, the strain pattern and the rock characteristics prior to making other calculations. An open pit never reaches its final shape until it is finished and during its life it is constantly changing and the rocks are always in a transient condition of stress and strain. The actual shape and size and the overburden ratio is controlled mainly by rock characteristics and not solely by the shape of the ore-body.

Because of their histories and their granular nature, rocks have stored residual strain energy which tends to be concentrated in patterns which consist of curvilinear surfaces of higher than average shear strains and these are planes of preferred shear because they are the planes along which the rock preferably will crack.

Again a rock will generally be intersected by a system of joints and bedding planes and though individual pieces of rock between joints and planes may be very strong these will provide surfaces of weakness and stability will therefore be dependent on the shear strength available along these surfaces, on their orientation in relation to the rock slope and on the water pressure acting across the surface. It follows therefore, that the main difference between the analysis of stability in soil and in rock slopes is that in the latter the possible failure surfaces are dictated by the orientation and position of joints.

The stability of slopes in both soil and rock can usually be improved if the pore-water pressure can be reduced. The majority of remedial measures for slides and slope failure consist of, or include, the improvement of drainage and in slope problems, consideration should be given to the part water pressures play either in soil, rock mass or in fissures and joint planes. It is possible to estimate the extent to which slopes may be steepened when drainage is provided and a comparison of costs made. Drainage may be by deep drainage to reduce water pressure within the rock or soil mass or shallow drainage to intercept surface water.

A modern example of the need for and care in planning of a large surface enterprise is the planning and development of the Kedia d'Idjii ore-bodies of SA des Mines de Fer de Mauritanie in NW Africa. Mauritania is bordered by Algeria, Mali and Senegal and the Atlantic Ocean. Planning took place from 1951 until the first ship was loaded with iron ore in 1963.

The mineralized area, principally high-grade haematite ore and banded haematite and magnetite quartzites (BHQ), of the Kedia d'Idjii has been classified into eight deposits of which two, Tayadi and I'Derik, have been explored in detail and another, Rouessa, partly explored. The nearest industrial centres are Dakar and Agadir and the majority of the population has had little contact with modern life. The plain in which the deposits occur consists of early Pre-Cambrian granites which dip westwards under recent sediments and eastwards below Cambro-Ordovician cliffs. The roots of the Pre-Cambrian formation consist of white quartzites, shales and banded haematite and magnetite quartzites (BHQ). Subvertical and contorted, these strata stand out as inselbergs, steep-sided eminences rising from an arid plain wherever they have withstood erosion, of which the Kedia d'Idjii is the largest.

Early in the twentieth century high-grade ore was found in the area but proved only of academic interest until the Bethlehem Steel Corporation examined the outcrops in 1948. In 1951 an Anglo-Franco-Canadian mission recommended exploration and Miferma was set up in 1952 as a prospecting company with a capital of £40,000. The object was to determine the probability of a reserve of 100 million metric tons of ore containing over 60% iron and an adequate supply of water to justify further expenditure.

The major difficulties encountered were lack of water, problems of supply by road or plane with heavy transport charges and losses, abnormal wear of equipment from abrasive dust, lack of local labour and heavy turnover of expatriate labour.

Two possible routes for the ore were investigated and the advantages of Port Etienne as a harbour were found to outweigh the disadvantages of the Congo railway route.

Some 14,000 ft of drilling and over 300 yd of tunnelling had been completed by 1957 at I'Derik and 5000 ft of drilling and 600 yd of tunnelling at Tayadit and 3300 ft of drilling at Rouessa, and Port Etienne had been chosen as the shipping port, expenditure to date having totalled just over £$\frac{1}{2}$ million.

The International Bank for Reconstruction and Development was approached for a loan, successfully, and the capital of the company was increased to £1,650,000 but it was realized that to meet strong competition the commercial grade of the ore would need to be raised to 64% iron, 3% SiO_2 and 1% Al_2O_3 and that constituents would have to be closely controlled. The staff in the field and at head office was expanded and planning completed by 1959.

Tayadit was chosen as the initial surface mining venture because it contained the largest proved reserves and was capable of producing an annual output of 4 million tons. I'Derik was to be equipped later for 1 million tons per annum and when producing Rouessa would be developed.

The Tayadit ore-body had been outlined to a depth of 500 yd and high- and low-grade material and proved tonnages had been established and the pit was designed as a V-shape cut down to 500 yd. Minable ore was in ample evidence below this but information was insufficient to extend planning at this stage any deeper. The pit shape was governed by the 60° dip of the ore-body and by its strike along a hog-backed ridge flanked on each side by deep ravines with several million tons of ore exposed over a height of 430 ft along its southern end.

The slope of the pit walls was decided empirically by reference to other mines, the hanging wall in massive banded haematite and magnetite quartzite dipping outwards was to be benched back 13 m in every 20 m, while the footwall, underlain by soft schists will be benched back 10 m every 10 m. Provision was to be made for service roads along pit sides.

The north end of the ore-body is partly covered by a 'bun' of BHQ on the hanging wall but once this has been levelled off, stripping will be reduced to benching back the walls of the pit and the average ratio of 2 of waste overburden to 1 of ore will be adhered to as closely as possible throughout the life of the pit. No problems were raised by the disposal of waste down to the 440 m level. The pit has direct access to a large valley beyond the hanging wall and the waste from the 440 m level can be hauled with a 60 m lift. Below the 440 m level the choice of means of waste disposal will be between a skip and a haulage road depending on the volume of materials to be transported.

In the choice of equipment the largest was to be used compatible with the selectivity required in order to reduce manpower requirements, powerful and strong enough to handle heavy, tough materials in difficult conditions. Three Ward-Leonard electric 8 yd^3 shovels were selected for stripping overburden and a Ward-Leonard 6 yd^3 for ore with pressurized and air-conditioned cabs. Three $2\frac{1}{2}$ yd^3 Diesel shovels are to be used on narrow faces and $1\frac{1}{2}\%$ of water is sprayed on the transport line ahead of the shovel buckets when digging. The benches and dumps are cleared by track and tyre-mounted bulldozers.

Haulage roads are to be 15 m wide with maximum grades of 8% and to be regularly sprayed with lignosulphonate. Ore and waste are to be hauled by twenty-two 60-ton trucks with oleo-pneumatic suspension and hydraulic retarders.

The drills to be used are down-the-hole type 6 in diameter to drill a 43 ft hole at any angle with a crawler-mounted chassis with a column 33 ft $7\frac{3}{4}$ in adjustable at any angle. Blasting with commercial slurries containing TNT or nitroglycerine placed in the bottom of the hole and the rest of the charge fine-grained ammonium nitrate mixed with 5 to 6% of fuel oil is being tested.

The lack of a background organization created early difficulties particularly in 1957 when it was necessary, when the final technical and financial programme finalized. The company relied greatly on consultants in order to reduce the number of personnel who would become redundant when planning changed to construction and this to operation. On the mining side was the Société Miniere et Metallurgique de Penarroya, who by contract agreed to provide the staff and organization to plan, develop and operate the mine during its first few years. The staff was divided into three sections—designers, constructors who negotiated contracts, and the planners. Peak contractors' personnel reached 535 Europeans and 3000 Africans. The project was broken down into eight 'jobs' and each 'job' was laid out on a Gantt chart.

At I'Derik it was finally decided to put in at the 420 m level a longitudinal drift carrying inclined diamond drill holes 100 m in length and 40 m drifter holes every 20 m, all holes being fully sampled. Trenching and surface drilling were continued and from the surface and internal sections thus established it was possible to demonstrate the structure and distribution of the first 10 million tons to be mined.

OPEN PIT PLANNING—COPPER

The most important factors to be taken into account in planning an open pit, particularly for copper, are the following:

(1) Assays of the deposit.
(2) Its geology and the topography of the area.
(3) The mining method to be adopted including height of benches in overburden and in ore and the machines and equipment required and their capacities.
(4) The reduction plant capacity and its location.
(5) Economic factors influencing operating costs.
(6) Capital expenditure required.
(7) Profit required and expected.

The sampling procedure must be sufficiently detailed to decide and classify the material to be removed from the open pit. For this purpose the assay values must be presented on plans and sections of the ore-body together with geological information. Assay sections are subsequently prepared with ore and overburden blocked out by bench intervals. Plans are prepared of each bench level of the pit. In a particular case block plans are provided in which each block is 100 ft^2 and represents 40,000 tons of rock.

A decision must then be taken to determine the pit slopes to be adopted and this is a crucial decision since too steep a high wall leads to slides that may be troublesome and even disastrous. On the other hand, a slope on an unnecessarily lower angle will increase the amount of overburden to be removed.

The ramp gradient to be selected must next be decided. The tonnage of overburden to be stripped for a grade of 6% to the base of an 800 ft deep pit amounts to some 21 million tons against 10·6 million tons for a 12% grade in the same conditions.

The boundaries of the pit must be decided before further planning stages are attempted.

If the ore-body has been completely drilled out with vertical holes on 200 ft centres this may be adequate to evaluate the ore-body but insufficient to locate its boundaries.

It would be quite possible with this layout to position the boundary 50 ft from its correct position. In such a case, a strip of strata 50 ft wide along the walls of a pit 7000 ft long and 3000 ft wide and 800 ft deep would amount to some 48 million tons, so that in this case this amount has wrongly been included in the limits of the pit or wrongly excluded from the limits of the pit. Pit boundaries are generally fixed in the first instance on sections. The decision whether strata along an open pit boundary should be accounted ore or waste depends upon three factors:

(1) The value of the material in per cent per ton.
(2) The amount of overburden requiring removal.
(3) The cost of winning it generally expressed per ton of ore.

Of these, the overburden to ore ratio to be removed and the value or grade of the ore are the parameters which need consideration. Arbitrary maximum overburden to ore ratios are often adopted to fix pit boundaries but the correct limits should be determined by correctly establishing the grade and overburden factors, for each section or incremental strip under consideration.

The following formula may be adopted or modified to suit particular circumstances. The rock is considered to be ore and should be mined if the grade is better than $(A + B) R$, where A is the break-even grade which should be mined and processed plus the grade to furnish the required profit, B is the grade to equal the cost of removing and R is the ratio of overburden to ore.

The following illustrates the method of using the formula.

If the copper ore is 1% copper and it can be mined and processed at 0.4% copper profitably and if the proceeds of 1.5 lb of copper pay for a ton of stripping; then the overburden ratio is:

$$1.00\% > 0.040 + (0.075 \times R)$$

$$0.60\% > 0.075\% \times R$$

$$\therefore R < \frac{0.60}{0.075}$$

$$< 8.$$

When the open pit boundaries have been fixed in this way and established on each section, such positions from each section are plotted on a master plan. A saw-tooth type of pattern generally results, which is equalized to give a smoothed-out boundary which can be worked up to without leaving pockets by the extractive machines.

The final master plan shows the toe of each bench and includes roads and ramps for transport. The positions of the bench toes are then transferred to individual bench block plans showing the areas to be mined.

An important factor which must be decided previous to actual planning is the sequence of mining the various grades of ore available. If in a given case ore is available at 1%, 0.7% and 0.4% grades, with an average grade of 0.7% copper, it may be decided to work the highest grade ore first to recover as quickly as possible the money invested in opening out and equipping the property. The 0.7% grade would probably be worked next leaving the inferior grade to the last. Alternatively, it could be arranged to produce at or near the average of 0.7% throughout the life of the mine. In many cases alternative schemes are worked out concerning the sequence of mining the various grades to give a combination of fairly quick recovery of capital followed by a period of steady costs and profits.

The immediate object of planning is to determine as exactly as possible what can be worked and the second to forecast as accurately as possible the results obtainable as mining proceeds and the third is that of ensuring control of mining operations, such that production will be maintained at the planned rate in the conditions and subject to the factors predicted. On the other hand, planning must remain flexible to cope with changes in metal prices, changes in production and other costs which may result from new methods or equipment becoming available, and increases or decreases of production required through market fluctuations. It should be realized that mine plans are not sacred and are not expensive to revise. Planning should be carried out for different periods of time. First a long-range plan for the complete extraction of the ore-body over a period. This plan will probably undergo considerable alteration before the ore-body is exhausted.

The plan is built up in stages, the first five or six at six-monthly intervals followed by another five, say at yearly intervals and then the remainder at five-year periods. Prints of the composite bench plan make convenient work sheets in formulating the mining plan. Bench toes representing the beginning of the stage are placed on the composite plan and on the one overburden tonnages to be mined are calculated, making sure that sound working conditions are provided throughout the stage. The composite plan is to the same scale and can be placed over the bench block plans making it easy to select areas and tabulate the ore and overburden tonnages or volumes, the tabulation being such that enough grade classifications are provided so that various cut-off values can be considered without retabulation. These tabulations are filed as part of the mining plan so that if it is required to know what the situation would be if the cut-off grade were raised the answer would be available quickly from the tabulations.

Each month a mining plan is made for the next two months, each month being considered separately. This is useful as it allows co-ordination with the long-range plan, and production to achieve its objective while providing a guide to daily planning.

This monthly plan shows the areas to be worked with tonnages of ore and overburden and the ore grades and provides direction for the mining operations. Together with the two-monthly plans a general forecast of production for the two-monthly period can be provided

Where the distribution of ore and overburden is irregular a daily mining plan or schedule may be used. Similarly if both oxide and sulphide copper ores are worked, a daily schedule may also be required.

The daily schedule shows the date, the shift, the number of the shovel to be used, the bench, a description of the material to be worked, the grade and the disposition of the material to be mined. Its function is to control the tonnage and grade, to facilitate sorting and to make the mining conform to the long-term plan and make the best use of men and equipment.

Every morning the new brows and blast holes are surveyed and plotted. Prints are made of the parts of the bench plans covering working areas and assays are posted. The schedule for the next day is then shown on the prints. Brows which have been surveyed are shown, areas to be mined are laid out for each shift. Blast hole numbers and assays are also shown. The shift foreman uses the schedule and sets of conveniently sized prints, say 17 in by 11 in, to direct the mining operations for the next 24 hours. Although departures from the schedule may be necessary to meet unexpected conditions, generally it is followed quite closely.

Computers are being increasingly used in open pit planning, particularly for fixing pit layouts and computing quantities and grades of ore and overburden. After the preparation of the fundamental data, the results of operating the open pit under several sets of assumptions can easily be obtained.

EXPLORATIONS FOR SAND AND GRAVEL

A map of the sand and gravel reserves showing for the first time the essential facts concerning the distribution and working of gravel and associated sands was published by the Ordnance Survey for the Ministry of Housing and Local Government in the autumn of 1965 and is the latest addition to the national planning series of 1/625,000 (10 miles to an inch) maps of Great Britain. It covers all of England south of the Lake District and the whole of Wales. From this area comes over 85% of the sand and gravel produced in Great Britain. A total of about 850 gravel workings are shown.

Among the minerals of Great Britain, gravel, the bulk of which is used for concrete aggregate, is second only to coal in terms of both output and value, and production has grown rapidly from 16 million yd^3 in 1937 to 90 million yd^3 in 1964 in response to the demands of the building and constructional industries.

The Waters Report on sand and gravel provides information of the location of such deposits in the United Kingdom but variations within such localities renders necessary geological and reconnaissance surveys to confirm that viable deposits are present when a particular site is under consideration.

In overseas undeveloped areas, particularly where dense vegetation prevails, there may be little or no surface indication of the presence of sand and/or gravel beneath. By reason of the modes of deposition of these deposits and the erosion and redeposition which may subsequently take place during their geological history, variation in thickness and quality are likely and present problems of investigation and evaluation seldom encountered in the exploration of other non-metallic and industrial minerals and rocks.

There are three main types of deposits:

(1) River deposits, which may consist of flood plains or terraces, are generally the most consistent in thickness and quality and most easily located, but buried channels of thick material, washouts and other troubles are difficult to discover by conventional methods, and geophysical methods can be of great assistance.

(2) Glacial and fluvi-glacial deposits are likely to be inconsistent in both quality and thickness and the overburden cover may also be considerable, again adding to the difficulty and expense of evaluation.

(3) Bedded deposits occur as pebble beds in sandstone as in the Bunter of the Trias and commonly occur in irregularly shaped lenses.

Sand and gravel are worked also from the sea, lake and river beds and may require special examination.

Of these modes of occurrence each type may present its own investigation problems but in terms of drilling and geophysical prospecting they present many common points.

In the United Kingdom the preliminary survey of a proposed site should proceed by reference to an ordnance survey of the district on the 1/2500 scale on which the boundaries of the prospect are marked. The topography should be studied and the relation of the area with reference to neighbouring cities and towns including any large civil engineering works under construction from which a demand for sand and gravel may be expected. The position of roads, rivers, canals, railways, electric power lines and water and gas mains should be noted, as they may affect the economical working of the proposed pit. The course of rights of way and bridle paths should be noted, particularly with a view to their diversion to permit economical working.

Enquiries should then be pursued on the supply of water and of manpower in the neighbourhood—both are essential. Where the site does not belong to the company interested, the time available to carry out the survey may be limited by reason of a limited option period, near approach of an auction date or other restrictions. Access to much of the site may be restricted by reason of possible damage to valuable crops, objections of tenants, or marshy or flooded ground. The cost of carrying out the survey may be restricted for commercial reasons or by agreement with the land owner who may be sharing the cost. Under all these considerations the application of geophysical survey techniques is relevant since the methods allow large areas of ground to be explored quickly and cheaply. Assuming the deposit is 20 to 30 ft deep, an average of 60 acres per day can be covered in reconnaissance fashion by the electrical resistivity methods giving one observation per 4 acres.

Light but easily operated robust equipment is used, not requiring the use of a vehicle. Either the seismic or the electrical resistivity method may be adopted. In the former method a portable shallow refraction unit includes the creation of a seismic shock wave by a tamper which is picked up by a geophone and displayed on a cathode ray tube or recorded on photographic paper. The method can be used on land or under water.

The cost of shallow seismic surveys of this type using a tamper as a wave source averages £25 to £30 per day during which readings at 10 to 15 stations with maximum tamper to geophone distances not exceeding 100 ft can be achieved. Seismic surveys take longer than the electrical resistivity method and the latter is to be preferred, particularly when the sand and gravel layer is overlain and underlain by clay, since the resistivity contrasts in this case should be well marked. On the other hand, the seismic refraction method is unlikely to be successful as there will almost certainly be a velocity inversion between the gravel and the clay beneath which will prevent detection of the interface. If the sand and gravel rests on hard rock such as sandstone, rock marl, limestone or chalk there may be very little resistivity contrast but a seismic contrast would be probable.

Surveys by the resistivity method cost about £10 per acre and though this gives, in a proper case, the extent of the deposit and to some extent its nature, it does not indicate the grading of the gravel or its sand content for which test boreholes are essential. Experience in the United Kingdom has shown that, except in the most variable situations, an average of one geophysical observation per acre coupled with one borehole or trial pit per 10 acres is a reasonable compromise between cost and accuracy for small and medium sized sites say up to 500 acres.

For larger sites, particularly if of several thousand acres in extent, half of these intensities may be acceptable but they should be considered the absolute minima.

It is necessary to obtain accurate and specific information of the overburden and of the deposit by test holes and deep boring, the former may be dug manually which probably causes least damage to crops or by 12 in or 18 in diameter augers. From the results the depth of overburden is determined and this may indicate that the deposit is not viable. When this is not the case drilling is next carried on to indicate the grading of the material in the deposit, also any impurities which would affect the quality of the product such as clay, chalk, shale, loam, coal, peat or vegetation. Conglomerate, ironstone, soft sandstone and soft-centred pebbles are also deleterious. Samples from the test holes are taken for sieve testing and silt analysis, each sample being carefully labelled and the size of the sample will depend on the grading of the gravel. If few stones over $1\frac{1}{2}$ in diameter are present then a sample of 14 lb is sufficient but if material up to 7 in diameter is found then the sample should be 5 cwt. The sieving sample is selected by successive quartering, care being taken to obtain a representative sample. Sampling is best carried out in accordance with British Standard Specification 882, Appendix B. Material passing $\frac{3}{16}$ in mesh is considered sand and that passing 100 mesh is silt and only a small percentage of this is permitted in the BS 822 specification.

The water requirements for normal washing may be taken as 5 gallons per minute for each ton of material put through in an hour. Where the water-table is high the use and disposal of washing water is easy, provided the waste water is not allowed to silt up the main workings. With a low water-table, where the deposit is worked dry after de-watering measures have been taken, sites are also generally without difficulties and after being worked out may be used as a silt pool.

Where deposits are dry they may require a series of settling pools capable of holding two or three days water supply, into which the dirty water is discharged at one end and clean water at the other. Artesian wells may be drilled to supplement the water supply from streams but such supplies are controlled under Section 6 of the Water Act 1945 by the Ministry of Health.

Information should also be obtained of the availability of electric power supply from the local Electricity Board, but if the site is an isolated one, a diesel-driven generator may be required for power supply. Sanitation and the supply of drinking water will also need attention.

The legal position will of course be examined by the organization's legal advisers, particularly with regard to the mining rights and the effect of the Town and Country Planning Act 1947. The results of these preliminary surveys and planning may be presented in a number of ways depending on the particular conditions.The report should incorporate a plan of the area with the position of boreholes, trial pits and geophysical prospecting stations with each hole or station clearly marked so that subsequently, reference can be made to any particular hole or station. The scale already mentioned, 1/2500, is suitable for the smaller areas but for those extending to thousands of acres, a scale of 6 in to the mile is more appropriate.

A detailed record of the strata penetrated in each hole together with diameter of hole and date of drilling, depth of water-table and depth at which samples were taken and tested should be filed for future reference. It is not sufficient to keep overburden and productive thicknesses. Geophysical data is not necessary in the report but all field book plottings and computations should be filed for future reference. Cross-sections showing overburden thickness, deposit thickness and water levels should be plotted along selected lines across the area. Where significant variations occur separate diagrams may supplement the cross-sections. Isopachytes or contours of equal thicknesses of overburden and of sand and gravel and contours of sand and gravel to overburden ratios should be available. The trends of the deposit are highlighted in this way and may be useful in obtaining planning permission. The report should also include volumes of sand and gravel and of overburden in the area, average, maximum and minimum thickness of sand and gravel and of overburden, average gravel and overburden production per acre and average sand and gravel to overburden ratios.

THE DEVELOPMENT OR PRE-PRODUCTION PERIOD OF QUARRIES AND OPEN PIT OPERATIONS

The period of development between the recognition of the viability of an ore-body or deposit to be exploited by quarrying or opencast methods is of importance in determining the period during which funds expended on the purchase of plant, machinery, buildings, construction and equipment are lying idle and unremunerative. It is important, therefore, that the phasing of the work should be arranged so that plant representing a heavy capital outlay, on which loan charges are accruing is not lying idle for a long period. The control of the execution and timely completion of a project lies principally in efficient programming and progressing, both physical and financial. For this, progress charts are used to phase out the work initially and, as it proceeds, the actual and the target progress of each job or portion of a job, can be compared and any difference indicated if any section of the work is behind schedule. The method of critical path analysis is being increasingly adopted for this purpose in an endeavour to reduce delays and the resulting extra financial burden which ensues.

In addition to the geological conditions of the ore-body and its geographical position with respect to potential markets and access to the site which have always been of paramount importance in determining the period and cost of the development of opencast sites, the economic climate at the time of development, the political situation in the particular country, the laws of land tenure there, the facilities, the markets and the amount of governmental backing are all important in relation to the period required for development.

An examination of the period of development of a number of important opencast sites in the past quarter of a century, and particularly in the past decennial period is given in Table II which is divided into three sections: (a) those sites taking two years or less, (b) those taking two to five years and (c) those taking five to seven years.

The sites in category (a) generally had very favourable physical and geological conditions with simple ores easily beneficiated and of high unit values, the minerals being comparatively easily won by conventional methods of working and processing with no large-scale transport problems, an assured market and only short-term finance required for development. In category (b) will be found the majority of new sites where the conditions may be described as 'average'. In category (c) are many of the large sites with large potential outputs of low-grade deposits, often requiring complex concentrating processes. In addition are those in a category requiring a development period of seven years or more and include those where difficult physical and geological conditions prevail, resulting in complex technological problems of winning and processing and difficult economic conditions. Western Deep Levels gold mine in South Africa, eight years, is a case in point.

TABLE II

1. SURFACE MINES WITH 2 YEARS OR LESS DEVELOPMENT PERIOD

Surface mine and minerals	Initial production (tons per day)	Commenced operation	Development time (years)
Palawan, Philippines, mercury	150	1955	2
Walton, Nova Scotia, barite	400	1941	1
Black Rock, Australia, copper	1,000	1963	1
Getchell, Nevada, gold	1,500	1962	2
Brynnor, British Columbia, iron	3,000	1962	2
Kidd Creek, Ontario, copper–zinc–silver	6,000	1966	2
Castle Dome, Arizona, copper	10,000	1943	2
Yerington, Nevada, copper	12,000	1953	2

2. SURFACE MINES WITH 3 TO 5 YEARS DEVELOPMENT PERIOD

Surface mine and minerals	Initial production (tons per day)	Commenced operation	Development time (years)
Mary Kathleen, Australia, uranium	1,120	1958	4
Gunnar Beaverlodge, Sask., uranium	1,250	1955	3
Brookfield, Nova Scotia, limestone	1,800	1965	4
Northgate, Ireland, lead–zinc–silver	2,000	1965	3
Carlin, Nevada, gold	2,000	1965	3
Coalinga Asbestos, California	2,500	1962	3
Zeballos, British Columbia, iron	3,200	1962	3
National Gypsum, Nova Scotia	6,000	1955	4
Endako, British Columbia, molybdenum	10,000	1966	$2\frac{1}{2}$
Esperanza, Arizona, copper	12,000	1959	4

3. SURFACE MINES WITH 5 TO 7 YEARS DEVELOPMENT PERIOD

Surface mine and minerals	Initial production (tons per day)	Commenced operation	Development time (years)
Pima, Arizona, copper	3,000	1957	5
Silver Bell, Arizona, copper	7,500	1954	6
Bomi Hills, Liberia, iron	10,000	1951	6
Quebec Cartier, Quebec, iron	10,000	1961	5
Mineral Park, Arizona, copper	12,000	1964	5
Lavender, Arizona, copper	12,000	1954	6
Palabora, South Africa, copper	1,750	1963	6
Weipa, Australia, bauxite	33,000	1960	7

The development periods in Table II are somewhat wider than is sometimes assumed in that they include a detailed evaluation exploration stage in which the existence of the mineral in marketable quantities at an economic, viable price is confirmed and the necessary finance and subsequent engineering plant and works determined and put in train. The examples quoted do not include opencast coal sites but in fact the factors which influence the length of the development period are similar to those where other minerals are concerned, but the period tends to be some 25% shorter.

The use of computers is becoming an integral part of planning the long- and short-range developments of open pits. The use of a computer allows planning variations to be explored in order that optimum results may be obtained. From drill-hole data the economic limits of an open pit may be determined. The computer programme relates the cost-value data for equal size blocks of ore and waste. Thus a three-dimensional filing system is produced in which any geological or technical data may be stored to be used to provide a profit figure per block, as input to the pit optimization programme for maximum profits which results are then fed back into the technical data file to give a pit outline and ore reserve statistics. Similarly truck haulage problems may be analysed.[1] The interaction of economic factors such as population, personal income levels, building activity, road building programmes and others may be used to analyse their effect on the production and demand for mineral materials such as sand, gravel, cement and aggregates, used in the construction industries. The analysis uses a multiple regression technique for use with a digital computer.

---

[1] Improvements in drilling technique have kept the cost of exploration drilling very stable over the past 60 years. Churn drilling in 1966 cost $7 per ft for overburden drilling and from $7 to $12 for diamond core drilling.

REFERENCES

'Geochemical Prospecting', D. H. Yardley. *Mining Engineering*, February 1964, p. 77.

'Elements of Gravel Pit Design', D. A. Webb. *Quarry Managers' Journal*, 1966, pp. 1–16.

'Location and Evaluation of Sand and Gravel Deposits by Geophysical Methods and Drilling', G. Vann. *Opencast Mining, Quarrying and Alluvial Mining*, Institution of Mining and Metallurgy, 1965, pp. 3–19.

'Computer Techniques in Mine Planning', T. R. Carlson, J. O. Erickson, D. T. O'Brian and M. T. Panu. *Mining Engineering*, May 1966. p. 53.

'Surface Minerals and Planning Powers', M. J. Hezzier. *Quarry Managers' Journal*, February 1966, p. 57.

'Planning a New Quarry', K. H. Goodacre and R. M. Farahar, *Quarry Managers' Journal*, April 1966, p. 131.

'Applications of Rock and Soil Mechanics to Surface Mining', A. C. Meigh and D. J. Henkel. *Opencast Mining, Quarrying and Alluvial Mining*, Institution of Mining and Metallurgy, 1965, p. 369.

'Groundwater Control in Opencast Mining', S. C. Brealey. *Opencast Mining, Quarrying and Alluvial Mining*, Institution of Mining and Metallurgy, 1965, p. 390.

'Estimating Reserves of Surface Mine Properties', N. Pundari. *Coal Age*, March 1967, p. 68.

'The Pre-Production Interval of Mines', W. C. Peters. *Mining Engineering*, August 1966, p. 63.

Planning and Mechanized Drifting at Collieries, J. Sinclair. Pitman, 1963.

'Some Considerations Involved in Opening up a Quarry', J. K. Mercer. *Quarry Managers' Journal*. April 1968, p. 143.

CHAPTER 4

REMOVAL OF OVERBURDEN

The two machines which occupy the major roles in the stripping of over-burden are the shovel and the dragline, although ancillary machines such as the bull- and calf-dozer, the shovel loader and the caterpillar and the four-wheel tractor-drawn scraper may also be used in a secondary capacity.

The capital investment required to install these individual items of equipment, particularly in large-scale operations, may run into millions of pounds sterling. Before investing such large sums, an extremely detailed analysis of the overall economies must be made in order to be certain that the right tool is to be obtained for the job in hand.

Coal, phosphate, iron and copper ores are four types of minerals obtained by opencast mining and more especially by stripping equipment removing the overburden. As long as the general geological characteristics exist, that an efficient application can be effected, any mineral can be mined economically by this method. The total reserves available provide the criteria for the size of equipment required and the resulting capital investment required.

With the general rule of thumb that the larger the unit the greater the saving and the availability of 200 yd^3 shovels and 220 yd^3 draglines, it might be considered that large units would be the best solution economically, but this is only the case when very large reserves of mineral are available in one site. Widely separated non-continuous types of deposits, divided ownership of large deposits, limited markets for a particular product and many other reasons may determine that smaller units should be adopted for maximum economy and efficiency. In every case the reserves must be sufficient to absorb the amortization cost of the machine without unduly inflating the price of the product.

In comparing the two types of machine available for overburden stripping the choice depends very much on the geological conditions of the deposit.

The particular advantages of the dragline are that longer reach and range are obtained. The dragline is more flexible with regard to variations in the configuration and general geology of the deposit. The greater reach and dumping radius enable deeper overburden to be removed, so that for the same capital outlay, considerably more depth of overburden can be stripped enabling deposits to be worked profitably which would be uneconomic for shovel operation.

The dragline is the more easily manoeuvred machine and is more versatile. It is located on or near the ground surface, thus eliminating certain problems. Slides in either the bank or spoil cannot block or impair its operation. Water also presents little difficulty, as the run-off can be easily controlled and seepage is less of a problem on the surface than within the confines of a pit.

Localized pitches and rolls are of little importance since overall planning can correct such variables and their effect on the required cut widths or spoil area and variations in the top of the mineral deposit can easily be negotiated by the dragline.

The dragline can 'chop down' a certain proportion of the overburden if necessary, although this is not as efficient as the normal operating procedure, it can be used on occasion, with advantage. Inequalities in ground topography can be eliminated by 'chopping down' and the dragline then provides itself a constant level horizon from which to work and an increase in the dumping radius can also be obtained in this manner.

It may be necessary in order to open up a new property to begin with a box cut when the greater reach and location of the dragline can be fully utilized. When the final cut or last phase of a stripping operation is achieved, the dragline can take advantage of dumping both into the spoil area and also on top of the ground surface, being thus able to expose the deposit to a greater depth.

For the larger sizes of shovels and draglines, considerably lower ground bearing pressures are possible with the dragline, for example, a 45 yd^3 shovel has a bearing pressure of 52 lb/in^2 whereas a 35 yd^3 dragline though somewhat heavier in working weight (approximately $3\frac{1}{2}$ million lb against $3\frac{1}{4}$ million lb), gave a bearing pressure of only 12 lb/in^2.

Although rehandling of spoil adds to the cost of overburden removal, it is a simple operation with a dragline and in some cases rehandling of some spoil may be a necessity.

If conditions change, the boom of a dragline can be lengthened or shortened fairly easily and the working range can be altered by changing the boom angle simply by raising or lowering the boom. Of course, simultaneously bucket capacities would change in conformity with allowable loads at varying radii.

The shovel, on the other hand, has the advantage of greater capacity than the dragline, particularly in the larger units. With equality of capital cost a shovel with a dipper of 45 yd^3 capacity would be equivalent to a dragline of 30 yd^3, or 30% more capacity in the shovel.

Since the dipper on a shovel is fastened to the driving mechanism by a solid connection, positive control results in better filling of the dipper, giving a higher loading factor, particularly in the harder or rockier types of overburden. Scooping up through the boom will also assist in loading the dipper better when compared with dragging the bucket of a dragline by cable through the same material. Blockier material will have less overall effect on the shovel's efficiency of loading and considerably less expense for drilling and blasting will be required in shovel operation.

As the material is loaded in an area directly in front of the shovel and is dumped within a relatively short radius of the loading point, the cycle time will be lower than for a dragline and the difference is increased where long dragline booms are used, when considerable 'chopping down' is done or in deep overburden.

An advantage of the dragline, that it works from or near the original ground surface, may on occasion, prove disadvantageous. Since the location of the shovel on top of the mineral deposit is not controlled by the original topography, then with the shovel, the possibly inefficient and expensive preparation of a working bench is eliminated, particularly in extremely undulating areas.

A certain berm width is usually maintained by the shovel in the pit, which makes a roadway available for haulage units. Where haulage cannot be carried on successfully on the beds below the mineral deposit (for example spavin or fireclay below a coal seam) the berm is a definite advantage.

The cost of overburden removal per cubic yard is lower with a shovel than with a dragline but the difference is generally marginal.

HARD OVERBURDEN

Where a large proportion of the overburden is hard, blasting must be resorted to in order to break it up before excavation. Such conditions obtain where the covering of a massive deposit is of considerable depth and lateral extent, the winning of which will extend over a considerable period so that most of the overburden must be dumped outside the limits of the deposit, also where the hanging and footwalls of veins and masses which must be stripped to prevent them falling in on the workings as the ore is removed. With hard overburden as occurs in bedded deposits of fairly shallow depth so that into a worked out area, overburden from adjacent unworked areas can soon be dumped, where the operation and the daily amount of overburden to be stripped is small, a tractor shovel may be sufficient and economical. Where hard overburden overlies a soft or disturbed bed it may be difficult or impossible to maintain a satisfactory haulage road on the soft bed. In this case a dragline working on the top of a bench in the hard overburden may provide a satisfactory solution, the longer digging cycle time than for the shovel can be compensated by the dragline carrying a larger bucket. More blasting is, however, required since the lump size must not be large enough to throw over the bucket by striking a lump, but excessive blasting may cause instability of the working bench. More spillage may result with draglines when 'spotting' the bucket over a truck or conveyor.

There is generally more scope for cut-and-fill working with bedded deposits than with masses or veins and cut-and-fill gives the shortest distance for the movement of overburden between cut and spoil bank. It is necessary that the dragline or shovel should have sufficient reach to dump the spoil into the worked out area, without the need for intermediate rehandling.

In selecting the excavator for a cut-and-fill operation, the ore deposit having been prospected and its limits defined, the rate of extraction will be decided after taking account of the market for the product over the life of the major machines, say 15 to 20 years or less if reserves are more limited. The required bucket capacity can be derived from the formula:

$$B = \frac{100RO}{FC}$$

where O is the output of ore in yd^3 required per hour (if in tons convert to yd^3, for ironstone divide by 1·7).

R is the ratio of depth of overburden to depth of ore.

F is the bucket factor, that is the actual yd^3 moved per hour divided by the nominal bucket capacity in yd^3 × cycles per hour (this in broken rock may be as low as 50%).

C is the average cycles per hour available for working. Cycle times rarely average better than 60 per hour and availability higher than 75% of manned hours is difficult to maintain.

B is the nominal bucket capacity in yd^3.

The dump radius is best determined by drawing pit sections to show the layout of stripping and loading operations. It is generally advisable to adopt the layout which shows the lowest requirements in stripping machine dump radius. Information which will be necessary to draw these sections is the angle of repose or slope to which the spoil will run, 5 horizontal to 4 vertical being a commonly acceptable figure for rock (39° from the horizontal falling to 32° when fines are present). If no local information is available, a tentative 35° should be adopted. The width of the cut and the spoil bank area will be controlled by the ore-loading machine cut width. Cut faces are not as near vertical as they seem and allowance must be made for their inclination; it is also dangerous for machines to travel too near the edges of open faces and room for cables and other services may have to be provided.

Records of operating times of excavators are useful for control and for the planning of future operations. These are available in a number of different types depending on the exact information it is desired to record. The control of excavators has improved and air-operated controls lighten the work of the operator and prevent the end-of-shift fatigue which often increases operating times towards the end of a busy shift. This falling off is also experienced at busy winding shafts at mines. Ward-Leonard control of excavators as with mine winding engines lightens control and gives faster response to control signals.

The front-end equipment on shovels tends towards single dipper handles with rope crowd operating from the machinery deck or the Marion system of positioning the crowd machinery on the A-frame, both having the effect of reducing the weight on the boom which must be counterbalanced.

Where a shipper shaft is used the boom may be braced back to the A-frame at this point and the boom beyond jointed to allow for some movement when digging. A wide range of walking draglines is being developed in the USSR ranging in bucket capacity from 5·2 to 105 yd^3 with boom lengths 125 to 390 ft, dumping radii 128 to 397 ft and weights from 186 to 5100 tons. Draglines are widely used for mining bedded deposits such as coal, iron, manganese ores and other minerals. The normal methods are cut-and-fill when the overburden material is handled by draglines on to spoil banks in mined-out areas. Some rehandling by draglines is found to be necessary.

In the Urals a coal bed 30 m (93 ft) thick is being worked under rock overburden, some of which is transported to spoil dumps by road and rail and the remainder spoiled into the worked out areas. After mining out coal or ore the dragline excavates overburden from the next cut, the dragline alternates removal of overburden and mining of mineral. If two or three draglines are used in an open cut, mining goes on continuously round the clock. This 'excavator quarry' system is used in winning brown coal and iron ore in the Tula region.

Two firms, Marion, and Bucyrus-Erie with its British subsidiary Ruston-Bucyrus, predominate in the excavator machine industry throughout the world, although Ransome and Rapier and Smith also hold a prominent place in the British market. It is proposed to describe the Model 110 RB Ruston-Bucyrus shovel and dragline and the Marion 8700 walking dragline.

Ruston-Bucyrus 110 RB electric shovel

The specification of the 110 RB electric revolving shovel equipped with crawler-type mounting (Figs. 16 and 17), and arranged to operate from an a.c. power source to drive the motor generator set supplying power to the

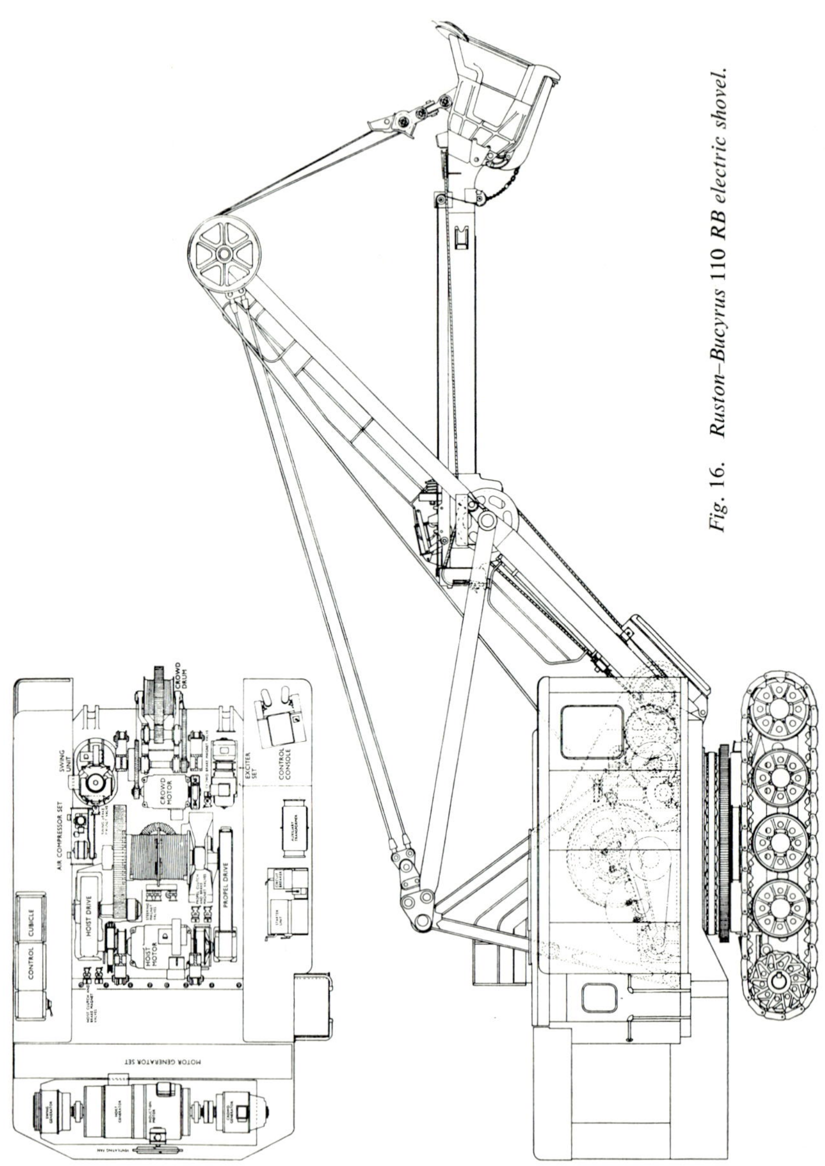

Fig. 16.   *Ruston–Bucyrus 110 RB electric shovel.*

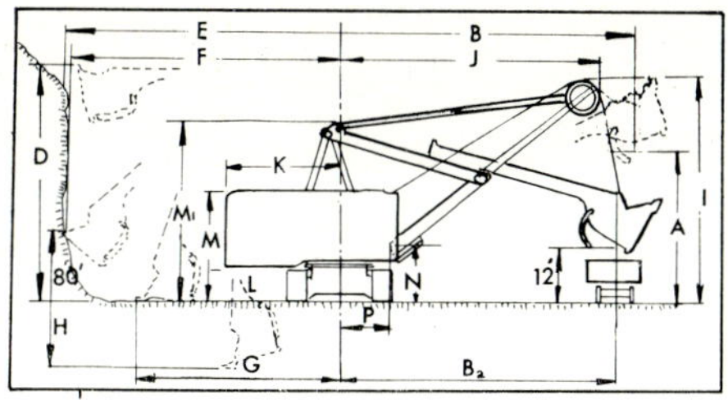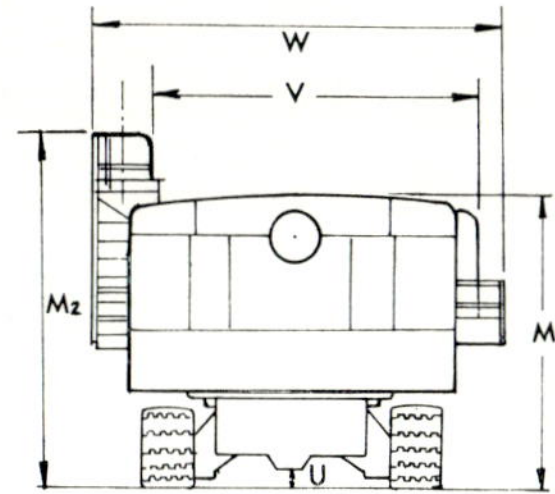

Fig. 17. Specifications of 110 *RB electric shovel. The lettering is identified in the text.*

dc operating motors, is fitted with Ward-Leonard variable voltage control and the machine is provided with boom, handle and dipper.

MAIN SPECIFICATIONS

Net weight, domestic, without ballast, approximate	310,000 lb	$138\frac{1}{2}$ tons	140·430 kg
Working weight, including ballast, approximate	340,000 lb	152 tons	154·020 kg
Ballast—supplied by purchaser	30,000 lb	$13\frac{1}{2}$ tons	13·590 kg
Shipping weight—prepared for export, no ballast, approximate	325,000 lb	145 tons	147·225 kg
Ships' option tonnage		260 tons	

Crawler mounting

Width of links—standard	3′ 0″	914 mm
Width of links—wide	3′ 6″	1·07 m
Overall width of mounting (3′ 0″, 914 mm links)	16′ 3″	4·95 m
Overall width of mounting (3′ 6″, 1·07 m links)	16′ 9″	5·11 m
Overall length of mounting	19′ 4″	5·89 m
Total effective bearing area (3′ 0″, 914 mm links)	100 ft^2	9·30 m^2
Total effective bearing area (3′ 6″, 1·07 m links)	117 ft^2	10·87 m^2
Diameter of idler rollers	39″	990 mm
Diameter of driving tumblers	$41\frac{3}{8}$″	1·05 m
Diameter of take-up tumblers	41″	1·04 m
Number and pitch of links	70—$14\frac{9}{16}$″	370 mm
Propelling speed—normal	65 fpm	19·81 m

Turntable

Diameter of roller track	9′ 10″	3·00 m
Number and diameter of rollers	40—$7\frac{1}{2}$″	190 mm
Diameter of swing rack	10′ 11″	3·33 m

Drums—Sheaves—Ropes

Diameter of hoist drum	30″	762 mm
Diameter of crowd drum	24″	610 mm
Diameter of shipper-shaft sheaves	39″	991 mm
Diameter of boom-point sheaves	48″	1·22 m
Diameter of hoist ropes—single part—twin dual	$1\frac{1}{8}$″	29 mm
Diameter of crowd and retract ropes—twin	$1\frac{1}{4}$″	22 mm
Diameter of bridge-strand suspension—4 single strands	$1\frac{5}{8}$″	41 mm

MAIN SPECIFICATION—*continued*

Electrical equipment—Ward-Leonard

Hoist motor (blown)	.125 hp	230 V, 75°C, cont.
Swing motor (blown)	. 44 hp	230 V, 75°C, cont.
Crowd motor	. 35 hp	230 V, 75°C, cont.

Generators for Ward-Leonard control are equivalent in capacity to their respective motors and are driven by a suitable induction motor.

WORKING DIMENSIONS

	Dipper capacity	$4\frac{1}{2}$ yd^3	3·44 m^3
	Length of boom	36′ 0″	10·97 m
	Effective length of dipper handle	21′ 6″	6·55 m
	Overall length of dipper handle	25′ 10″	7·87 m
	Angle of boom	45°	
A	Dumping height—maximum	23′ 3″	7·09 m
A1	Dumping height at maximum radius . . .	15′ 9″	4·80 m
B	Dumping radius at maximum height A . . .	38′ 6″	11·73 m
B1	Dumping radius—maximum	40′ 3″	12·27 m
B2	Dumping radius at 12′ 0″, 3·66 m, elevation . .	40′ 0″	12·19 m
D	Cutting height—maximum	34′ 9″	10·59 m
E	Cutting radius—maximum	46′ 3″	14·10 m
F	Cutting radius at 8′ 0″, 2·44 m, elevation . . .	44′ 0″	13·41 m
G	Radius of level floor	31′ 3″	9·53 m
H	Digging depth below ground level—maximum . .	9′ 3″	2·82 m
I	Clearance height of boom-point sheaves . . .	36′ 0″	10·97 m
J	Clearance radius of boom-point sheaves . . .	34′ 9″	10 59 m
K	Clearance radius of revolving frame . . .	18′ 3″	5 56 m
L	Clearance under frame to ground level . . .	5′ 3″	1·60 m
M	Clearance height with boom and A-frame lowered .	16′ 9″	5·11 m
M1	Height of A-frame	24′ 0″	7·32 m
M2	Height over stairway	20′ 0″	6·10 m
N	Height of boom foot above ground level . . .	8′ 5″	2·57 m
P	Distance from boom foot to centre of rotation . .	7′ 4″	2·24 m
U	Clearance under propelling gear case to ground level	1′ 3½″	394 mm
V	Width of superstructure with platform and stairway removed	17′ 9″	5·41 m
W	Overall width of superstructure	20′ 10½″	6·36 m

Mounting

The machine is equipped with crawler-type mounting, composed of two belts, one on each side of the truck frame, driven by tumblers and running around rollers mounted on shafts firmly held in heavy, box-section cast-steel girders forming the crawler frames. These girders support the truck frame by means of heavy integral lugs. The driving tumblers are alloy-steel castings, with lugs to engage the link castings. They are keyed to hammered-steel shafts which run in bronze-bushed bearings in the frames. This shaft also carries a driving gear on a splined section which meshes with a pinion on an intermediate shaft. Both gear and pinion are enclosed in an oil-tight, dirt-proof casting.

Steering clutches

Both crawler belts may be driven simultaneously or independently through multi-jaw clutches. These are on the afore-mentioned intermediate shaft and also serve as couplings between the propelling machinery on the truck frame and the gears in the crawler frames. They are manipulated by air cylinders controlled from the operator's station. Locking of either crawler belt is similarly controlled.

Crawler

Each crawler belt has three idler rollers of large diameter which support the machine and run free on hammered-steel shafts securely held in the frames. A sealed, grease-tight closure is installed on the outer ends of the bronze bushings in the idler rollers and take-up tumblers, and a dirt seal ring protected by machined grooves is provided on the ends of the hubs adjacent to the crawler frames. The tumblers and idler rollers are differentially hardened.

The links are long-pitch, heavy, alloy-steel castings of a patented design and are connected by pins of special heat-treated steel. The narrow roller path gives lateral flexibility to the links, to reduce strains caused by uneven ground. The roller path and connecting-pin holes are flame hardened. A screw-type jack is provided to take up the slack of the belts by adjusting the position of the take-up axle which rides in guides formed in the crawler frames.

Truck frame

The truck frame is a structural steel unit having members of generous proportions solidly welded together to provide strength and rigidity. To ensure alignment and solid foundation for machinery parts attached to the frame, machined surfaces are provided for the mounting of these parts.

On its upper surface provision is made for the attachment of the large circular swing rack which carries the roller rail. The rail sections are alloy-steel castings set in a machined groove to maintain them on centre and are held in place with tapered chock bars. At the centre of the truck frame a large, bronze-bushed journal receives the cast-steel centre sleeve. This sleeve, which is rigidly attached to the revolving frame, has provision for take-up and forms the connexion between the truck frame and the revolving frame. This sleeve is bored to provide bearings for the vertical propelling shaft.

Roller circle

The roller circle is composed of a complete ring of rollers which roll against the cast rails on both the revolving frame and truck frame. The roller pins and roller frame serve only as spacers to position the rollers.

The roller frame is made of heavy bar rings outside and inside, which are separated by the large-diameter pins. The inner and outer rings of the frame are made in sections for easy removal of a group of rollers or a section of roller tail. The rollers are heat-treated, high-carbon alloy-steel forgings. Means for lubricating the rollers is provided on each roller pin.

Revolving frame

The revolving frame consists of an annealed-steel casting having integral lugs for the boom feet and the front and rear A-frame legs. On the upper machined surface is mounted the machinery.

This casting is bored for the centre pintle and for the vertical swing shaft. The underside is machine-grooved for the cast alloy-steel roller rail sections which are held in place with tapered chock bars.

Fastened to the rear of this main casting is a welded-steel structure with side wings which carries the motor generator set, etc., on the upper face and the ballast inside.

A frame of column construction

The forward and rear legs are fabricated from parallel H-beams with cross members to ensure rigidity and are pin-connected to the revolving frame. The forward legs and the upper portion of the rear legs form a welded truss over the hoist machinery which can be pivoted forward for travelling the machine when overhead clearance requires it; they also serve as a rigid support for the strut to the shipper shaft. The lower portion of the rear legs is pin-connected to the truss at a point below the roof line. The apex of the A-frame is high and well back, minimizing boom suspension loads and compression in the boom.

Twin-dual single-part hoist

There are two independent hoist ropes. Each is anchored to the hoist drum; passes over a groove in the corresponding boom-point sheave, through one of the equalizing sheaves attached to the side of the dipper, and is doubled back over another groove in the boom-point sheave to be anchored again to the hoist drum. This arrangement is referred to as the twin-dual, single-part hoist and steadies the dipper in the bank when digging. By doubling each rope in this manner, it is possible to use a smaller diameter rope, thus increasing the ratio of the drum and sheave diameters to rope diameter. The machinery is designed to facilitate conversion from shovel to dragline.

Regenerative lowering

All of the hoisting machinery is engaged while the dipper is being raised or lowered, the motor functioning as a motor when hoisting and acting as a regenerative brake when lowering. This design eliminates the necessity of operating a clutch or brake while digging; the dipper being at all times under the control of the hoist controller lever.

The motor is connected to the primary pinion shaft by a smoothly engaging, electrically controlled, air-actuated clutch, which also serves as a flexible coupling and slipping clutch for overloads. The bearings for this shaft are of the anti-friction type, and are carried in an oil-tight gear case, integral with the left-hand side frame. The primary pinion, integral with its shaft, drives the intermediate gear, keyed to its shaft, which turns in anti-friction bearings in the left-hand side frame. The hoist pinion is integral with the intermediate shaft and drives the hoist gear, keyed to the hoist shaft, which runs in renewable, babbit-lined bronze shells in the cast-steel right- and left-hand frames mounted on the revolving-frame casting.

The drum is a split lagging with machine-turned grooves for the twin-dual ropes and is securely bolted to the hoist gear and a flanged hub at the right-hand end of the shaft. When converting to dragline, the shovel hoist lagging is replaced with a drag lagging.

Brake for power failure

The hoist brake, mounted on the primary pinion shaft is spring-set, air-released and electrically controlled and is used to hold the hoist drum in a fixed position when desired, and sets in case of failure of power supply.

All shafts are hammered-steel forgings, all pinions are alloy-steel forgings, and all gears are alloy-steel castings. All gears and pinions have machine-cut teeth. The first gear reduction has double-helical teeth.

Propelling machinery

The crawler belts are driven by the hoist motor through an electrically controlled, air-actuated clutch, which is a duplicate of the one used for hoisting. The motor is connected, through the clutch, to a combined shaft and driving sprocket running in anti-friction bearings in a cast-steel base, and drives a sprocket on the intermediate shaft by means of a roller chain in an oil-tight, welded structural chain case. A spur gear on this shaft drives a gear on the upper horizontal propelling shaft which runs in babbit-lined bronze bushings. This shaft drives the vertical shaft, longitudinal shaft and centre drive shaft at the rear of the truck frame through three pairs of bevel gears. The spur gears and sprockets above deck are machine cut. All the bevel gears and the spur gears in the crawler frame are made of heat-treated alloy-steel castings, with heavy-pitch teeth. All shafts run in bronze bushings. All shafts are splined except as described otherwise. All gears in the truck frame are enclosed to retain lubricant and exclude abrasive material.

Safety brake

An easily accessible band brake is provided in the truck frame to prevent movement of the machine when operating or should power fail when moving on a gradient.

Swing machinery

The swinging of the revolving frame and its attachments is accomplished by a single unit, located at the front and to the left of the centre line of the machine. It is built into a base casting securely bolted to the revolving-frame casting. The base casting contains the bearings for the shafts, the lower bearing for the second intermediate shaft being machined on the outside to fit into a bored recess in the revolving-frame casting to act as a shear plug in maintaining alignment and assisting the foundation bolts. The vertical shaft on which the rack pinion is mounted has two babbit-lined bronze-bushed bearings in the revolving-frame casting the shaft having square ends to fit in its rack pinion and gear.

All gears are alloy-steel castings, all pinions are alloy-steel forgings, and all have cut teeth including the rack and rack pinion. All gears with the exception of the rack and rack pinion, are fully enclosed in the base casting and run in oil. The first and second reductions are mounted on anti-friction bearings.

Power is provided by a vertical-type motor mounted on the base casting cover. The motor is provided with a spring-set, air-released brake which operates to prevent swinging of the machine in case of failure of the power supply. The brake is not used to retard the swing as the motor is designed to perform this function by plugging.

Sectional boom construction

The boom is constructed in two sections. The lower boom section is a single, welded unit, consisting of two side girders and two cross girders with cast-steel boom feet, and shipper-shaft bearings welded into the structure. The side girders are widespread at their lower ends for pin connexions to the revolving-frame centre casting and tapered at their upper ends to accommodate the shipper-shaft bearing structure.

The upper boom section is composed of two fabricated H-beams, a boom-point sheave pin welded thereto and suitable cross members. It has provisions for a hinge connexion to an extension of the shipper-shaft bearings at the lower end and for attachment of the bridge-strand suspension at the upper end.

Permanent strut and bridge-strand suspension

Two boom struts, consisting of fabricated H-beam members, have bearings on the shipper shaft at their outer ends and pin connexions to the A-frame structure at their inner ends. The inner-end connexions on the struts provide suitable connexions for the attachment of the bridge-strand suspension.

Boom machinery

The shipper shaft is a large-diameter, hammered steel forging supported in large solid bearings in the lower boom structure and carrying the saddle block.

The saddle block is a single alloy-steel casting with bronze-bushed hubs and has full-circumference segmental-type, renewable liners in either end which guide the dipper handle.

The saddle block is equipped with a device to retard handle rotation should slight slack develop in the hoist lines when starting a digging stroke under the boom and close to the crawlers.

A key on the lower side of the dipper handle engages a keyway in the liners and these tend to rotate with the handle. Liner segments at the forward end of the saddle block have gear teeth which engage with a pinion that provides resistance to handle rotation through a spring-loaded friction disc.

On the saddle-block hubs, large-diameter sheaves for the crowd and retract ropes are mounted, so that the ropes are actually outside of the saddle-block casting. These sheaves are steel castings with deep machined twin grooves, and run on large-diameter, bronze, flanged bushings that are secured to the saddle-block hubs.

Air-operated dipper trip

The saddle block provides a mounting for an air-actuated cylinder which operates the dipper latch by means of a rope.

The large-diameter point sheaves are steel castings with machined twin grooves and bronze-bushed hubs which turn on the boom-point pin.

Crowd machinery

The dipper handle is crowded out or retracted by a positive rope crowd, actuated by the crowd machinery located at the forward end of the revolving-frame casting. The crowd machinery is driven by an independent motor. The drum is a steel casting with machined grooves for the twin crowd and retract ropes and has bronze-bushed hubs. Rotation of the drum in one direction forces the handle and dipper out, and rotation in the opposite direction retracts it.

The ends of the crowd rope are wedge-anchored at the rear of the handle, on the stop casting. Prior to this, each line passes over a half-sheave segment positioned by a large-diameter screw, thus providing adjustment for the crowd and retract ropes. The bight of the crowd rope passes over a half-sheave at the middle of the crowd drum under the drum gear rim.

The ends of the retract ropes are wedged to opposite ends of the crowd drum with single-taper wedges, which eliminate bending over of rope ends. The bight of the retract rope passes over a large half-sheave on top of the handle at the dipper end of the handle. Horizontally mounted deflection sheaves for the retract lines are provided on the lower boom section.

Air-actuated slipping clutch

The motor drives the drum by means of a primary pinion in the armature shaft meshing with a gear on the first intermediate shaft and through two additional spur-gear reductions. The intermediate shafts run in anti-friction bearings. An air-actuated overload slipping clutch is provided on the first intermediate shaft.

All the shafts are mounted in a single, cast-steel frame, mounted on the revolving-frame casting, which is removable as an assembled unit.

Automatic brake

The motor is provided with a spring-set, air-released brake, which holds the dipper in a fixed position should power fail, or may be operated by a switch in the operator's cab.

Dipper handle

The dipper handle is a single large-diameter tube of heavy section. Structural members and a casting welded to the lower end form a connexion for the dipper body. The two large-diameter screws on the rear handle stop that provide adjustment for the rope can be reached from the cab roof. The handle stop arrangement is mounted on a large vertical pin, to provide equalization for the crowd lines. A fixed half-sheave is fitted on top of the handle at the dipper end for the bight of the retract rope and provides equalization of these lines.

A key is provided on the bottom of the handle tube which engages the keyways in the saddle-block liners, through which connexion is made to the handle-stabilizing device.

Dipper

The standard dipper is of welded construction. A single manganese-steel casting forms the front and the lip, and has integral sockets for the inserted-type, renewable teeth. The latch-bar keeper is cast integral with the dipper front. Annealed-steel castings, having provision for attachment of triple-pin connected twin-rope equalizers, are welded to the front and connected to the dipper back which is an integral part of the dipper handle. Connexions for the high-carbon steel, heat-treated hinge pins, are provided in lugs on the dipper back. These lugs are provided with hardened-steel bushings.

The door is an alloy-steel casting with the hinges and the latch-bar guides integrally cast.

Cab

The cab is constructed entirely of steel and is self-supporting without auxiliary framing. Removable panels are provided over the motor generator set, the swing machinery and to the sides and rear of the A-frame. The cab completely encloses all the machinery and the operator, and suitable doors and windows are provided. The operator's compartment is isolated from the main machinery.

A large motor-driven fan in the rear of the cab provides ventilation.

Ropes

One complete set consisting of hoist, crowd, retract, dipper trip ropes and suspension bridge strands is included. Each pair of bridge strands has a load-equalizing link at the A-frame head anchorage.

Operating levers

The master switches for operating the machine are located at the forward end of the revolving frame, to the right of the boom, giving the operator a clear, unobstructed view of the work. A panel with switches is located at the operator's position for the control of the auxiliary functions of the machine. An adjustable, padded seat is provided for the operator.

Ballast

Space is provided in the rear section of the revolving frame for the ballast which will be supplied by the purchaser. Close-lying scrap and pig iron or punchings or a combination of these materials is suitable for this purpose. Volume of ballast boxes in rear end 320 ft^3 (9·05 m^3).

Motor generator set

The machine is equipped with a motor generator set composed of an alternating-current motor driving generators which furnish direct-current power for the hoist, swing and the crowd motors. A separate motor-driven set is provided for excitation. The motor generator set is arranged for auto-transformer starting and may be stopped from the operator's position by a conveniently located switch, 'killing' all generator fields and applying motor brakes in an emergency.

D.c. motors

The hoist, swing, and crowd motors have separately excited field windings and are of a type suitable for high-peak duty. They are designed with small flywheel effect, have anti-friction bearings and class B insulation.

The swing motor is of the vertical type and the hoist and swing motors are force-ventilated by externally mounted blowers driven by separate squirrel-cage motors.

Contactor control

At the operator's position are master controllers operating on magnetic contactors which vary the resistance in the generator fields, so establishing the voltage of the generators and thereby the speed and direction of rotation of the various driving motors. The generators are designed to limit the current in each motor circuit to a value which will develop the maximum torque required.

Regenerative braking

The hoist motor, through gears and rope, is connected at all times to the dipper when the hoist is engaged. The motor delivers power to the machinery when required and automatically acts as a generator when overhauled by the descending dipper. This is the regenerative feature and is made practical by the use of separately excited motors; it gives complete control of the motion without the operation of mechanical clutches and brakes. In the same way the regenerative feature is utilized in the swing and crowd.

Transformer

The various auxiliary motors for exciter, generator, air compressor and main motor blowers are supplied at medium voltage by an auxiliary transformer which also provides a low-voltage supply for lighting and heating.

Collector rings

Large-diameter collector rings are attached underneath the revolving frame, taking current from heavy shoes located on the base. The rings are of ample capacity, simple design, well insulated and protected, and are accessible for inspection.

Dipper trip

The mechanical device for tripping the dipper is controlled by a magnet valve operated by a thumb latch switch on the crowd master controller.

Lighting and heating equipment

The lighting equipment consists of floodlights at the front of the machine to light the digging operations, together with internal house lights and inspection lamp sockets. Heating equipment is fitted in the operator's compartment.

Air compressor

An independent motor-driven air compressor provides air for the operation of clutches and brakes.

Trailing cable entry

A suitable high-voltage cable coupler is provided with a bell-mouth gland to suit the trailing cable.

Ruston-Bucyrus 110 RB electric dragline

The specification of the 110 RB electric dragline (Figs. 18 and 19) is similar in many of its item specifications to that of the 110 RB electric shovel. Given below are those items which are not common to the two machines. The machine is provided with a boom and bucket equipment to suit the operating conditions from the combinations available.

MAIN SPECIFICATIONS

	minimum			*maximum*	
Length of boom					
80′	24·40 m		120′	36·60 m	

Net weight, domestic, with bucket, without ballast, approximate

298,200 lb	$133\frac{1}{4}$ tons	135·100 kg	302,200 lb	135 tons	136·895 kg

Working weight, including ballast, approximate

363,200 lb	$162\frac{1}{4}$ tons	164·545 kg	367,200 lb	164 tons	166·340 kg

Ballast—furnished by purchaser

65,000 lb	29 tons	29·445 kg	65,000 lb	29 tons	29·445 kg

Shipping weight—prepared for export, no ballast, approximate

313,200 lb	$139\frac{3}{4}$ tons	141·900 kg	317,200 lb	$141\frac{1}{2}$ tons	143·695 kg

Ships' option tonnage

305 tons	340 tons

The above weights include buckets of the size and type shown in the table on p. 78.

Crawler mounting

Width of links—standard	3′ 0″	914 mm
Width of links—wide	3′ 6″	1·07 m
Overall width of mounting (3′ 0″ links)	16′ 3″	4·95 m
Overall width of mounting (3′ 6″ links)	16′ 9″	5·11 m
Overall length of mounting	19′ 4″	5·89 m
Total effective bearing area (3′ 0″ links)	100 ft²	9·30 m²
Total effective bearing area (3′ 6″ links)	117 ft²	10·87 m²
Diameter of idler rollers	39″	990 mm
Diameter of driving tumblers	$41\frac{3}{8}″$	1·05 m
Diameter of take-up tumblers	41″	1·04 m
Number and pitch of links	70—$14\frac{9}{16}″$	370 mm
Propelling speed—normal	70·40 fpm	21·45 m

MAIN SPECIFICATIONS—*continued*

Turntable

Pitch diameter of roller track	9′ 10″	3·00 m
Number and diameter of rollers	40—7½″	190 mm
Diameter of swing rack	10′ 11″	3·33 m

Drums—Sheaves—Ropes

Pitch diameter of hoist drum	26″	660 mm
Pitch diameter of boom-point sheaves	24″	610 mm
Pitch diameter of padlock sheave	20″	508 mm
Diameter of hoist rope (2 or 3 part)	⅞″	22 mm
Diameter of boom hoist rope—8 part	1″	25 mm
Diameter of bridge-strand suspension—used with mast— 2 strands	1⅝″	41 mm

Electrical equipment (Ward-Leonard)

Hoist motor (blown)	.125 hp	230 V, 75°C, cont.
Drag motor (blown)	.125 hp	230 V, 75°C, cont.
Swing motor (blown)	. 44 hp	230 V, 75°C, cont.

Generators for Ward-Leonard control are equivalent in capacity to their respective motors and are driven by a suitable induction motor.

WORKING DIMENSIONS

K	Clearance radius of revolving frame	18′ 3″	5·56 m
L	Clearance under frame to ground level	5′ 3″	1·60 m
M	Clearance height with boom and A-frame lowered	16′ 9″	5·11 m
M1	Clearance height of A-frame	24′ 0″	7·32 m
M2	Height over stairway	20′ 0″	6·10 m
N	Height of boom foot above ground level	8′ 5″	2·57 m
P	Distance from boom foot to centre of rotation	7′ 4″	2·24 m
T	Distance from boom-point pin to pin for attaching bucket	5′ 6″	1·68 m
U	Clearance under propelling-gear case to ground level	1′ 3½″	394 mm
V	Width of superstructure with platform and stairway removed	17′ 9″	5·41 m
W	Overall width of superstructure	20′ 10½″	6·36 m
F	Throw of bucket—depends upon the ability of the operator		

BAX buckets

Capacity		Weight		Dimension 'R'		
yd³	m³	lb	kg	ft	in	m
5½	4·20	9800	4445	20	6	6·25
5	3·82	9200	4175	20	6	6·25
4½	3·44	7700	3495	18	6	5·64
4	3·06	7000	3175	18	4	5·59
3½	2·67	6400	2905	17	6	5·33
3	2·29	5700	2585	17	3	5·26
2½	1·91	4600	2085	16	1	4·90

Single drag rope

The drag machinery is designed to handle the bucket on a single rope attached to the bale on the bucket and to the drum. All of the machinery

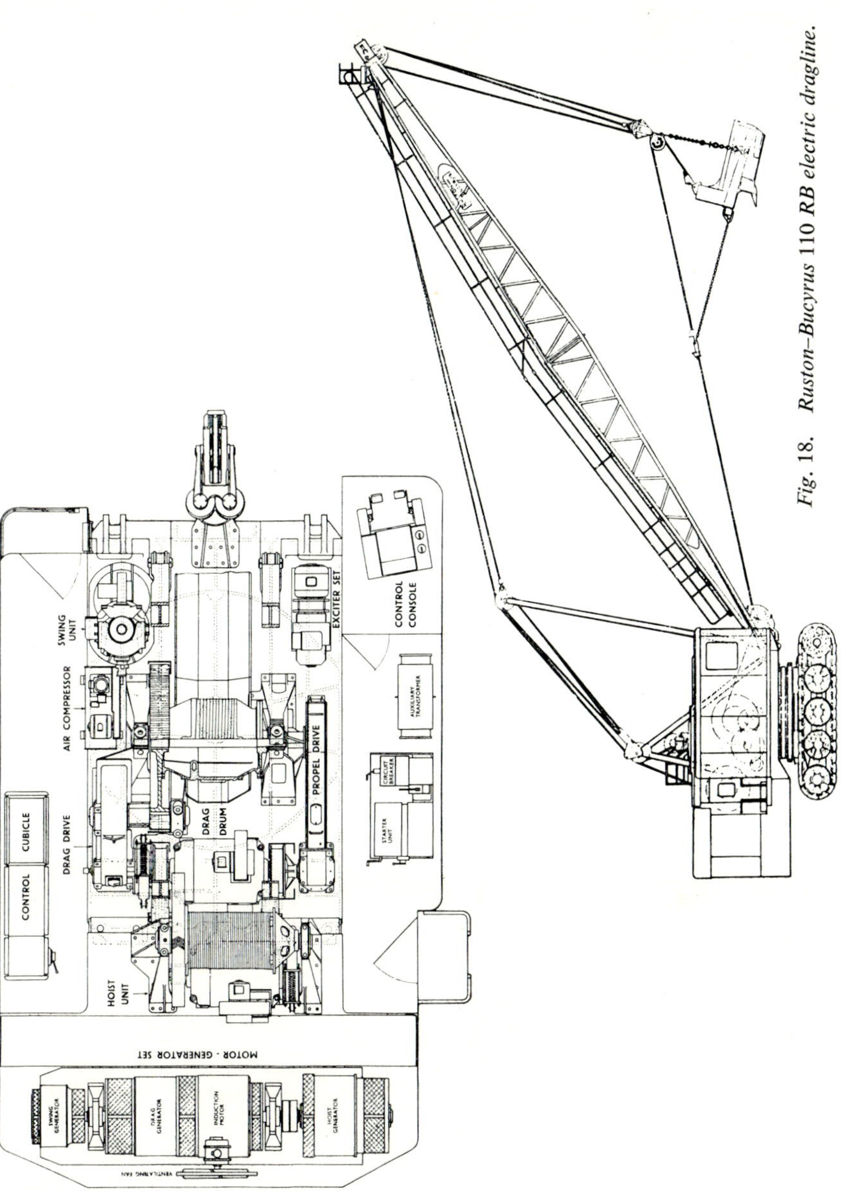

Fig. 18. *Ruston–Bucyrus 110 RB electric dragline.*

is engaged during the entire cycle, thus placing the bucket at all times under the control of the drag controller lever.

Independent drag motor

The motor is connected to the primary pinion shaft by a smoothly engaging, electrically controlled, air-actuated clutch, which also serves as a flexible coupling and slipping clutch for overloads. The bearings for this shaft are of the anti-friction type, and are carried in an oil-tight gear case integral with the left-hand side frame. The primary pinion, integral with its

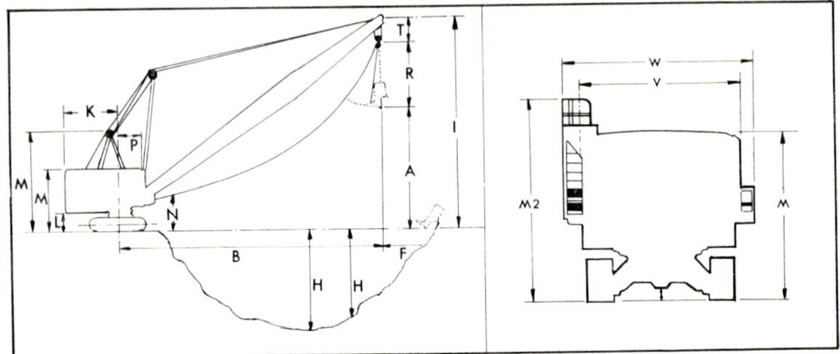

Fig. 19.   *Specifications of* 110*RB electric dragline. The lettering is identified in the text on p.* 78.

shaft, drives the intermediate gear, keyed to its shaft, which turns in anti-friction bearings in the left-hand side frame. The drag pinion is integral with the intermediate shaft and drives the drag gear, keyed to the drag shaft, which runs in renewable, babbit-lined bronze shells in the cast-steel right- and left-hand side frames mounted on the revolving-frame casting.

The drum is a split lagging with machine-turned grooves for the drag rope and is securely bolted to the drag gear and a flanged hub at the right-hand end of the shaft. When converting to shovel, the drag lagging is replaced with a shovel hoist lagging.

Brake for power failure of drag motor

The drag brake, mounted on the primary pinion shaft, is air actuated and electrically controlled and is used to hold the drag drum in a fixed position when desired and sets in case of failure of power supply.

All shafts are hammered-steel forgings, all pinions are alloy-steel forgings, and all gears are alloy-steel castings. All gears and pinions have machine-cut teeth. The first gear reduction has double-helical teeth. The drag rope leads from the drum, through the fairlead, to the bucket.

Hoisting machinery

The hoisting machinery is designed to handle the bucket on a two- or three-part hoist depending on the weight of the loaded bucket.

Independent hoist motor

A pinion on the extension of the motor armature shaft drives a cast alloy-steel gear bolted to the semi-steel hoist drum and keyed to its hammered steel forging shaft. This shaft runs in anti-friction bearings in the cast-steel side frames mounted on planed surfaces on the welded rear end of the revolving frame. The hoist reduction gears, which have double-helical teeth, are suitably guarded.

Brake for power failure of hoist motor

The brake, which is mounted on the motor armature shaft opposite the pinion is spring-set, air-released and electrically controlled and is used to hold the hoist drum in a fixed position when desired and sets in case of failure of the power supply. The hoist rope leads from the drum, over a sheave at the A-frame apex, over a sheave at the top of the mast when fitted, and over a sheave or sheaves at the boom point, to the padlock to form a two- or three-part hoist.

Fairlead

The universal fairlead is mounted at the front of the revolving frame. It has two horizontal, cast-steel sheaves, keyed to pins turning in bronze bushings in a cast-steel pivoted frame, adjustable for different sizes of drum, and provides a direct lead from the fairlead itself to the drum. Two vertical cast-steel sheaves, keyed to pins turning in bronze bushings, are mounted in a cast-steel swivelling frame which swings in line with the drag rope regardless of the position of the bucket.

The frame is fitted with renewable cast-steel guards for leading the rope into the sheaves.

Boom

The latticed and cross-braced boom is of welded construction and has structural steel chord members and lacing. It is designed to combine light weight with required strength and is constructed to permit lengthening or shortening by the addition or omission of boom sections. The lower end has a widespread connexion to the revolving-frame casting.

The cast-steel sheaves at the boom point are equipped with anti-friction bearings mounted in a universal swivel frame to provide proper lead for the ropes. The cast-steel sheave in the padlock is provided with sealed, babbit-lined bronze bushings.

Boom suspension
The boom is raised and suspended by a multiple-part rope suspension system operated by a motor-driven, worm-geared drum. The compact boom-hoist unit is mounted on a structural support on the roof in the truss formed by the upper part of the A-frame.

Motor-driven boom hoist
The motor is reversible and locally operated. For the shortest length of boom the suspension rope is reeved between sheaves at the apex of the A-frame and at the boom point.

Mast suspension for long booms
Longer booms are supported by an auxiliary mast mounted at the foot of the boom and having bridge-strand guys between the upper end and the boom-point pin. The suspension rope is reeved between the A-frame apex and sheaves at the top of the mast.

Drag bucket
A Ruston-Bucyrus drag bucket of the size and type specified will be supplied. The bucket is constructed to provide good digging and filling qualities. Hoisting and drag chains and a dump rope and sheave are supplied with the bucket.

Operating levers
The master switches for operating the machine are located at the forward end of the revolving frame to the right of the boom, giving the operator a clear unobstructed view of the work. A panel with switches is located at the operator's position for the control of the auxiliary functions of the machine. An adjustable, padded seat is provided for the operator.

Cab
The cab is constructed entirely of steel and is self-supporting without auxiliary framing. Removable panels are provided over the motor generator set, the swing machinery and to the sides and rear of the A-frame. The cab completely encloses all the machinery and the operator, and suitable doors and windows are provided. A large motor-driven fan in the rear of the cab provides ventilation.

Ropes
One complete set of ropes for hoist, drag and boom suspension is included. Bridge-strand guys are included when a mast is supplied.

Ruston-Bucyrus 110 RB diesel-electric shovel or dragline

Where a suitable electricity supply is not available in undeveloped, difficult terrain demanding maximum mobility the capacity and characteristics of the electrically driven 110 RB shovels and draglines are available

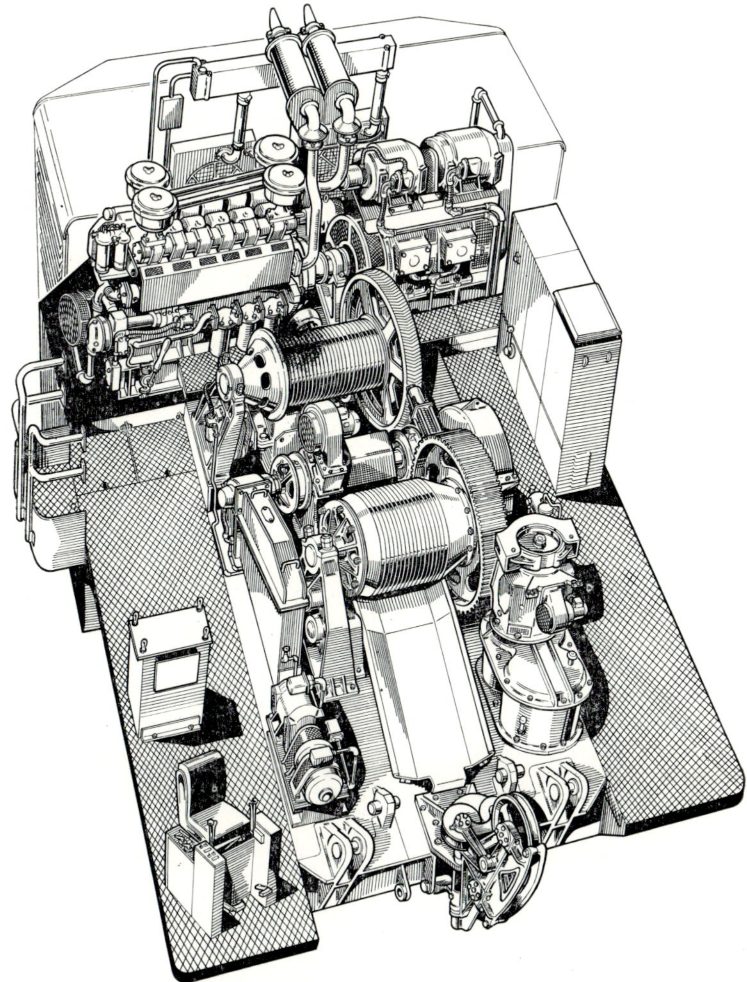

Fig. 20. *Diesel-electric drive for shovel or dragline.*

in the diesel-electric models. The induction motor and generator of the electric model is replaced by a self-contained power unit consisting of a twelve cylinder vee-form compression-ignition diesel engine of 528 hp and a generator assembly which supplies to the dc driving motors (Fig. 20). Engine cooling is by water circulated through large capacity radiators arranged to operate in parallel, having shutters to permit adjustment of the cooling to suit ambient operating conditions. Sections are also incorporated into the radiators, the engine lubricating oil, air, lubricating oil and fuel oil are filtered, and pressure and temperature gauges are fitted on the engine and alarm equipment gives the driver visible and audible warning of high water temperatures or low lubricating oil pressure. The fuel and lubricating oil systems each have a primary pump to ensure an immediate supply when starting.

The three generators are equivalent in capacity to their respective motors and the alternator supplies power for the ac auxiliary motors and through a transformer for the main lighting and heating equipment.

The generators and alternator are installed in a compact two-tier assembly. The fuel tank provided has a capacity sufficient for 24 hours' operation. Highly rated batteries are provided for stationary and emergency lighting and a battery powered fuel–oil pump for easy filling of the tank.

Marion Type 8700 walking dragline

In the giant class of draglines is the Type 8700 of the Marion Co. (Fig. 21); which is of the walking type. The machine is equipped with an 85 yd^3 bucket and is used to uncover and remove coal from the famous Mammoth Vein for the Jeddo-Highland Coal Co. of Hazelton, Pennsylvania.

The boom is 300 ft in length and the total working weight is $9\frac{1}{2}$ million pounds (4241 tons). It can dig to a depth of 200 ft below ground level and spoil overburden to a height of 148 ft. The machine is powered by eighteen dc motors totalling 9750 hp with six hoist motors each of 625 hp, four drag motors of 625 hp each, four swing motors of 500 hp each and four propelling motors of 375 hp. The working cycle of digging, swinging dumping and swinging back occupies less than 60 seconds.

As the term walking dragline implies, the machine is propelled by doing a walking step. The pair of walking shoes are 55 ft long and 13 ft 6 in wide and the machine takes an 8 ft 9 in step. When the dragline is in operation it rests on a large circular base 65 ft in diameter. This base or 'tub' provides a bearing area of 3318 ft^2 for the weight of the rotating frame and the boom.

The operation of the dragline is controlled by one man from a station located at the front of the upper frame and permits comfort and maximum visibility.

Figure 22 shows a Marion dragline bucket of 130 yd^3 or 195 tons of overburden capacity. It is for the Type 8900 Marion dragline for the Thiess-Peabody-Mitsuz Coal Pty. Ltd of Moura, Australia. The boom will be 275 ft long.

Fig. 21.   Marion type 8700 walking dragline removing overburden at an opencast coal site. Power shovel loading the bared coal.

Bucket wheel excavators

Bucket wheel excavators are finding increasingly a place in surface mining where large tonnages of relatively soft minerals in thick beds have to be worked or disposed of as overburden. They have proved particularly suited to opencast coal, including German brown coal and the Demerara bauxite deposits in Guyana and are used to mine to a height of over 130 ft and a width of 75 ft. Hourly quantities of 200 to 10,000 tons can be handled and it can also be used for stacking and reclaiming of bulk materials.

A range of small bucket wheel excavators are being built under license from Weserhutte Otto Wolff of Germany by Strachan & Henshaw Ltd (Table III). The range of models are capable of handling material from loose gravel to soft consolidated rock at a rate of 100 to 2500 yd^3 per hour, at a cost that can be as low as 1d per yd^3.

Fig. 22.   Marion dragline bucket of 130 yd^3 (195 tons) capacity.

The digging head comprises two rotating flanges between which are mounted eight buckets, and as these rotate the head is impelled forwards or downwards into the ground causing the buckets to be filled. Between the revolving flanges at the forward part of the wheel a blanking plate serves to retain the material which has been dug in the buckets and as these approach a horizontal position the plate is cut short and is replaced by an inclined chute. The material falls down the chute and passes to the conveyor sited in the forward boom of the excavator.

Guide boards on either side of the conveyor permit occasional overloading as well as preventing fine materials from being scattered by winds. Impact idlers are fitted on the conveyor belt below the wheel chute loading point and on the delivery conveyor at the transfer point. The tail boom which carries the delivery conveyor can be slewed through a 180° arc relative to the bucket-wheel boom and can be raised or lowered for loading into vehicles, onto conveyors or stockpiles.

TABLE III

DETAILS OF WESERHUTTE–STRACHAN & HENSHAW
BUCKET WHEEL EXCAVATORS

Model	*AR45*	*AR125*	*AR220*	*AR380*	*AR600*
Discharge, min	32·80	32·76	32·72	32·68	32·64
Bucket capacity ft^3	1·59	4·25	7·77	13·25	23·0
Material output					
heavy material	130	325	585	910	1,640
light material yd^3	195	520	910	1,560	2,660
Tangential bucket force in					
tons at $K = 30$	2·6	3·5	4·5	5·3	6·5
Installed motor power, kW	56	106	170	245	340

A hydraulic system is provided for the lowering and raising of the booms, simplifying the operation of the machine and permitting a low compact construction with no superstructure for mounting, hoisting ropes and winch units required.

The excavators are self-propelled through crawler tracks which have independent electric motor drives and excavators of special design are available both crawler and rail mounted.

LOADING AND TRANSPORT OF OVERBURDEN

The most convenient method of loading overburden when the waste must be transported, when the cut-and-fill method of disposal is not appropriate, is the close-coupled shovel particularly for hard rock overburden. Draglines can and do load into vehicles or hoppers delivering to belt conveyors but the free-swinging bucket and movement of the bucket away from the machine when dumping, makes 'spotting' the bucket over the truck or hopper difficult and spillage is likely to be increased.

Waste may be transported from the excavator to the dump by rail, belt conveyor or by rubber-tyred trucks.

Rail

Although rail haulage was used extensively in the past it cannot now, in many cases, compete economically with the other methods available and no new installations are to be expected. The initial capital cost of track, rolling stock and locomotives and the high cost of maintenance of track, which needs frequent relaying in new positions in following the working face are major considerations.

Belt conveying

If the tonnage of material to be dealt with is sufficient to justify the high capital outlay involved, the belt conveyor is probably the cheapest method of transport over short distances. Hard rock overburden needs to be broken small so that it does not damage the belt when loading and the material should not be wet and sticky so that it adheres to the belt at transfer and discharge points. The rate of face advance may be decisive as some overburden is suitable for conveying when freshly wrought but weathers quickly to a sticky clay and there are seasonal effects.

To prevent damage to belts by oversize material, portable grizzlies may be necessary and portable crushers to deal with the oversize may be installed; the impact or hammer mill type of crushers are most suitable for this purpose because of their moderate weight and ability to handle large reduction ratios in a single stage.

Trucks

The most flexible method of transport suitable to most conditions is that by rubber-tyred trucks. Two types are available: those conforming with statutory highway requirements, particularly as to width and axle loadings, known as 'on highway' trucks and necessary where a portion of the journey to the dumping site is over a public highway; and those which do not conform with the legal requirements and are suitable only for sites with access from the quarry or open pit, not involving public highways and known as 'off-highway' trucks. Some, however, come within the legal dimensional and weight limits for highway running but the intrinsic load-carrying capacity must be reduced when running over the highway so that maximum permitted axle loadings are not exceeded.

The capacity of the dump truck required is related to the size of the excavator. The most economic size for average haul distances is such that the trucks are not standing waiting to be filled longer than three or four complete cycles of the loading machine and that part-filled bucket loads are not required should the excavator ever be waiting for an empty truck. The loading capacity of the excavator having been decided on the daily output required, the truck size for most economic working should not exceed four times the bucket capacity. The round journey to the dumping site and back, including loading and dumping time, will determine the number of trucks required. For long hauls the cost of transport may outweigh the cost of loading and larger capacity trucks may be economic. Trucks up to 240 tons capacity have been used for opencast coal haulage in Ohio. It has a 1000 hp engine mounted at each end and an electric drive with a loaded weight to power ratio of 380 lb. It is 96 ft long, 15 ft wide and 15 ft 7 in high and has a turning radius of $42\frac{1}{2}$ ft. The brakes are multi-disc oil-cooled, road irregularities are compensated by eight nitrogen-over-oil suspension cylinders, and dual controls are provided in the steering system.

The rear dump truck is generally the best when the dumping point is over the edge of a bank and advances as the bank is built up. For fixed dump stations and small material side and bottom dump types are most suitable.

For steep gradients, soft or slippery surfaces and poor haulage roads, four-wheel drive (4×4) or six-wheel (4×6) or (6×6) may be required but if possible gradients should be reduced and road surfaces improved so that rear-wheel-drive four-wheel trucks (2×4) may be used with a reduction of capital, and running maintenance costs. Where conditions of surface and gradient are good semi-trailers up to 100 tons and more capacity can be used. If gradients are less than 1 in 10 they can be negotiated by rear-wheel-drive (4×4) trucks under most conditions.

The electric wheel drive has an electric motor and reduction gear built into the centre of a rubber-tyred wheel and power units of one or more diesel engines driving through mechanical transmission systems with or without hydraulic torque converters. Electric and hydraulic transmissions with gas-turbine units are in use.

Speeds up to 35 mph are adopted as at N'Changa copper open pit but good road construction and surface is essential.

LOADING AND TRANSPORT OF MINERAL

The nature of the deposit, from which the overburden has been removed, will govern the succeeding operations. The fragmentation of the useful material will depend upon the drilling and blasting technique adopted if the use of explosive is required to excavate the material and reduce it to a condition for easy loading in the rock pile so produced which should be as uniform as possible.

The capacity of the loading and transport equipment required, neglecting ancillary equipment such as bull- and calf-dozers, caterpillars, scrapers, rippers and tractor shovels depends either on the actual production required from these operations or where processing, such as crushing, is required the capacity of the processing equipment (e.g. a crusher) will affect not only the capacity of the face loading equipment, and thus their number and capacities of the individual machines and vehicles, but also the number of loading points in operation at any one time The largest piece of rock which the bucket or dipper can handle should also be the maximum size which can be conveniently passed through the crusher. The crusher also affects the capacity required of the transport unit and its type since it is essential that the load can be fed smoothly and quickly into the crusher without delay.

Since the most economical method of transporting large quantities of material of the correct size from the quarry or open pit is belt conveying, fore-crushing of the material at the face to a size suitable for conveyor transport is being increasingly adopted. This of course requires a mobile crusher preferably installed as near as possible to the rock pile so that only conveyor transport is required.

Mobile primary crushers

Although the use of field conveyors for the transport of material from the working face to the processing plant has become standard practice in the sand and gravel industry, they have not yet found favour in the hard rock quarrying industries because the larger size of the rock handled at the quarry face is usually unsuitable for direct loading on to a normal sized conveyor.

For some years the quarrying industry in Germany has been developing the use of mobile primary crushers in conjunction with a fixed secondary crusher and subsequent stages. The mobile plant consists of the primary crusher, usually of the impact breaker type, and a feeder which is loaded directly by the face excavator and the product is of a suitable size to be taken by conveyor to the fixed processing plant.

A mobile completely self-contained crusher unit by GEC (Process Engineering) Ltd is carried by a standard 33 ft long Merriworth trailer (Fig. 23) adapted to accommodate a crusher and a fully equipped laboratory where crushing tests conducted on the site can be quickly evaluated and probable throughput rates and power consumption can also be assessed.

Any machine from the range of reversible impactors and single and double-rotor hammer mills can be mounted on the trailer which incorporates a 118 KVA diesel-alternator to drive the crusher motor.

It is fitted with triple air brakes and performs to Ministry of Transport and Continental requirements.

An advantage claimed for the mobile crushing plant is that a larger proportion of the installation runs as a continuous process. When an excavator, dump trucks and a static plant is installed continuous operation does not start until the feeding of the primary crusher is continuous. With a mobile crusher, however, the only unit that is not continuous is the excavator whose operation is dictated by the requirements of the plant and not by the availability of dump trucks. This, in many cases, results in increasing excavator output by as much as 60% in some cases, with a corresponding reduction in loading costs.

An 80 ton per hour closed circuit mobile crushing and screening plant known as the Screen Loader 1131 by Frederick Parker Ltd of Leicester will screen aggregate into four sizes and feed direct to lorries or to a stockpile as required. The plant is provided with three conveyors to feed lorries and these are folded when the unit is on tow. Screening is by a 48 in by 120 in Rapide four-deck vibrating screen the oversize passing to

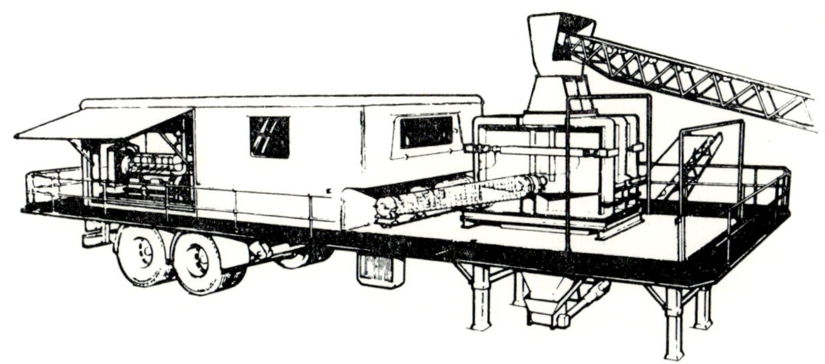

Fig. 23. Mobile self-contained crusher unit (GEC).

Stonesizer 36 in by 6 in granulators for crushing; these giving high outputs of well-shaped stone. The granulators are powered by a 105 hp Dorman diesel engine built into the plant the other items are driven by individual electric motors. Travelling height is 13 ft 6 in, length 43 ft and minimum width 8 ft 6 in and dual air brakes are fitted.

Interrelation of loading and transport facilities

Interdependence between loading and transport arrangements are essential to economic handling of the deposit between the face and the processing plant or the consumer.

The first item to be decided is the capacity of the bucket or dipper and the transport to achieve the required production with the particular loading system adopted. The main alternatives are: (a) the trucks back up to the shovel on one side of it so that there are pauses in the loading by the loader as the trucks manoeuvre to change places; (b) where the trucks drive up or back up to the shovel on both sides of it or two-way steering is provided so that there are no pauses in loading as the trucks change over.

It is also necessary to decide how many trucks are required with regards to the length of the transport run, go and return to loading, truck changing and tipping times.

In the formulae which follow it is assumed that the loading and transport is in constant use but allowance must be made for interruptions occurring in the quarry or open pit or at the processing plant due to breakdown or other emergency. This will depend on the particular circumstances of each installation. Average conditions in this respect have been assumed in the formulae.

Shovel dipper and truck capacity required for target production
   The hourly output Q in tons at a loading point is

$$Q = T \times N \times \frac{C}{R}$$

Also
$$N = \frac{R}{L_t + t_b}$$

and
$$L_t = \frac{T}{D} \times D_c$$

where T = truck capacity in tons
   N = number of trucks employed
   C = conversion factor, 60 (min/hr)
   R = time for transport cycle (min)
   C/R = number of transport cycles/hr
   L_t = time for loading truck (min)
   t_b = time for backing up truck to shovel (min)
   D = dipper load (contents) tons
   D_c = time required by dipper for performing loading cycle.

By combining these, the following are obtained:

$$Q = \frac{T \times C}{\dfrac{T \times D_c}{D} + t_b}$$

and
$$D = \frac{T \times D_c}{\dfrac{C \times T}{Q} - t_b}$$

   To calculate the dipper capacity D_f from the dipper load D this must be divided by a factor $F = 0\cdot8 \times 1\cdot6 \text{ tons/m}^3 = 1\cdot28 \text{ tons/m}^3$ (efficiency factor for dipper filling $= 80\%$; one cubic metre of rock pile $= 1\cdot6$ tons)

$$D_f = \frac{T \times D_c}{\left(\dfrac{C \times T}{Q} - t_b\right) \times F}$$

If the truck changing time t_b be dropped then

$$D_f = \frac{T \times Q}{C \times F}$$

To calculate the number of trucks required for a predetermined output; the transport system must deal with the hourly output Q,

$$Q = T \times N \times \frac{C}{R}$$

In addition $R = t_c + L_t + t_b + t_{cr}$

$$t_c = \frac{2d}{v}$$

$$L_t = \frac{T}{D} \times D_c$$

where t_c = time for travelling to and returning from crusher (min)

t_{cr} = time spent at crusher (min)

d = distance from loading point to crusher (km)

v = average speed of truck travelling empty and laden (km/hr).

Combining the above:

$$Q = \frac{T \times C \times N}{\dfrac{2d}{v} + \dfrac{T \times D_c}{D} + T_b + t_{cr}}$$

$$N = \frac{Q\left(\dfrac{2d}{v} + \dfrac{T \times D_c}{D} + t_b + t_{cr}\right)}{T \times C}$$

If the trucks back up to the shovel on one side of it only so that there are pauses on loading then

$$D_f = \frac{T \times D_c}{\left(\dfrac{C \times T}{Q} - t_b\right) \times F}$$

If there is no lost time the equation is

$$D_f = \frac{Q \times D_c}{C \times F}$$

Dump trucks

The primary requisites in a modern dump truck for surface mining work are simplicity of control as trucks working overseas are often in the control of relatively unskilled native operators. Simplicity of design is also important so that trucks may be understood and serviced by plant mechanics. Comfort and ease of control reduce driver fatigue, increase productivity and reduce the risk of accidents when working in confined areas at high speeds. Body heating by engine exhaust gas was first introduced in cold climates to prevent freezing of payload on long hauls but is now used to prevent wet sticky material building up in the front of the body. Aluminium bodies are used to reduce tare weight and increase payload.

Reliability and ease of maintenance is essential to withstand continuous rugged work under arduous site conditions.

Power-to-weight ratio is of the order of 7 to 8 hp per ton of gross weight and with the use of high-tensile steels and more compact high-powered diesel engines robust 'off-highway' trucks with a ratio of tare weight to payload of 70 to 80% are available with even lower chassis weights in the ultra-large dump trucks.

While body capacity should be three to four shovel dippers for short hauls this can be increased to five or six for long hauls.

Normally aspirated engines are used on small and medium sized trucks but turbo-charged engines are used to keep down engine size on large trucks with aftercooling of such engines to increase power and engine life. A flat torque curve is required for torque converter applications since the torque absorption is a function of the square of the speed.

Hydrostatic transmission gives a variable stepless hydraulic drive from maximum tractive effort to maximum road speed.

Positive and reliable service brakes are necessary with heavy payloads down steep gradients and either assistance or full power is applied to the wheel brakes. The parking brake is between the gearbox and the rear axle taking advantage of the axle reduction to give a boost equal to the gear ratio. Hydraulic retarders are available for controlling the speed of trucks on long, steep down gradients.

Tipping gear is provided with double-acting hydraulic rams working at 1500 lb/in^2. Power-assisted steering is essential with front axle loads greater than $3\frac{1}{2}$ tons and on the larger trucks provides practically the entire steering effort. Shuttle trucks working in restricted areas may have two-way steering allowing the driver to face forward or reverse with the steering and all controls acting in the correct sense.

The overall height of the truck should be such that it makes a good target for the shovel dipper. Length should be a minimum and a short wheel base with ability to turn round in cramped situations in a minimum of time and get under the dipper also in a minimum of time prevents congestion and improves productivity. Tachographic recorders mounted in the cab record engine speed and stoppages and total engine running time. Safety devices provided include low air pressure and oil pressure or engine temperature too high, when an automatic cut-off stops the engine.

In the USA a comparison of running costs between 90-ton trucks powered by V12 turbo-charged diesel engines with 40 and 50 ton units showed a saving of 47%. Driving costs are halved and maintenance and fuel costs per ton-mile are reduced but the capacity of trucks should be related to size of the shovel dipper, from three to six times the capacity of the dipper depending on the length of the haul.

The major requirements of tyres for 'off-highway' dump trucks include strength, wearing properties, toughness, traction and heat resistance and in some respects some of these are incompatible.

Heavy construction is needed to provide strength but this makes the dissipation of heat a problem. Selection therefore involves size and ply rating suited to the load and then the best design for the conditions, particularly tread depth. Although a deep tread offers maximum resistance to cuts, hauls should not exceed 2 miles each way and speeds should be below 30 mph. A shallow tread is suitable for hauls over ten miles when heat is liable to be a problem. Steel cord $29 \cdot 5 \times 29$ and $29 \cdot 5 \times 35$ truck tyres have been used on 65 ton coal trucks in Illinois with thin walls and shallow treads to dissipate heat quickly in high ambient temperatures. In some limestone quarries medium-pressure tyres (3 to $4\frac{1}{2}$ atmospheres) have been found to give best results.

To obtain maximum tyre life maintenance should include: systematic tyre pressure testing and inflation, valve caps on all valve systems, systematic tyre inspection to detect trouble or the need for retreading, analysis of the causes of tyre wear or damage, accurate records of tyre life and training of maintenance men and drivers in care of tyres.

Diesel-electric trucks
Electric drives are available in the larger dump trucks of 6 tons payload and above.

Gas turbine-electric trucks
Large turbo-electric 100 ton capacity 'off-highway' dump trucks of the M100 Lectra-Hauls manufactured by Unit Dig and Equipment Co. have been operating successfully at Kennecotts China Division powered by a 1200 hp LM100 General Electric gas turbine and the Berkeley open pit of the Anaconda Co. with a 1100 hp Solar International Harvester gas turbine. The use of the gas turbine as the power unit is a logical advance on the success of electrical transmission and the motorized wheel since the turbine offers simplicity, reliability and reduced weight for a given power output.

Caterpillar tractors

These machines are constructed from three basic assemblies, two truck frame assemblies, the engine and the transmission. The arrangements vary depending on the particular design adopted by the different manufacturers. Generally a truck assembly consists of a truck frame resting on track rollers and at the rear is a large sprocket wheel driving the track and at the front the idler around which the track travels. A number of supporting rollers are provided to prevent the track sagging and the idler is adjustable to tighten the track.

A dead axle running through the final drive casing ties the two tracks together and the sprockets are mounted so that the tracks can each move independently in a vertical plane at the front of the machine and a diagonal bracing extends backwards from midway along the tracks to a hinge at the back of the machine.

Rigidly connected and supported by side beams running forward from the transmission housing is the engine clutch housing, transmission and steering clutch housing. Driver comfort results from independent track oscillation and this contributes to accuracy in dozer work but where high loads are carried as when front loading attachments are used oscillation between tracks and the engine assembly may be unsafe, and a rigid bar may be used to tie the assemblies together.

Diesel engines are well suited to the work of the larger units because of their economy and ruggedness and their thermal efficiency is twice that of petrol engines which are, however, used in the smaller models because of their high output to size and weight ratio. Diesel running costs for fuel are one-quarter to one-sixth those of a comparable petrol engine since diesel fuel is much cheaper than petrol. Both two- and four-cycle diesel engines are used and each has its particular advantages. The four-stroke cycle has more complete combustion and a lower fuel consumption for a given power but a less even torque throughout the cycle and less carbonization of piston rings and exhaust passages. The disadvantages are that the variable crank-pin effort requires a heavier flywheel, the engine is heavier per rated hp and exhaust and inlet valves and their motions are necessary.

The advantages of the two-stroke engine on the other hand, are no exhaust valves, inlet valves or valve motion and the power developed per unit volume of piston displacement is greater and there is a more uniform torque requiring a lighter flywheel, but combustion is inferior giving a smoky exhaust and better lubrication and more cooling water is required and also increased fuel consumption.

The transmission is generally of one of three types—the fluid coupling or clutch, the torque converter and the variable-speed coupling.

Most tractors are steered by declutching the drive to the inside track and if necessary, braking it so that the power is applied to the outside track only, no differential being provided.

Hydraulic operation of the steering clutches is usual except on small machines, and planetary gearing may be used instead of clutches. The sprocket driving wheels usually have an odd number of teeth and the tracks an even number to distribute wear evenly. Three common types of track are used: full grousers with a single high cleat running across to give good traction and minimum slipping and skidding; semi-grousers which have one to three low cleats running across them; and flat shoes when machines carry rather than push and when the ground must not be cut up. Sideslip can occur with these. Conversion shoes can be bolted on to convert full grousers to flat shoes when running on roads and vice versa.

Tractors are used in conjunction with bulldozing and angledozing blades, scrapers and rippers and the present trend is to prefer the four-wheel-drive rubber-tyred tractors with their higher travelling speeds; caterpillar tractors for these jobs are still used extensively particularly in harder strata.

Bull- and angledozers are used for cleaning up work in conjunction with excavators, digging softer strata, grading and short-range transport and the tractor is fitted with the appropriate sized blade weighing one to six tons which is raised or lowered by hydraulic or cable control. The dozer blade normally has a straight cutting edge with a curved mould board but special shaped blades are used for particular jobs. Back ripper teeth may be fitted which dig in when the dozer is reversed but float when moving forward. Blades on cable-controlled machines may have their cutting edge set forward to provide a sharp cutting angle giving good penetration from blade weight alone, while with hydraulic operation downward pressure can be applied and less forward set is required. Angledozer blades are mounted on a C-frame so that they can be angled without fouling the tracks.

When bulldozing, the pushing run is usually made in bottom gear and limited to some 400 ft and the use of a scraper should be considered for hauls exceeding about 100 ft depending on conditions. The push-arms mounting the blades are mounted outside and pivot on the track frames and the blades on hydraulically operated dozers are raised or lowered positively by double-acting rams at the rear of the tractor acting through a crank, the hydraulic pump being powered from the front power take-off and the blade can be lifted, forced down or floated under its own weight or locked at any point by the hydraulic control valves. Alternatively the rams may be mounted at the front of the tractor on the central assembly and the tractor springs then cushion the movement of the blades and rams. The front of the tractor is raised considerably before the tracks are raised by the tractor springs when the blade is pushed down and the front of the tractor is pulled down when the blade is lifted thus reducing the blade lift. The track frames are relieved of twisting strains by these characteristics, shocks are cushioned, the blade is kept level in uneven ground and the tracks are kept on the ground when downward pressure is exerted.

Another design has the blade attached to the front of the tractor frame and the central assembly mounted on hydraulic rams instead of springs. The advantages of hydraulic control include the facts that the blade is always under positive control, there are no fast-wearing parts, the blade can be used to help free a bogged machine and the operation is smoother. On the other hand, regular inspection of glands and the hydraulic system is essential, operating speeds are lower and are lowered further by low temperatures.

When cable operation is adopted the blade is raised by a cable reeved through a sheave mounted at the front of the tractor and the control unit is driven from either the front power take-off point or the rear take-off point, the former is generally used for the blade only but the latter may have two or three drums and may be used for scraper work and other work. The blade is usually more quickly raised by cable than by hydraulic control.

Rippers, hydraulically or cable controlled, suitable for towing behind a tractor are used to dig up hardpan, pavements and soft rock. The hydraulic type use double-acting hydraulic cylinders to control the ripper and down pressure can be applied to the teeth if required. The maximum depth of ripping is commonly 12 to 24 in and in the USA rippers are being used in opencast coal as the production tool in seams up to 6 ft in thickness, making three cuts.

The cable-operated type consists of a beam mounting at one end a tooth or a number of teeth held in a cross frame. The beam can be lifted directly from the tractor by the cable or supported on wheels which are mounted on a triangular frame and by pivoting one corner of the triangle on the beam, attaching the wheels to the second and the cable to the third corner, the wheels can be forced down by the pull on the cable to lift the teeth clear of the ground and control the ripping depth.

Shovel loaders

Shovel loaders are employed in surface mining as primary loaders and also as ancillary plant for clearing up after excavators such as draglines and power shovels.

A typical modern appliance of this class is the $1\frac{1}{4}$ yd^3 capacity Type 125 crawler loader of the International Harvester Co. This incorporates double-reduction planetary, final drives, a new type of heavy-duty mounting frame, wide track gauge and an advanced control layout.

The engine is a 60 hp BD 281, four-cylinder direct-ignition diesel engine which develops its maximum horsepower at 1600 rpm and produces its maximum torque of 220 lb ft at 1200 rpm.

The engine drives through a 13 in diameter single plate, overcentre clutch, a sliding gear type transmission which gives four forward speeds of up to 4·66 mph and two reverse speeds of 1·68 and 2·75 mph respectively.

The first forward speed at rated engine speed is 1·42 mph and is ideal for both heavy-duty digging and precision 'inching'. The final drive is a gear, three planetary gears and a ring gear at the sprocket which is the point of application. By keeping the torque low throughout the entire power train a reduction of up to 25% is obtained on the transmission steering clutches and shafts with a corresponding reduction of maintenance while the tractive effort, measured by maximum drawbar pull is 14,100 lb.

Two rigid frame crossbars transmit all stresses direct to the track frames isolating the engine and transmission. The track guage is 54 in and the width of the triple bar grousers is 13 in. These features, combined with the $75\frac{1}{4}$ in length of track, placed on the ground by the five roller frames give a high degree of flotation, stability and traction. The centre of gravity is such that sufficient weight is provided at the rear of the machine to offset the bucket load when digging in heavy conditions enabling a break-out force of 14,070 lb to be fully used.

A standard four-in-one bucket of $1\frac{1}{4}$ yd^3 capacity has a width of 77 in and weighs 1485 lb. A general purpose bucket of the same capacity but having a width of 71 in and a weight of 710 lb is also available. The hydraulic system which operates at 1500 lb/in^2 is maintained by a front-mounted gear type pump and is controlled by a three-valve control block.

An International Model B-100 provided with a four-in-one bucket for a variety of jobs is shown in Fig 24. It is driven by an indirect injection diesel engine developing 50 hp at 1450 rpm and is provided with a one yd^3 bucket 68 in in width and speed variation from 1·5 to 5·4 mph.

The articulated-frame design incorporated into the rubber-tyred bulldozers has also been applied to the larger rubber-tyred loaders and these are proving well capable of the most arduous quarry work and economical and are available with buckets of 3 to 10 yd^3 capacities, the latter for light materials. Many of them can be converted to bulldozers by removing four hinge pins.

The Caterpillar Tractor Co. manufacture a range of wheel shovel loaders as follows:

Model	HP	Bucket capacity
922B	80	1 –3 cu yd
944	105	2 –4 cu yd
950	125	$2\frac{1}{4}$–4 cu yd
966B	150	$2\frac{1}{4}$–5 cu yd
980	235	4 –5 cu yd
988	300	5–$6\frac{1}{2}$ cu yd

Ancillary equipment in surface mining

An indispensable adjunct to the face shovel in modern surface mining to supplement its work by collecting scattered material and building up

the toe of the rock pile is carried out by ancillary equipment. This work
was previously performed by the face shovel but was liable to cause breakdown and delay; the machine can also be used for road grading. Previously
most of this ancillary equipment was crawler mounted, but rubber-tyred
machines fitted with non-skid chains present advantages in some conditions.
Such machines fitted with a $2\frac{1}{4}$ yd^3 bucket have been used by British

*Fig. 24.   International B–100 crawler loader with four-in-one bucket for scraping, loading,
bulldozing and clamshell work.*

Gypsum Ltd at Cropwell Bishop in conjunction with Aveling-Barford
SL dump trucks, by the Hoveringham sand and gravel firm and working
as an ancillary to face shovels to load granite at the Enderby and Stoney
Stanton quarry.

The rubber-tyred tractor shovel is speedy and easily manoeuverable.
Repairs are cheaper than for crawler-mounted tractor shovels since the
travelling gear of a 150 hp crawler shovel takes some 300 man-hours to
renew against 8 hours to renew the tyres of a wheeled shovel.

Where reclamation of a site is required by regulations or lease conditions
it is necessary to remove, and store separately from the overburden, the
top soil and this work is often carried out by crawler diesel, tractor-drawn
or the speedier rubber-tyred scrapers which are also used for overburden

removal. They are used among other applications at the French bauxite open pit of SA Bauxite & Alumines de Provence where two 30 yd^3 caterpillar tractor scrapers are employed each moving 50,000 yd^3 of laminated sandstone and calcareous strata per month. The scraper consists of a blade similar in action to that of a carpenter's plane, carried in a steel rectangular truck fitted with oversize rubber tyres. The skimmer plate is adjustable and skims off the soil or loose overburden to a variable depth of some 12 in according to the type of material being removed. They are drawn by diesel tractors and follow each other in line along the site to the soil spoil heap and back to the site again. A push-loaded scraper will load faster and carry greater payloads because the load is packed tighter.

Bull- and calf-dozers are used like tractor shovels to clean up after the face shovel.

A combination often used for stripping softer overburden is a small shovel and one or two bulldozers. In these conditions the bulldozers usually work together and remove either soft material which does not need blasting or handle part of the overburden which has been blasted and it is not uncommon for the bulldozers to remove 24 ft of overburden.

Land reclamation really begins as the overburden is removed since bulldozers may level spoil banks as mining proceeds.

Because of the increased speed and power of some modern bulldozers they are being increasingly used for overburden stripping where little or no blasting is required, generally working in pairs. For efficiency a limit of 35 ft of overburden should be moved and the terrain should be gently rolling or hilly to assist in the movement of the material. Pushing should be at 90° to the outcrop after the initial cut is made along the outcrop and the bulldozers should work together one following the other and slightly overlapping the path of the loading machine to pick up side spillage.

The development of the hydraulically operated ripper mounted on the rear of a large bulldozer has increased the range of the scraper. The shales and soft rocks that previously resisted handling by scrapers are now successfully loosened. In general ripping should proceed as deeply as possible.

A large bulldozer with a hydraulic ripper may be used to cut down and level 15 to 20 ft of overburden to provide a level bench for a large dragline. A rubber-tyred tractor scraper may be used to make an initial cut and a working bench for a dragline.

A typical large bulldozer is the International & Hough TO25B developing 230 hp, the engine being turbo-charged and has a full range countershaft power shift transmission for matching the power to the load demand. Planetary power steering is provided and four speeds both forward and reverse.

A machine designed as a heavy motor grader but which by attachments can also function as an 11 tine V-type scarifier or as a 9 ft bulldozer, both being hydraulically controlled, is the Super Heavy 500 Grader by Aveling-Barford. It has a Rolls-Royce E6N 170 hp diesel engine running at 2100 rpm or a Leyland AU/600 163 hp at 2200 rpm. The transmission is a Clark 15 in single stage torque converter or an Allison CRT3531 power-shift transmission as an alternative. The total weight is $14\frac{1}{2}$ tons.

Road building

Roads should be planned well ahead of requirements. Main roads should have wide beds and good alignment to permit safe high speeds, with all curves super-elevated. Hard core material should be compacted and then a top layer of crushed rock should be laid down and compacted. This top layer should be applied in several layers, each layer separately compacted. Roads should be sprinkled in dry dusty weather to maintain visibility and to keep dust out of engines and other moving parts. It is usually possible at quarries to avoid special expenditure on road construction, except the loop road round the crusher or other road sections with heavy traffic where hard surfacing is advantageous. Where large trucks run at high speed special surfacing may be worth while, as at N'Changa open pit copper mine where 65 ton trucks run at 35 mph.

Transport accidents at surface mines

The report for 1965 of the Chief Inspector of Mines and Quarries shows that of 31 fatal accidents (12 at sand and gravel quarries) and 91 seriously injured (25 at sand and gravel quarries), nine (six at sand and gravel quarries) fatal and 25 (11 at sand and gravel quarries) seriously injured occurred due to haulage and transport operations. Although a reduction of 18 occurred with combined totals compared with 1964, it is clear that if safe routines are to become established at quarries, management must make rules appropriate to the operating conditions which should be as nearly ideal as good planning can achieve. Drivers should be fully competent and their operational behaviour frequently checked; their duties and that of their machines should be well within their competence.

Proper appreciation of operational hazards is the first step to higher standards of safety.

A large block of limestone weighing 4 tons was being towed by a light tractor up an uneven road rising 1 in 12. The driver made a sharp turn on to a section rising immediately at 1 in 6 when the tractor fell backwards on to the stone and the driver was killed. The short towing chain was connected to the wrapping chain at the bottom of the block and by a pin to the tractor 32 in above ground level and made an angle of 30° to the vertical. The blocks are now broken *in situ* by explosives and loaded into trailers.

In two fatal accidents drivers lost control of their vehicles which then ran off the road and into ponds, the steering being defective in one case. A foreman employed by an outside firm of construction engineers was killed by a vehicle being reversed up a ramp on a gradient of 1 in 12. He and a senior quarry engineer, who was knocked aside, were engrossed in their discussion and failed to see the lorry as they stepped on to the ramp.

A driver, after tipping a load, took a short cut and drove the lorry with the body raised under an elevated conveyor not permantly secured which collapsed onto the cab when hit by the lorry body. Forgetfulness of the position of the body appeared to be the cause and some indication in the cab of the position of the body should be fitted as standard.

The collapse of the tip edges is a hazard and two accidents occurred from this source, one being fatal, the other driver was saved from death but was seriously injured, being saved by the vertical hydraulic rams behind the cab.

The driver of an articulated dumper reversed it towards the rear of the excavator and trapped and fatally injured the driver standing by the machine. The customary practice for dumper drivers to await acceptance signals from excavator drivers was not followed in this case.

A driver with an instructor was driving an articulated lorry under instruction but after a few hours' tuition and in response to a signal from a driver standing by an excavator, attempted a close turn to the excavator for loading but misjudged both the distance and the amount of lock required. The excavator driver was trapped and seriously injured. Drivers should only be allowed to operate in areas remote from production until fully competent.

An experienced tyre fitter was preparing to change the rearside outer wheel of a pair on a 24 yd^3 dump truck and was kneeling in front of a wheel when the rim of the inner wheel broke and the sudden release of air at 70 lb/in^2 from the inner tyre projected the free outer wheel which struck and seriously injured him when he was crushed against a wall. Examination revealed fatigue cracks due to high stress concentration. Other rims in the transport fleet were then tested for cracks by a detector and this led to an improved rim.

Plant maintenance

In order to prevent loss of time and production through breakdown of plant, planned maintenance of equipment and systematic inspection and lubrication is a necessity, particularly overseas with relatively unskilled operatives. At some surface mines each operator is required at the beginning of each shift to check his machine oil and water levels and to grease up at prescribed points. The major part of the daily servicing, however, is carried out from a mobile servicing and fuelling unit during meal breaks, wherever possible, especially if it is an excavator loading trucks or lorries. On a plant working three shifts daily the organization is as follows: a mobile grease truck equipped with two air-operated grease guns, oils, gear compound, kerosene and waste, is manned on the day shift by a shovel operator and a truck driver, both of whom do the shovel greasing. The operators on the shovels are required to assist and it is their responsibility to service the boom, rack arm and ropes, as well as cleaning the shovel, the operation taking 20 to 30 minutes to complete. The order of greasing the shovels is decided between the foreman mechanic and the operating foreman to fit in best with the production schedule. On afternoon and night shifts the shovel operators are responsible for greasing the important sections of the truck equipment and the high-speed portions of the shovel machinery. Fifteen minutes a shift is allowed for this.

For fuelling servicing, a 3-ton truck is equipped with a 500 gallon tank of diesel fuel, a petrol-driven fuel pump and an accurate flowmeter. The truck is also equipped with an air cleaner and transmission oils and hand-operated grease guns.

The truck driver is responsible for fuelling the truck and for noting fuel consumption and hour-meter readings. The bulldozer driver is responsible for greasing, oil checking and oil cleaner servicing as well as for the general cleanliness of his machine.

REFERENCES

'Stripping of Rock Overburden', T. M. Dover. *Opencast Mining, Quarrying and Alluvial Mining*. Institution of Mining and Metallurgy, 1965, p. 551.

'Economic Comparison and Detailed Discussion of Dragline and Shovel for Removal of Overburden', R. S. Zeindler. *Opencast Mining, Quarrying and Alluvial Mining*. Institution of Mining and Metallurgy, 1965, p. 565.

'Factors Affecting Dump Truck Design', P. J. Guy. *Quarry Managers' Journal*, October, 1966, p. 395.

'Loading and Haulage in Quarries', P. Flachsenberg. *Opencast Mining, Quarrying and Alluvial Mining*, Institution of Mining and Metallurgy. 1965, p. 299.

'High Capacity Mobile Crushing Plant', *Mining and Minerals Engineering*, April 1967, p. 142.

'Construction and Use Regulations for Commercial Vehicles', R. E. Armstrong and G. T. Leggett. *Quarry Managers' Journal*, December, 1966, p. 494.

'Dust Control in Quarries', J. M. Hodgson. *Quarry Managers' Journal*, May 1967, p. 191.

'The Use of Belt Conveyors', A. C. Low, *Quarry Managers' Journal*, June 1967, p. 213.

'Applying Planned Maintenance in Quarries', P. C. M. Bathurst, *Quarry Managers' Journal*, June 1965, p. 244.

CHAPTER 5

THE USE OF EXPLOSIVES IN SURFACE MINING

In the hard rock deposits in particular, changing from a somewhat haphazard system to a modern one is no trivial undertaking. Ultimate conditions depend very largely on the results of primary blasting and if this is to be really efficient it must be properly designed from a detailed assessment of all the factors.

The first factor to be considered is the burden on the shotholes. The burden and the spacing will depend on the type of rock to be blasted and its joints or other characteristics. Good fragmentation and efficient blasting with a good blasting ratio will only be obtained if the explosive is used in a well-balanced shothole such that the distance which it is required to break out to a free face, the 'burden', is such that the explosive can detonate, build up pressure in the shothole and break out the rock to the free face in a few thousandths of a second, usually leaving half of the barrel of the shothole visible on the quarry face. If the burden is excessive the breakout of the gases produced by the firing of the explosive is delayed and this gives poor blasting with considerable 'back break'. The latter may bring down more rock at a particular blast but it opens up and strains the rock behind the shothole and thus gives poor fragmentation in the subsequent blast. More time must also be spent plucking the face to make it safe for men to work at or near the face.

Careful trials and observation of results are the only certain method of determining the correct burden for any rock for it depends to a large extent on the strength, density and the diameter of the cartridges of the explosive used but for a given explosive the burden is directly proportional to the diameter of the explosive cartridge, so that $D/D_1 = B/B_1$ in which D and D_1 are the diameters and B and B_1 the burden in feet.

Thus if a 2 in diameter cartridge of a gelatinous explosive such as polar ammon gelignite in a particular rock will blast efficiently a burden of 8 ft, if the diameter of the cartridge is increased to 5 in it will blast efficiently a burden of $B_1/B = 5/2$, $B_1 = 8 \times 5/2 = 20$ ft.

An important factor to which attention must be given is the position and the inclination of the shothole whether vertical or inclined. The toe or bottom of the hole is particularly important as it is here that most work has to be done by the explosive to leave a clean face. Where shotholes are drilled to blast to quarry floor level more energy is required to shear the rock at right angles to the shothole and this must be provided for in the drilling and charging to ensure the 'toe' is blown out. The usual method of ensuring this and getting the extra charge at this point is to drill below the level of the quarry floor for a few feet, the exact distance being determined by careful testing with different sub-floor distances. It will depend on the particular rock concerned but with 6 to 9 in diameter holes it is generally 3 to 5 ft. A 'toe' is less likely to be left when angle drilling is adopted.

The spacing of shotholes is again determined largely on the diameter of the explosive cartridges, the larger the diameter the greater the radius of its disruptive force and consequently the greater the required spacing. The spacing will vary with the strength, nature and jointing of the rock for a given diameter of cartridge but generally the spacing will be approximately equal to the burden on them. With good, approximately vertical jointing and the quarry face being worked on the plane of the rock, spacing can be increased to about one and a quarter times the burden and, consequently, where the quarry face is being worked 'on end' at right angles to the main jointing the spacing may be reduced to about three-quarters of the burden to ensure good breaking between shotholes.

The main factors governing the amount of explosive charge required per shothole are the hardness, the resistance to shearing and the nature and degree of jointing of the rock. Generally the amount of rock broken per pound of high explosive will be between 4 and 6 tons and when blackpowder is used three to four tons. Taking the cost of high explosive as 1s 6d per lb the cost per ton of rock broken is $3\frac{1}{2}$ to 4d per ton. Only by trial can the best blasting ratio for a particular quarry or part of a quarry be determined but by charging fairly heavily a greater overall economy is generally obtained; a blasting ratio of the order of 4 to 5 tons per lb for the primary blast is generally about right and proper spacing with moderately heavy charges ensures good fragmentation of the rock and reduces or eliminates the amount of secondary blasting while minimizing wear and tear on the loading machine.

When rocks are extremely well jointed blasting ratios as high as 15 tons per lb of explosive may be obtained.

The volume of rock on each of a row of shotholes is approximately a prism, with dimensions equal to the burden, the spacing and the depth of the shothole, *i.e.* volume $= B \times S \times D$. The weight of rock is usually expressed in tons per yd^3 which for the usual quarry rocks is two tons per yd^3 so that weight of rock broken per hole is $N = B \times S \times D \times 2/27$ tons. Dividing the weight of the rock by the blasting ratio gives the amount of explosive per hole, *i.e.* charge per hole, $\text{lb} = B \times S \times D \times 2/27 \times 1/4{\cdot}5$

$$= \frac{BSD}{60{\cdot}75} \text{ lb}$$

so that the charge per hole can be obtained by multiplying the burden by the spacing by the depth all in feet and dividing by 60 or, if less heavy blasting is required, by dividing by 70 which gives a blasting ratio of 5·2 tons per lb.

The blasting ratio is expressed in kilograms of explosive per cubic metre of rock where the metric system is used. The charge of explosive required in kilograms is simply the blasting ratio in kilograms per cubic metre multiplied by the burden, spacing and depth in metres, so that for a blasting ratio of 4·5 tons per lb, this becomes

$$4{\cdot}5 \text{ tons/lb} = 0{\cdot}265 \text{ } kg/m^3$$

$$\text{Volume blasted} = B \times S \times D \text{ } m^3$$

$$\text{Charge} = 0{\cdot}256 \text{ } B \times S \times D \text{ } kg$$

and the figures corresponding to 5 to 6 tons per lb would be 0·20, 0·25 kg/m^3 respectively.

PRIMARY BLASTING EXPLOSIVES

The explosives used for primary blasting fall into four categories:

> Conventional high explosives
> Blackpowder
> Factory mixed blasting agents
> Compositions mixed 'on site'

Conventional high explosives

Conventional high explosives must be readily initiated by normal detonators or detonating fuse. The sensitizing agent is usually nitroglycerine or trinitrotoluene (TNT) together with oxidizing salts, fuels and absorbent materials. A wide range of explosives of this type are available with different densities and velocities of detonation.

The main conventional high explosives used in surface mining are the following:

Submarine blasting gelatine

This is only used in quarries in the hardest rocks because of high cost. It is the strongest commercial explosive available and is a nitroglycerine explosive designed for use under high hydrostatic pressures in underwater work. It has a high density and high velocity of detonation.

Polar ammon gelatine dynamite

A high-strength gelatinous explosive whose main ingredients are nitro-glycerine and ammonium nitrate. It also has high density and good water resistance and is recommended where a powerful shattering explosive is required. It is often used as the base charge in deep holes.

Opencast gelignite

Similar to polar ammon gelatine dynamite being also a gelatinous explosive but it is slightly less powerful. It was first formulated for blasting overburden in opencast coal mining but is now widely used in surface mining generally. It is manufactured only in cartridges of $3\frac{1}{4}$ in diameter and upwards.

Opencast gelignite 'Q'

A modified form of the opencast gelignite introduced for use in $3\frac{1}{4}$ in diameter primary blast holes and is supplied in 5 lb cartridges $2\frac{3}{4}$ in diameter.

Polar ammon gelignite

Slightly less strong than the preceding explosives but is of the same gela-tinous type with similar density and water resistance and is an 'all purpose' explosive.

Belex and polar rockite

These are lower density nitroglycerine-based explosives. They contain less nitroglycerine than the gelatine explosives and are not so resistant to water but are useful for blasting medium and soft rock where there is no water problem and where a greater 'spread' of the effect of the charge is required.

Trimonite

TNT and ammonium nitrate type explosives in which TNT replaces nitroglycerine as the sensitizer generally have lower inherent water resis-tance and lower densities which limits their utility but they are popular where conditions are suitable and a range of four are available known as Trimonite.

Blackpowder

This is the oldest known explosive, also known as gunpowder and black blasting powder. It does not detonate but produces a slow, heaving action instead of the shattering effect of high explosive and this is advantageous in the quarrying of building and monumental stone. Blackpowder is very susceptible to moisture and can only be used in dry conditions unless enclosed in a waterproof container. It is supplied in granular form and is glazed with graphite to make it free-flowing to facilitate the charging of shotholes.

Factory mixed blasting agents

Factory mixed compositions based mainly on ammonium nitrate and containing no sensitizer such as nitroglycerine are known as blasting agents and have been used in surface mining extensively over the past ten years, particularly in the USA. They cannot be initiated directly by a detonator or detonating face and require a primer of high explosive. By incorporating power-boosting ingredients and special processing during manufacture the blasting properties can be varied to suit conditions. These blasting agents are generally known as AN–FO explosives (Ammonium Nitrate Fuel Oil) after their main constituents. In this country Nobelite and Nobelite H have densities slightly over 1. Nobelite is supplied in 10 in packs weighing 50 lb for pouring directly into shotholes. Nobelite H is supplied in similar packs and in 5 in to 8 in diameter polythene and chipboard cartridges. Blasting agents poured loose are not suitable for wet conditions and are normally used in the upper part or 'column' of the borehole after loading a base charge of conventional high explosive or slurry explosive.

In cartridge form they have reasonable water resistance although the energy concentration is reduced compared with filling the full diameter of the borehole, generally of large diameter, with loose explosive. It is claimed that AN–Sodium Nitrate–FO (40% sodium nitrate) is 60% more powerful than AN–FO alone and bulk $NaNO_3$ is as cheap as bulk ammonium nitrate.

'On-site' mixed compositions

When mixed with a combustible material such as fuel oil, ammonium nitrate becomes an explosive of low sensitivity but adequate for use in large diameter boreholes. The mixture can be prepared on site often subject to strict regulations and is poured directly into the shotholes. The ammonium nitrate must be in free-flowing form. Nobel Agent No 128 is of this type.

Explosive slurries

These blasting agents have been developed during the past six years. They are insensitive to detonators or detonating fuse, water resistant, and have high density and are good in wet conditions. Compared with AN–FO explosives they have a 50% higher detonation rate and double the density. They are powerful and are used for bottom charges and in hard rock. The addition of metallic powder such as aluminium chips 5 to 30% enables the explosive to reach a high strength. Research by Du Pont and others indicates that future metallic slurries will enable higher tonnages per foot of shothole to be obtained and the desired degree of fragmentation to be obtained. They are used in cartridges but the gelatinous consistency enables them to be pumped into the borehole. There are several compositions of explosive slurries, one being a mixture of AN, TNT and water with a small amount of quargum. They can be bought ready for use or mixed by the mining company.

Of importance in hard rocks where drilling costs are high, the increased power and density of slurry explosives permits smaller shothole diameters and greater burden and spacing can be adopted. The Iron Ore Co. of Canada found that spacing of holes initially of 8 ft by 8 ft pattern could be increased to 12 ft by 12 ft with a 55% decrease in drilling costs. Although slurries cost 68% more than AN–FO overall blasting cost decreased by 42%.

In the USA slurry explosives cost from 7 to 18 cents per pound and AN–FO average between 4 and 5 cents; other commercial explosives of the same strength are two to three times more expensive. These new explosives have allowed American and other mineral industries to achieve considerable economies in costs. Table IV shows that twice the amount of

TABLE IV

RELATIVE STATISTICS OF EUROPEAN AND USA
SURFACE MINERAL INDUSTRIES

	Europe	USA
Bench heights (ft)	50–100	30–50
Shothole diameter (in)	3–4 (6)	6–15 (4)
Cost per man-hour (shillings)	7–10	17–45
Relative price of explosive (%)	100–260	32–160
Manshifts per 1000 tons/day	2–3	0·4–0·6
Relative explosive consumption	50–100	100–200 (400)

Figures in brackets indicate outside limits.

explosive is used per ton of mineral compared with European consumption with better fragmentation, reducing also secondary blasting, loading and haulage and crushing costs. It will be noted that American surface mining prefers low benches compared with the higher benches of the corresponding European industries where the bench height burden and shothole diameter are selected to minimize explosive consumption because of the high price of explosive in relation to labour costs. As the latter rises annually with inflation while the cost of explosives is much more stable, lower benches are being adopted in Europe which involves greater safety from falls of ground and smaller rock piles and perhaps increased efficiency of shovels and other loading equipment. Shothole diameters are less in Europe since a larger burden can be blasted but with increased explosive cost per ton or less fragmentation and hole spacing.

Where reduction of wet hole diameter occurs normal cartridges or bags cannot be used but these blasting agents can be bought in special expandable bags which are pleated to allow expansion and lined with polyethylene for waterproofing When dropped into the hole they expand from 7 in to $10\frac{1}{2}$ in without bursting.

With the increasing use of AN–FO blasting agents users are becoming more particular about the quality of the nitrate with which they are being supplied as well as the subsequent mixing and handling. Such properties as grain size, porosity, moisture content, whether flaked or prilled, are receiving attention. The economy of bulk buying of the nitrate as in the Pennsylvania anthracite field strip mines is also being considered where a service of bulk delivery in trucks and controlled mixing on site of nitrate and fuel oil and its final detonation has been established.

Although the constituents of AN–FO blasting agents are comparatively safe to handle, stringent rules should be established for storing, mixing and handling. Among these should be the following:

1. Special plants should be used for mixing AN and FO under closely controlled conditions.
2. The mixing plant should be sited in isolated areas away from houses and other buildings and should be built of fireproof materials.
3. As quickly as possible after mixing, the product should be removed from the plant.
4. Welding and cutting should be forbidden in the plant until all material, mixed or unmixed, has been removed.
5. Priming cartridges, detonators and detonating cord should be transported in separate containers and vehicles to the AN–FO and only brought together at the shothole.

Blasting using delay or short-delay detonators or detonating relays may increase the efficiency of explosives, improve fragmentation and reduce concussion. Reduction of vibration and shock is important near populated areas, and is a problem when overburden becomes thicker and requires heavier blasting charges. Reduction of noise is also important. There is a tendency to equate noise and air blast with ground vibrations, and undue noise may lead to complaints even where damage would not result. The noise from Cordtex detonating fuse lines can be deadened by covering them with sand, earth or drill cuttings. Where complaints are liable to arise it is advisable to blast during the working day and avoid periods of quiet, such as the evenings or week-ends, when noise and blast tend to be more noticeable. Air blast is much less obvious in wet humid atmospheres and under low cloud, and its effect is less apparent at higher altitudes. Conditions to be avoided if possible:

1. Relatively high atmospheric pressure.
2. Wide daily temperature variations at ground level.
3. Poor visibility and light winds early in the morning.
4. Light surface winds and low relative humidity at ground level and stratus clouds below 10,000 ft.
5. Clear somewhat hazy days with little wind and fairly constant temperature.

Normal buildings should not be damaged by vibrations with amplitudes less than 0·008 in if in reasonable state of repair. For quarry buildings and plant where occasional minor damage can be tolerated 0·016 in is a satisfactory limit. The maximum for other buildings in reasonable repair of 0·008 in should be reduced in the case of ancient monuments, and for buildings where vibrations might produce complaints or even possible litigation, smaller limits should be adopted.

The smallest ground vibration which can be detected by a man depends on the frequency of the vibration, the posture and whether he is engaged in some employment or is awaiting the vibration. In the latter case, with the range of frequencies associated with blasting, 0·001 can be detected.

A vibrograph study of the site must be made, e.g. by Nobel Vibrograph of the vibrations produced by blasting for an accurate forecast and the records obtained require expert interpretation. Afterwards amplitudes elsewhere on the site can be estimated from the formula:

$$A = \frac{K\sqrt{E}}{d}$$

where A is the maximum amplitude in thousandths of an inch

E is the explosive weight in pounds

d is the distance between blast and building

K is a constant depending on the site

Where no vibrograph study has been made, an approximation of the ground movement to be expected can be obtained by using $K = 100$ for hard rock and $K = 300$ for wet or clay sites.

In AN–FO mixed-on-site blasting agents, the optimum fuel oil content is $5\frac{1}{2}\%$ at which the mixture develops optimum explosive energy. In mixing, a wooden tray or trough of suitable size is used to hold the ammonium nitrate about 1 cwt at a time, to which the fuel oil is added as evenly as possible and mixed thoroughly and evenly with the nitrate using small wooden spades. The quarry management must obtain a license from HM Inspector of Explosives to prepare blasting agents in this manner.

The mixture is poured loose into the shothole to fill the full cross-section of the hole, this being necessary because the sensitivity to propagation is at maximum when the charge fills the hole.

AN–FO mixtures require to be primed, usually by cartridges of a gelatinous explosive, as they lack the reliability of conventional explosives in sensitiveness to initiation by detonator or detonating fuse. In addition it is usually best to use a substantial base charge of a gelatinous explosive as well to ensure a good concentration of energy at the toe of the hole and minimize the risk of desensitization of the AN–FO by water in the bottom of the hole. A good rule is that the length of gelatinous explosive base charge should be equal to the burden on the toe of the hole. After the base charge has been loaded the shothole is filled to the appropriate depth with AN–FO mixture and a further single cartridge inserted at the top of the AN–FO column as an additional primer. Cordtex detonating fuse is led to the base charge and the AN–FO is thus initiated at both the top and bottom.

Such multiple primers are considered in the USA to increase the apparent velocity of detonation. By starting the chemical reaction at several points, the time required for a charged hole to detonate is reduced and there is also a better chance of detonating the charge.

INITIATING EXPLOSIVES

Detonators

Plain detonators
For use with safety fuse a plain detonator of No 6 strength is specified and in most cases this detonator is adequate, but where an additional margin of power is required a more powerful detonator, the No 6 'Star' type, should be used. In secondary blasting using plaster gelatine slabs the use of the No 6 'Star' offers advantages.

Electric detonators
These are suitable for firing single shots or a number of shots simultaneously in a round. The fusehead assembly of a detonator (Fig. 25) fitted with insulated leading wires and sealed in position with a neoprene plug, provides the means of igniting the detonator composition which is lead azide and lead styphnate with a little flake aluminium as the priming mixture, and tetryl as the detonating composition. Electric detonators are normally No 6 strength but specially powerful No 6 'Star' detonators are also available.

Standard detonators

Those for use under normal conditions are fitted with 25 swg (British Standard wire gauge) plastic-covered tinned-copper leading wires. 'Hydrostar' detonators can be used under reasonable heads of water. They are provided with 23 swg tinned-copper plastic-covered leading wires and are tested to withstand high hydrostatic pressure.

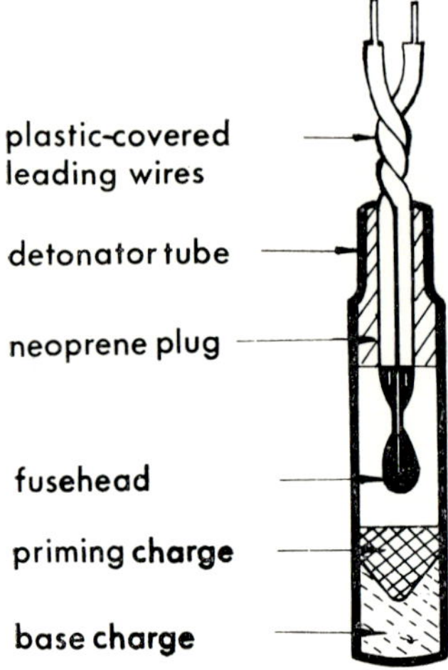

Fig. 25.   *Sectioned standard electronic detonator (ICI).*

Delay detonators

For initiating a number of explosive charges in a predetermined sequence at known time intervals, a simple and convenient means are delay detonators. Eleven delay detonators are available numbered 0 to 10. No 0 gives instant initiation and there is a standard time interval of approximately half a second between successive numbers. Each delay detonator consists basically of a standard electric detonator with a delay element interposed between the fusehead and the detonator composition (Fig. 26). They are compact varying in overall length from $2\frac{1}{4}$ to 3 in according to the delay period, the number of the delay being clearly marked on a tag attached near the end of one of the plastic-covered leads. They are made in No 6 strength in copper tubes and are provided with 23 swg plastic-covered tinned-copper leads.

Short-delay detonators

The range of short-delay detonators numbered 0–15 provides a series of explosions with very short time intervals between each. No 0 gives instant initiation and the intervals between succeeding numbers up to 4 is 0·025 second while from 4 to 12 the interval is 10·05 second and from 12 to 15 the interval between delays is 0·07 second. Short-delay detonators are made in No 6 strength in copper tubes and are provided with 23 swg plastic-covered tinned-copper leads. They have an important application in practically all types of rock breaking. They are used to reduce ground vibration due to blasting, which is important when blasting may have to take place near property. In certain conditions short-delay firing gives an improvement in fragmentation and reduction in back or overbreak of the rock.

Safety fuse

This is used for initiating blackpowder or for setting off plain detonators. Safety fuse consists of a core of blackpowder wrapped with layers of textile yarn or tape and with waterproof coatings to provide protection against

Fig. 26. Sectioned delay detonator (ICI).

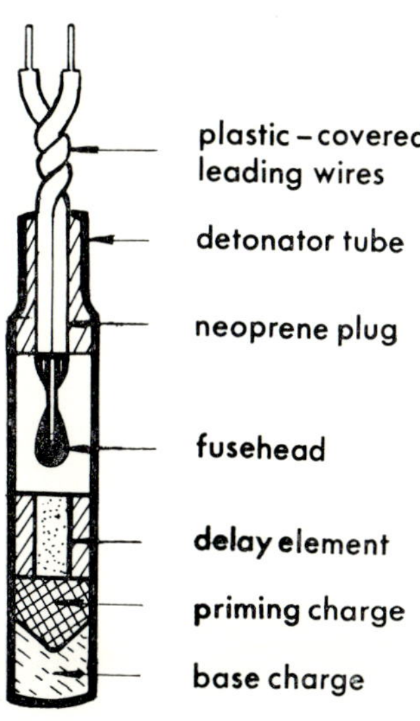

water penetration and mechanical damage. Careful testing pays particular attention to regularity of burning speed which must not vary more than 10 seconds above or below 90 seconds a yard and to water resistance of types supplied for use under damp or wet conditions. Capped fuses are lengths of safety fuse with plain detonators crimped in position before they are taken to the shothole. In the explosives factory specially designed crimping machines are used to fasten the detonators to the lengths of fuse, which ensures a weather proof seal between the detonator and the fuse, thus eliminating a main cause of misfires in safety fuse firing.

Capped fuses are supplied in various lengths in bundles of ten in cardboard packs, and are also supplied fitted with connectors for use with plastic igniter cord. Standard fuse lengths are 2 ft 6 in, 4 ft and 6 ft. For secondary blasting, such as breaking up large stones, 2 ft 6 in fuses are normally used.

Cordtex detonating fuse

This consists of a core of pentaerythritol tetranitrate (PETN) enclosed in a tape wrapped with textile yarns and then completely enclosed in a tubular cover of white plastic material (Fig. 27) which gives a strong

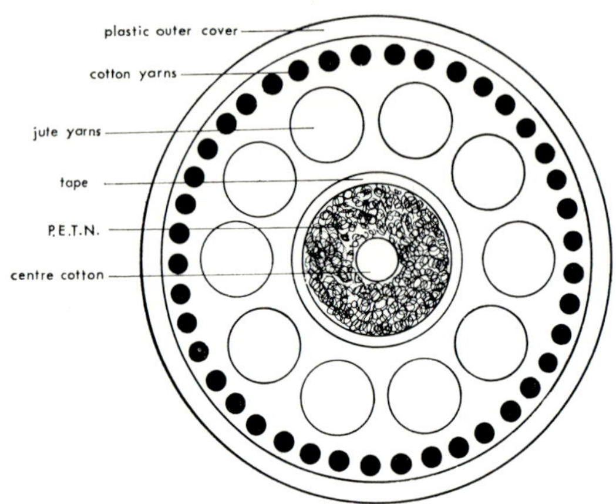

Fig. 27. Structure of Cordtex (diagrammatic) (ICI).

and flexible fuse which is waterproof, light and reliable. The external diameter is 0·195 in and it weighs 0·015 lb/ft, each foot containing 0·0007 lb of explosive. Cordtex does not readily deteriorate and will remain in good condition in storage for several years. The velocity of detonation is 20,000 ft (6500 m) per second and its propagation characteristics ensure the initiation of commercial explosives. It is very safe to handle and unless it is crushed in some way and the spilled PETN core accidently initiated by a second blow, detonation by impact is practically impossible but in spite of its relative insensitivity Cordtex can be initiated with certainty by a No 6 detonator. It has a high tensile strength of over 100 lb and resistance to abrasion which reduces the chance of failure caused by rough handling. It is particularly useful in shothole and heading blasts. In the former it makes possible deck loading and the placing of the explosive charge in the most advantageous position in the shothole with reference to the strata being blasted. In heading blasts it is used to link up the charges and provides a simple and efficient method of initiating the separate charges.

Detonating relays

Where Cordtex fuse is used for initiation, an excellent means of obtaining short-delay periods in blasts is provided by detonating relays. The method avoids the long electrical connections which are required when using delay detonators. The detonating relay is an assembly of two delay detonators in a thin rigid aluminium tube constricted in the middle and open at each end to receive Cordtex detonating fuse (Fig. 28). The delay for each detonating relay is nominally 20 milliseconds, the Cordtex line being cut where the delay is required and the detonating relay is then crimped between the

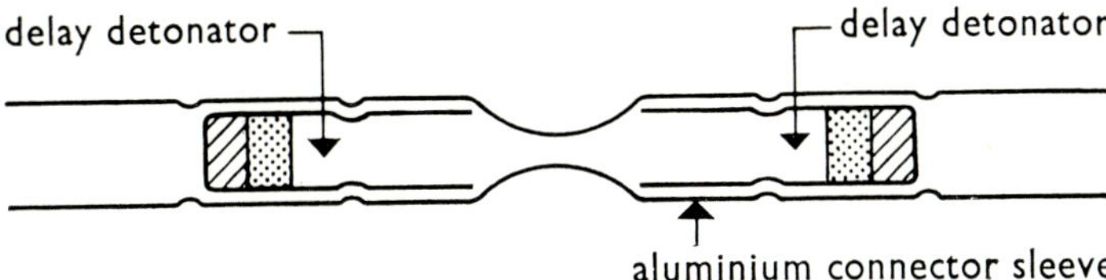

Fig. 28. Detonating relay.

two cut ends of the line, so that by judicious selection of the points at which the detonating relays are inserted any required blasting sequence can be provided.

Plastic igniter cord

For lighting a number of safety fuses in series plastic igniter fuse, an incendiary cord, has been developed of which two types (fast and slow) are available. When ignited an intense flame which will ignite the blackpowder core of ordinary safety fuse, passes at a uniform rate along the cord, but for ease of connection and certainty of ignition, special connectors for plastic igniter cord are supplied for attaching the cord to the ends of the safety fuses being lit.

The burning speed of fast plastic igniter cord is not less than $1\frac{1}{2}$ seconds per foot and it is enclosed in a thin outer plastic covering of overall diameter approximately 0·10 in. This fast cord is useful for lighting secondary blasting shots.

The burning rate of slow plastic igniter cord has a nominal burning speed of 10 seconds per foot and consists of a plastic incendiary composition extruded round a 25 swg copper wire. Both cords have excellent water resistance and good storage properties and can withstand reasonably rough handling. Their burning speeds are reliable and consistent even under adverse conditions as when burning in the opposite direction to a strong wind or when burning under water when, however, burning speed will be reduced if moisture penetrates to the centre of the fast cord. In order to ensure that the fuses will be lit with certainty under all circumstances it is essential that the trunk line of plastic igniter cord should be attached to each safety fuse and special 'beanhole' connectors have been developed to ensure this. They consist of small aluminium tubes closed at one end and containing a plug of incendiary composition. An oval aperture is cut through the tube and the incendiary composition so that a small loop of plastic igniter cord can be slipped through the tube and the incendiary composition which is securely closed round the cord by a special tool, which constricts the aperture into a figure-of-eight shape without crushing the igniter cord, making it impossible to dislodge it.

Electric powder fuse

By using electric powder fuses instead of safety fuses to initiate the explosive charges in blasting with blackpowder, advantages in control, safety and convenience can be obtained. The electric powder fuse consists of an electric fusehead assembly, similar to that used in an electric detonator, sealed into a cardboard tube containing a charge of blackpowder.

Exploders

The circuit should be tested in electric shotfiring before any attempt is made to fire the shots to make sure there is no open or short circuit and it is important that the current passed during testing should be limited so that there is no possibility of accidental explosion of the detonators. In addition, all testing must be done from a safe place and at a safe distance from the blast, and never, in any circumstances, at the face. Special ohmmeters are available for testing shotfiring circuits but only qualified electricians should maintain or adjust them since the safety characteristics can be destroyed by incorrect reassembly. At regular intervals exploders should be tested to ensure their ability to fire the required number of shots and special fuses are available for these tests consisting of fuseheads sealed into paper tubes and fitted with 12 in leads supplied in boxes of 50, sealed to prevent deterioration. Sealed, they remain in good condition for a year from the date of manufacture but unsealed for three months only, after which, tests may be unreliable. They should be kept free from contamination by oil or moisture and should not be exposed to high temperatures.

Single-shot exploders, such as the 'Little Demon', are magneto exploders operated by twist-action of a detachable key.

A typical six-shot magneto exploder is the 'Drake' which is operated by a twist-action detachable key.

A dynamo-type exploder for series firing up to 30 shots by a twist-action detachable key is the ME30. A device is incorporated by which no current is passed through the firing circuit unless the armature has a speed which ensures that the round does not partially misfire due to faulty use of the exploder by the shotfirer.

The 'Beethoven' dynamo condenser exploder comprises a dynamo which is operated by turning a bundle when the ac voltage generated is stepped up by a transformer, rectified and used to charge a condenser to a potential of not less than 1200 V. When the firing button is pressed the condenser is discharged through the firing circuit firing up to a maximum of 100 shots in series.

For circuits with multiple shots a range of Schaffler exploders is available.

Maximum efficiency and safety in blasting depends on choice of the right explosive for the task in hand, the correct application of the explosives and the requisite accessories using the best technique, and strict adherence to safety rules and practices and particularly by ensuring that

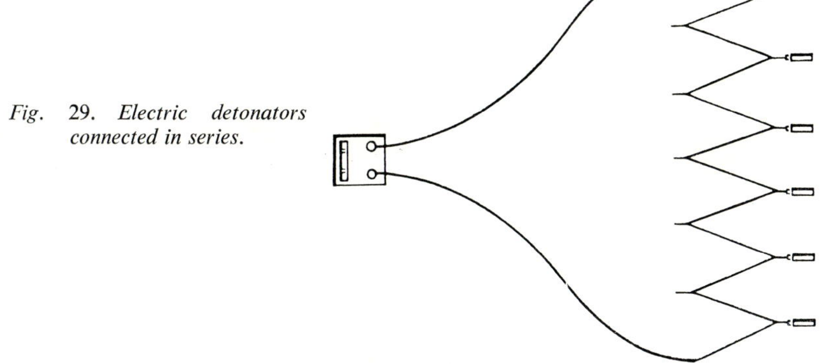

Fig. 29. Electric detonators connected in series.

'familiarity does not breed contempt'. To get best results the largest possible diameter cartridges should be used, provided they never need forcing into the hole. To initiate an explosion, one cartridge in each charge must be fitted with a detonator and safety fuse, an electric detonator which may be a standard, a delay or a short-delay detonator or a length of Cordtex detonating fuse. This cartridge is known as the 'primer' and special care is necessary in preparing it and placing it in good contact with the main charge since it reduces the risk of misfired shots. When several shots are to be fired simultaneously, electric detonators in a series circuit (Fig. 29) are recommended. The circuit is completed by connecting one leading wire of each detonator to one of the next until a continuous circuit is formed, the two free leading wires being connected to the shot-firing cable. Any fault in the connexions can be detected by a circuit tester since there is only one path for the current but care must be taken that the exploder has a capacity in excess of the number of shots to be fired. Good shotfiring cable is essential and all electrical connexions should be well made. Bare joints can cause misfires owing to earth leakage in damp conditions and they should be insulated by joint insulators or insulating tape.

Bare joints are suitable in dry conditions but they should be kept free from contact with the ground.

The following procedure should be followed if the firing circuit is found to be defective. The circuit tester should be disconnected and the connections in the firing circuit examined. If the fault is not located, the shot-firing cable is taken out of the circuit and tested separately. If the cable is sound, the fault must be in the circuit at the face. The face circuit is divided into two and the shotfiring cable connected to one half and re-tested from the firing point. This will show in which half is the faulty shot. The half circuit containing the fault is again divided into two and re-tested. This process is repeated until the faulty shot is located. It is *essential* that such circuit testing is done from a safe place, preferably the firing point.

DRILLING

Good preparation of material for loading is essential for good shovel or dragline performance and involves not only breaking overburden for easy removal but also breaking it at the lowest possible cost which involves maximum fragmentation with the least drilling and the most efficient use of explosives.

In deciding the best drilling and blasting procedure the benefits of fragmentation must be weighed against the cost of explosives required to break the rock smaller.

The seismic method of overburden analysis enables the overall consolidation of sub-surface materials to be measured quickly including rock hardness, stratification, fracturing and the degree of weathering. This method also makes it possible to determine quickly whether overburden can be ripped and where this is the case economies of 12 to 60% over drilling and blasting have been reported.

Factors of importance to the surface mining industries are the most economical drilling pattern, bench height, shothole diameter, burden and spacing and the important items determining these parameters are drilling and explosives handling costs, the final decision, of course, depending upon the determination of overall minimum costs and the summation of costs for drilling and blasting with those for loading, transport and crushing.

Recent developments have indicated that in many countries the trend is towards angle drilling rather than vertical drilling, The following advantages are claimed for angle drilling:

1. It is safer for men and equipment.
2. A considerable improvement in fragmentation and a reduction of secondary blasting.

3. The toe problem becomes insignificant.
4. Backbreak and its problems are practically eliminated and the face left shear with little or no overhang.
5. Explosive consumption reduced by 1% per degree of inclination from the vertical.
6. Less drilling per ton of overburden due to wider spacing of holes and/or larger burden and less sub-drilling.
7. Less ground vibration.
8. Overall costs, including drilling blasting loading and crushing reduced, in some cases, as much as 10%.

In the United Kingdom and Europe where shothole diameter and burden are small and benches are high the angle-drill method is important since it is difficult or even impossible to place the drilling machine for vertical drilling sufficiently close to the bench face-line to obtain the required burden which is often too big for the shothole diameter because of breakage and fall of rock from the upper edge of the bench and widespread fracturing of rock. Poor fragmentation, toe problem and backbreak result. Angle drilling eliminates all these problems, is safer and reduces costs.

In the report of the Chief Inspector of Mines and Quarries for 1964 it was stressed that it is necessary (*a*) to obtain geological knowledge of the formation and strength characteristics of the deposits to be worked and (*b*) to determine the angle of stability. Not only would knowledge of this kind be a great aid to productivity but it would undoubtedly have a marked effect on safety.

Except for a few undertakings with unusual geological features or where it is necessary that the mineral is won, shaped or examined by hand, quarries are now highly mechanized. In many hard rock quarries the adoption of modern techniques has reduced almost to a minimum the number of experienced men necessary to maintain the faces, but if full advantage is to be obtained three points are essential:

1. A face should be of modest height, say from 50 ft to 60 ft with, where necessary, correspondingly wide benches.
2. A reasonable backward hade of 25° to 35° should be maintained.
3. Full-face drilling should be established.

These practices, generally both reasonable and practicable, are to be found at an ever-increasing number of quarries and should result in a marked increase in overall efficiency, which, of course, takes into account standards of safety.

The necessity for securing face stability is illustrated by a fall of about ¼ million tons of slate and overburden which occurred at a large quarry where the mineral was being won from a series of benched faces giving an overall height of about 1200 ft. Indications of an impending fall had appeared in the overburden after heavy rains, but it was nearly three months afterwards before further heavy rains caused the collapse which left conditions that will be extremely difficult to rectify. Failure to have sufficient backward hade on the faces and benches of an adequate width were contributory factors to the incident.

Most new drill layouts have been designed for angle drilling and larger shotholes up to 15 in in diameter can now be drilled and are used particularly in the USA. Several drills can be placed upon a carrier to save labour costs and are operated by one man. For small units four machines may be used and for larger units drilling holes up to 9 in diameter, two machines with a hydraulic positioner which permits change of spacing, that is the distance between the drilling machines, and consequently the holes, may be used.

In the USA rotary dry drilling units are the most commonly adopted method for drilling vertical or angled holes although some improved vertical augers are used and some squeezing of drill holes in clay or soil has been solved by augering through the soft material and then changing to the regular tri-cone drilling bit.

A wide range of rotary dry-type machines is available to meet the needs of large and small surface pits and quarries in either deep or shallow overburden. Heavy-duty machines are generally mounted on crawlers as the Bucyrus-Erie blasthole drill (Fig. 30) which has variable pull-down pressures, variable rotary drive speeds and rapid levelling with hydraulic jacks. Vertical or angle holes can be drilled and the power supply is either diesel or electric. The hole diameters range from 6 in to 15 in and bit loadings from 32,000 lb to 90,000 lb. Smaller units may be mounted on truck frames. Hole diameters range from $5\frac{1}{2}$ in to 15 in. Modern heavy-duty machines can drill larger holes faster than smaller ones can be drilled by less powerful units. In field trials a unit designed for a $7\frac{3}{4}$ in bit penetrated overburden at the rate of 70 ft per hour, a $10\frac{5}{8}$ in machine drilled the overburden at the rate of 90 ft per hour, a $12\frac{1}{4}$ in unit at 100 ft per hour and a 15 in unit at the rate of 110 ft per hour. Bit life also increased from 6000 ft of drilling for the $7\frac{3}{4}$ in bit to 10,000 ft for the 15 in bit. Cubic yards of overburden prepared for blasting per shift increased from 7620 for the $7\frac{3}{4}$ in drill to 50,274 for the 15 in drill and drilling cost per cubic yard of overburden fell by 67%.

Large diameter holes permit more effective loading which concentrates explosives in the harder layers resulting in better fragmentation.

Where flexibility is extremely important and the undertaking is in the medium or small size range smaller, lighter and less expensive vertical rotary machines are available both truck and crawler mounted, and usually equipped to drill $4\frac{5}{8}$ in to $9\frac{1}{2}$ in holes.

Several horizontal rotary dry-type overburden drills are available to drill hard rocks at high speed. They have single or dual heads and can drill up to 12 in diameter holes to a depth of 116 ft.

Fig. 30.   Bucyrus–Erie blasthole drill.

When the overburden is not too thick or exceptionally hard, a single row of properly charged horizontal holes is usually sufficient to break the overburden. It is usual when blasting with a single row of horizontal holes to space them so that the burden or resistance of the material above the holes is approximately equal to the resistance of the material between the holes. If the holes are spaced too close together there will be a tendency to shear between holes rather than strike upwards and break the overburden.

It has been found economical to adopt what is termed explosive casting of overburden in some cases, that is the overburden is blasted on to a spoil bank. Large volumes of low-cost AN–FO mixtures are charged into medium diameter shotholes in the ratio of more than one pound per cubic yard of overburden and detonated with short-delay detonaters. When the shot is fired it is blasted away from the high wall on to the spoil bank where it attains the angle of repose. Removal of 30% to 50% of overburden by this method has been reported.

In the United Kingdom the surface mining industries cover a wide range of materials but average drilling requirements apply to a large proportion of them and for primary blasting, holes are 4 to $6\frac{1}{2}$ in in diameter and 50 to 200 ft deep. Since cable or churn drilling has been largely superseded, the choice vertically lies between the down-the-hole and the rotary methods, drag bit and heavy rotary.

Down-the-hole drilling

The down-the-hole method is suitable for both igneous and sedimentary rocks and gives a medium diameter hole but drilling speed is relatively slow so that, for a given output, drilling labour costs tend to be relatively high. A pneumatic percussive hammer drill is used down the hole so that the mounting need not be heavy and the capital expenditure low. From a given rig the range of hole diameter is small but this is often unimportant. In badly broken rock, however, jamming commonly occurs with the possibility of hammer failure.

The rock is fragmented by repeated impaction presented directly to the rock. Attempts to solve the problems associated with down-the-hole drilling include 250 lb/in² air pressure in the USA. In the UK air at 150 lb/in² has been used and has the advantage that standard type compressors can be used and many other users are using compressed air at 120 lb/in².

In broken strata performance may be improved by a larger mounting, the heavy stable rig and increased rotary torque and greater hoisting capacity and a better supply of compressed air for flushing reduces the danger of hole deflection, improves ability to free jammed rods and gives cleaner holes through better flushing with less chance of wedging rods at the same time. Large down-the-hole hammers can be accommodated, greater depths achieved and higher drilling speeds attained, but the rig is no longer small, portable and inexpensive.

High-pressure down-the-hole drilling rigs

Holman Bros. of Cambourne have added to their types of down-the-hole hammer drills two models, the Universal Vole HP drilling rig (Fig. 31) using a four-wheeled trailer and powered by a Rotair RO 37 HP portable

Fig. 31.   *Universal drilling rig (down-the-hole hammer) here being used on high-pressure operation, 150 lb/in².*

(four wheeled) rotary screw compressor designed to give an output of 375 ft³/min at 150 lb/in² and a crawler-mounted type, the Voletrac HP (Fig. 32) with the same compressor for power supply. Both models use $3\frac{1}{4}$ in or 4 in down-the-hole hammer units for drilling down, breast or lifter holes.

The Universal Vole (Fig. 31) has a 6 ft tube change while the Voletrac will take 10 ft tubes. Both drill units are mounted on a chain feed cradle with feed pressure supplied by a reversible pneumatic feed motor which can be adjusted to suit the nature of the ground being drilled, while full power is available for withdrawing the tubes.

Drill rotation is powered by a second pneumatic motor mounted on the platform which travels on the drill cradle. Pneumatically operated centralizers facilitate coupling and uncoupling of the feed tubes.

 Quarrying, Opencast and Alluvial Mining

Fig. 32.   Holman Voletrac drilling rig—without guard.

The Universal Vole HP rig (Fig. 31) has a U-shaped main frame of heavy-gauge steel tube with two large pneumatic tyred wheels at the rear and with a dual wheel assembly and tow bar at the front. The chain feed cradle is carried on a swing frame which is raised and lowered by a hand-operated worm and screw gear.

The Voletrac HP rig (Fig. 32) has the drill unit mounted on a crawler-type vehicle the tracks of which are driven by powerful pneumatic motors through reduction gearing. The control is of the 'dead man's' type and automatic brakes are fitted. The drill cradle is bolted to a steel boom and the elevating, lowering and slewing of the boom or the swinging, dumping or crowding of the cradle are all accomplished hydraulically and controlled from a single panel. Hydraulic pressure is supplied by a pump driven by an air motor.

With the down-the-hole HP machines a number of modifications have been made to the standard 80 lb/in^2 compressed air powered Holman Universal Vole and Voletrac designs to adapt them for high-pressure operation, including the following:

1. Large control valves are fitted in circuit with the rotation and feed motors and a pressure reducing valve has been incorporated in the air supply line to some of the accessories.
2. A three-pint oiling unit is fitted which is double the capacity used on the standard rigs.
3. The drilling control of both Universal Vole and Voletrac have been resited to give improved accessibility.

Both models use a new Mark IVa hammer unit (Fig. 33) which is now the standard for use with Vole drills operating at pressures from 80 to 150 lb/in^2. The Mark IVa has been designed to take maximum advantage of the higher air pressure without sacrificing reliability, the chief modification being the fitting of a heavy-duty piston.

The Rotair RO 37 HP compressor is basically a hybrid of two existing compressors, the RO 37P and the RO 60P, both of which deliver air at 100 lb/in^2.

Table V

FIELD TRIAL RESULTS WITH HOLMAN HP VOLETRAC

Rock type	Holes	Depth Max	Min	Speed ft/hr Max	Min	Aver.	Conventional rig ft/hr	Total ft
Elvan	34	60 ft	10 ft	46	24	31	14	1156
Hard granite	20	82 ft	50 ft	31	22	24·6	12	1211
Limestone	30	63 ft	15 ft	50	23	36	20	1541
Elvan and apeite	28	71 ft	15 ft	35	19	25	12	1006

Both rigs are supplied with a tube rack, a safety guard, a non-return valve for wet drilling, reaming washers, hammer unit servicing clamp, a magnetic fishing tool and tungsten carbide bit grinder.

Table V is a comparison of results between the Voletrac and a conventional rig.

Drag-bit drilling

This is a method of pure rotary drilling in which a wing-type bit is given a steady thrust with a strong rotary torque such that rock fragments are broken off by a combination of wedging action and shearing. It is the fastest method of drilling in friable non-abrasive rocks, but its use is

Fig. 33.   *Mark IVa Vole hammer.*

limited in harder rocks by the high thrust required resulting in heavy bit wear in abrasive rocks. In friable oolites speeds of 7 ft/min can be achieved provided the diameter of the hole is related to the power available at the drill and adequate air flushing is provided. Sandstones can be drilled if the bonding material between the silica grains is weak and the airstream is sufficient to remove cuttings from the bit as quickly as they are produced to prevent bit abrasion. Rounded sand particles cause less abrasion than sharp marine sand.

Drag-bits are of a large number of types and include fishtail, three- and four-bladed bits and combinations of these types. They have no moving parts and drill by a shovelling action of the blades which are dressed with tungsten carbide. The water courses through the bit are of the jet type and so placed that the fluid is directed on to the blades keeping them clean without erosion. The courses provided should be sufficient to handle the volume of fluid (mud) required to be circulated and are usually nine in number in a four-bladed bit. They are used to drill soft formations and require the circulation of large volumes of mud. In drilling for oil roller bits are replacing them.

Heavy rotary drilling

The first bit of the roller type was designed by H. R. Hughes and with subsequent improvements this led to the increasing adoption of rotary drilling. The most popular and suitable type for practically all strata is the tricone bit. For soft formations the cones have long, widely spaced teeth, but for hard formations they have shorter, closely spaced teeth.

Tooth length, spacing and pattern are related and arranged to give the fastest possible penetration rate with minimum wear.

In heavy rotary drilling, heavy thrusts are imparted to a tricone bit sufficient to overcome the compressive strength of the rock and for this the drilling rig must be heavy to counter the downthrust. Larger shothole diameters and deeper holes with higher penetration rates are available with this method.

The most important factor in drilling with roller bits is the provision of the required applied thrust, that is, the weight on the bit. Penetration speed increases with increase of weight after the critical weight required to crush the particular strata being drilled has been reached, until clogging of the bit occurs, the bit bearings fail or the hole begins to wander. If one tooth only in each cone is imagined to be in contact with the rock at any one time, it becomes apparent that the weight is concentrated on to a very small area. The teeth are lineal so contact area increases with bit diameter and not bit area to rock. The compressive strength in the solid of a hard rock may be assumed to be 30,000 lb/in^2. If drilling with an 8 in diameter bit with a contact area of $1\frac{1}{2}$ in^2 the down pressure must be 45,000 lb. The actual contact area of the bit will, of course, depend on its design and the amount of wear. Increase of wear will require increase of thrust because the contact area increases. For soft rock, bits with large, widely spaced, sharp-pointed teeth are used, large teeth spread the available thrust over a larger area which is easily accomplished since only a small total thrust is required to overcome the compressive strength of the rock, larger fragments are produced, less clogging and faster drilling results. On the other hand, in hard rock, finer teeth concentrate the available thrust. Bits for soft rock have the cones offset so that when rotating, each tooth is twisted slightly to impart shear, improving penetration without requiring extra thrust which might cause the hole to wander.

Thrust requirements, expressed in thrust required per inch of bit diameter, for various types of strata are:

Very soft friable rock 1000–1500 lb/in bit diameter
Soft sediment   2000–3000 lb/in bit diameter
Hard sediment   3000–4000 lb/in bit diameter
Very hard rock   5000–6000 lb/in bit diameter

The strength of roller bits varies as the square of the diameter, the surface area of the bearings virtually determines the strength so that an 8 in diameter bit will accept about four times the thrust that can be taken by a 4 in diameter bit before it fails. There is, therefore, a minimum theoretical bit size for each type of rock, the softer the rock the smaller the minimum size of bit which can be used. A bit $4\frac{3}{4}$ in in diameter, therefore, can be used only in the softest rock with a thrust of 4750 lb at 1000 lb/in diameter. Thus hard sedimentary rock probably has a minimum bit size of $6\frac{1}{4}$ in and the thrust needed would be about 20,000 lb. This explains why rotary drilling rigs at quarries need to be substantial.

Rotary speed depends upon the hardness and the condition of the strata with regard to fissuring, hard and fissured strata imposing shock loads on the bit bearings and requiring slower rotational speed of say 40 rpm compared with 80 rpm in soft rocks.

For flushing out cuttings by compressed air a velocity of 3000 ft/min up the hole is regarded as the optimum figure. With lower rates of air flow chippings are not cleared immediately and energy is absorbed comminuting the rock fragments and penetration rate suffers. In siliceous rocks, however, too high a velocity leads to bit abrasion. In many cases where the compressor is integral with the drilling frame this adds to the stability of the unit though a separate compressor or the quarry compressed air supply may be utilized. The air pressure required depends on the depth of the hole especially in water-bearing strata since the static water head must be overcome. For holes 100 ft deep a pressure of 40 lb/in^2 is required, deeper holes may require pressures up to 100 lb/in^2 in order to clear the bit.

Water flushing at a speed of 120 ft/min may be used for rotary, but not for down-the-hole drilling.

Hoisting capacity must be adequate to handle the heavy string of stems and lifting capacity in adverse conditions may be invaluable in freeing a clogged or wedged string.

Broad conclusions on the spheres of interest of the different types of drill in surface mining may be summed up as follows: in very soft rocks there is no satisfactory substitute for rotary drilling, in medium to hard sedimentary rocks rotary or down-the-hole drilling are equally applicable but much depends on the size of hole required while in the hardest rocks down-the-hole drills are more suitable. However, in cavernous or fissured rocks or where alternating hard and soft layers of rock occur or rocks with clay backs, the rotary drill is generally more successful. Rotary drills give a high rate of penetration but have a high capital cost. On the other hand it may be possible to replace four down-the-hole drill and compressor units with no loss in drilling since the penetration rate may be four times with a rotary drill and as a large-diameter hole is also possible burden and spacing may be increased to give improved production per footage drilled. The higher drilling speed might allow a group of amalgamated neighbouring quarries, and these days when rationalization is very much practised to husband financial resources, one expensive machine may service such a group with saving in labour and maintenance cost.

The St Ives Sand and Gravel Co. Ltd have adopted such a system of primary drilling at their three quarries in South Wales, namely Craig-yr-Hesg at Pontypridd, Trehir at Llanbradach and Sennybridge in Breconshire which is 33 miles from Pontypridd, but instead of maintaining separate bench or wagon drilling equipment at each of these quarries it was decided to use one crawler-mounted rig complete with air compressor to serve all three quarries.

An Atlas Copco BVB 61 Trac-rig powered by a screw compressor, also by Atlas Copco, delivering $600 \, \text{ft}^3/\text{min}$ at $100 \, \text{lb/in}^2$ is used, the compressor being permanently mounted on a lorry chassis and the Trac-rig being transported from one quarry to the next on a tipping lorry but a trailer is being designed to carry the rig behind the compressor lorry.

Blue Pennant sandstone, aggregate crushing value 17·7, crushing strength 25,000 to 34,000 lb/in^2, resistance to abrasion 15·7 and polished stone coefficient 0·74 is won at Craig-yr-Hesg and is used chiefly for aggregate, monumental stone and large square blocks for sea defence works, crazy paving and garden stone. Blocks of up to 8 tons are produced regularly but large block beds predominate in the lower main face and specimens up to 90 tons are available, the largest block produced to date weighed 180 tons.

At Trehir, Blue Pennant sandstone is also won, the aggregate crushing value is 22·0, the crushing strength is 28,000–30,000 lb/in^2, resistance to abrasion is 17·7 and polished stone coefficient is 0·68, so that the stone is similar to that produced at Craig-yr-Hesg and has similar uses.

At Sennybridge, granite is produced with an aggregate value of 17·0, a crushing strength of 32,000 to 37,000 lb/in^2, a resistance to abrasion of 9·2 and a polished stone coefficient of 0·71.

The drill rig has to be flexible to meet the varying conditions at each quarry. At Craig-yr-Hesg in particular the number of shots in a round has to be limited because of the proximity of houses. Originally this quarry was worked from top to bottom by means of small-diameter holes to a maximum depth of 180 ft. The masonry block stone was worked in a similar manner. It was then decided to split the 180 ft face into two levels with a down-the-hole drill with 4 in diameter bits. It was found that 3 in diameter holes were more economical since the drilling speed is increased by 16% and the fragmentation by 20%. Secondary blasting is rarely necessary as a 19 RB shovel fitted with a 30 cwt drop-ball is used. Similar results were obtained at Trehir and Sennybridge except that at Sennybridge 60 ft horizontal holes have been drilled for road construction and after this only 60 ft vertical holes will be drilled. Drilling speeds at all quarries vary from 45 to 52 ft per hour. A bit life of only 700 ft is obtained in the abrasive Blue Pennant. The rig allows the direction of drilling to be adjusted within wide limits, the boom is movable in the vertical plane and has a floating connection to the platform at its lower end. A swivelling bracket is at the upper end of the boom for the feed retainer of the hydraulically controlled feed beam and the bracket is fitted with pivots arranged so that the feed beam can be moved both longitudinally with and across the Trac-rig. All movements are effected by means of four hydraulic cylinders controlled by levers on the left-hand side of the rig.

Results have proved that a group of quarries not too far apart can use a 'commuting' drill rig economically if the drilling speed is high and it can be operated by one man.

The large rig designed for rotary drilling may be used also for large down-the-hole hammer drills but a specifically designed heavy down-the-hole rig cannot be used for rotary drilling except drag-bit drilling in certain circumstances.

The components of a rotary drill (Fig. 34) are a power source and transmission, diesel or electric, rotary system, pull-down hoist system, drilling mast, self-propelled mounting generally tracked, air compressor and drilling stem handling equipment. In addition dust extraction fans are usually provided.

Diesel power enables the rig to be self-contained and not dependent on electric power lines and sub-stations and has increased mobility. Most rigs at quarries are diesel powered. If the rig is truck mounted the prime mover may be the truck engine with a power take-off or an auxiliary may be provided.

The power may be transmitted for the various operations of the rig by mechanical, pneumatic or hydraulic means. The former is little used for blasthole drilling. Although compressed air is required for flushing it is notoriously inefficient as an agent of power transmission. To rotate the drill-string of a 6 in rotary drill may need a hydraulic motor developing 45 bhp but an air motor for the same job would require some 800 ft^3 per min of compressed air at 100 lb/in^2 which would require 200 bhp of primary power. When the other operations and their power requirements such as the pull-down system are taken into account the sole use of compressed air appears impracticable.

This objection does not apply to down-the-hole drilling as the torque and pull-down demand is less but, of course, hydraulic transmission is as or perhaps more suitable.

Fig. 34.   Crawler-mounted C650 Reichdrill (Consolidated Pneumatic).

Rotation is obtained by a top drive by which a hydraulic motor imparts rotary motion direct to the drill-string and at the end of each pass or stem the lead is disconnected and raised up the mast so that a new length of stem can be added directly to the string. The drill stems used in heavy rotary drilling weigh 260 lb for a 20 ft stem of $3\frac{1}{2}$ in diameter and 380 lb for a 20 ft stem of $4\frac{1}{2}$ in diameter and are therefore too heavy to be man-handled conveniently. Two methods available for doing this mechanically are known as the magazine and the continuous loading systems. Of the first method is the barrel loader which is attached to the mast and carries five or six stems in a rotating and swivelling rack. The loader is swung in at the end of each pass so that a new stem is positioned for coupling to the drill-string. A six-stem 20 ft barrel loader enables holes up to 140 ft to be drilled without recharging the barrel, there being one in the mast and six in the loader which is hydraulically operated and is indexed so that a fresh stem is offered each time the loader is swung in. It is charged by means of a winch.

In the second method a continuous system is employed which is better for deeper holes and one example is the side-arm loader in which stems are stored on the side of the rig horizontally. A single stem is lifted by a winch during drilling on to a side arm which is then swivelled to the vertical. At the end of each pass this stem is swung into position.

To prevent deviation of the hole in both vertical and angle drilling the diameter of the drill rods should be increased to give extra rigidity in the hole. A heavy rotary drill of 6–7 in diameter with a full range of equipment to drill to a depth of 200 ft costs some £30,000.

The lighter down-the-hole rigs cost £1500 to £5000 but the heavier rigs of this type to drill a 6 in hole at about 30 ft per hour in hard limestone and about 150 ft per hour in soft limestone would also cost £30,000. To justify a £30,000 drill of either type a production of at least three or four thousand tons must be visualized. The weight of a medium-sized rotary drill is about 15 tons, that of a heavy rotary drill about 30 tons.

PRIMARY BLASTING

Blasting operations in surface mining are divisible into two: primary blasting breaks the rock from the solid, secondary blasting reduces any large blocks to sizes convenient for loading.

Large-diameter holes

It is often preferred to use large-diameter drills where quarry faces are very high. The shotholes may range from 5 in to 9 in in diameter but $6\frac{5}{8}$ in is a fair average. The holes are usually drilled to floor level but in some cases it is better to drill 2 to 3 ft below floor level to make sure the rock breaks out satisfactorily at the toe.

Most high explosives and blasting agents are suitable for large-diameter hole blasting. Provided that normal care is taken in charging and stemming shotholes there is no risk of accidental explosion. In view of the large charges involved, however, it is a wise precaution to stack the explosives at a sufficient distance to prevent sympathetic detonation in the event of a charge detonating in any hole. Thus a 100 lb stack should be at least 35 ft from the nearest hole to be charged and a 500 lb stack at least 60 ft away. Cases of explosives should be opened as required and carried one at a time to a spot at least 6 ft from the hole. From these the cartridges should be passed singly.

Medium-diameter holes

These holes are from 3 in to 5 in in diameter and although the optimum depth of the faces on which they are used is 60 to 70 ft they are also used on faces up to 100 ft in depth. On such faces the general practice is to adopt full-face drilling.

Inclined holes, medium and large diameter

The potential hazard from slides and loose stones on high vertical faces, particularly when the rock is extensively jointed, can be reduced by working the face with a sloping or angled profile.

The face line and inclination with properly controlled blasting will conform to the geometric pattern of the shotholes as successive rounds are fired and to establish a suitable slope of the face, the shotholes have to be drilled at the required predetermined angle. The inclined drilling technique is often of advantage in improving breakage of the toe in addition to the advantage of safer working conditions. However, each case must be considered on its merits as the direction and nature of the jointing in the rock formation may make angle drilling impracticable. In any case accurate drilling in respect to angle of inclination and horizontal direction requires that the setting up of the drill in both planes is correct and the angle of inclination of the hole should be checked as soon as penetration has been established since an error of 2° will increase or decrease the burden by 3 ft 6 in in a 100 ft hole. Poor blasting and unbalanced shots result, and an uneven face line will be produced from variations in the drilling angles and convergence of holes. Holes should be straight without appreciable deviation and should be examined for deflection before charging and this requires that suitable drilling equipment should be selected which ensures accuracy of drilling. Both rotary and down-the-hole hammer drills have proved satisfactory in most deposits but the choice must depend upon the type of strata and its condition, the quantity and method of compressed air supply and, particularly, the skill of the operator. The optimum drilling angle must be determined empirically by careful experiment. Although angles of 45° are possible, the greater the angle of inclination from the vertical, the greater the problem of setting out and drilling the holes; angles between 10° and 25° from the vertical are sufficient to prevent overhang and present the minimum difficulty in drilling.

The explosives of the gelatinous type most in use are opencast gelignite or opencast gelignite 'Q' but combinations of two or more explosives may prove useful for certain conditions such as where the bottom bands of stone in a succession are particularly hard. Polar ammon gelatine dynamite is sometimes used as the base charge with opencast gelignite forming the main explosive or 'column' charge. The 'column' or upper part of the charge in a shothole is mainly to assist fragmentation throughout the face. A lower concentration of energy here may be adequate and a

TABLE VI

TYPICAL CHARGES AND BURDENS FOR PRIMARY BLASTING BY SHOTHOLE METHODS

Minimum finishing diameter of hole (in)	Cartridge diameter (in)	Depth of hole (ft)	Burden (ft)	Spacing (ft)	Rock yield (tons)	Explosive charge (lb)	Blasting ratio*	Tons of rock per foot of drilling
1	$\frac{7}{8}$	4	$2\frac{1}{2}$	3	2·3	5 ctgs†	4·0	0·6
1	$\frac{7}{8}$	5	3	3	3·3	7 ctgs†	4·2	0·7
$1\frac{1}{4}$	$1\frac{1}{8}$	8	$4\frac{1}{2}$	$4\frac{1}{2}$	12·0	$2\frac{3}{4}$ lb	4·4	1·5
$1\frac{3}{8}$	$1\frac{1}{4}$	10	5	5	18·5	4	4·6	1·9
$1\frac{3}{4}$	$1\frac{1}{2}$	12	6	6	32·0	7	4·6	2·7
$2\frac{1}{4}$	2	20	8	8	95	21	4·5	4·8
		25	8	8	120	26	4·5	4·8
		30	8	8	140	32	4·5	4·7
		40	8	8	190	42	4·5	4·9
3	$2\frac{1}{2}$	30	9	9	180	40	4·5	6
		40	9	9	240	55	4·5	6
		50	9	9	300	70	4·5	6
		60	9	9	360	80	4·5	6
$3\frac{1}{4}$	$2\frac{3}{4}$	50	10	10	370	85	4·5	7
		60	10	10	450	100	4·5	7
		70	10	10	520	115	4·5	7
		80	10	10	600	130	4·5	7
4	$3\frac{1}{4}$	40	12	12	425	95	4·5	11
		60	12	12	650	140	4·5	11
		80	12	12	850	190	4·5	11
		100	12	12	1100	230	4·5	11
$6\frac{5}{8}$	5	50	20	20	1500	330	4·5	30
		60	20	20	1800	400	4·5	30
		80	20	20	2400	530	4·5	30
		100	20	20	3000	660	4·5	30
$6\frac{5}{8}$	6	60	22	22	2200	475	4·5	36
		80	22	22	2800	630	4·5	36
		100	22	22	3600	800	4·5	36
9	8	70	25	25	3300	725	4·5	46
		80	25	25	3700	825	4·5	46
		90	25	25	4200	925	4·5	46
		100	25	25	4600	1050	4·5	46

* Tons of rock per pound of explosive.
† 9–10 cartridges to the pound.

lower density explosive such as Trimonite No 4 or the free-flowing Nobelite H can be used in place of part of the opencast gelignite or opencast gelignite 'Q'.

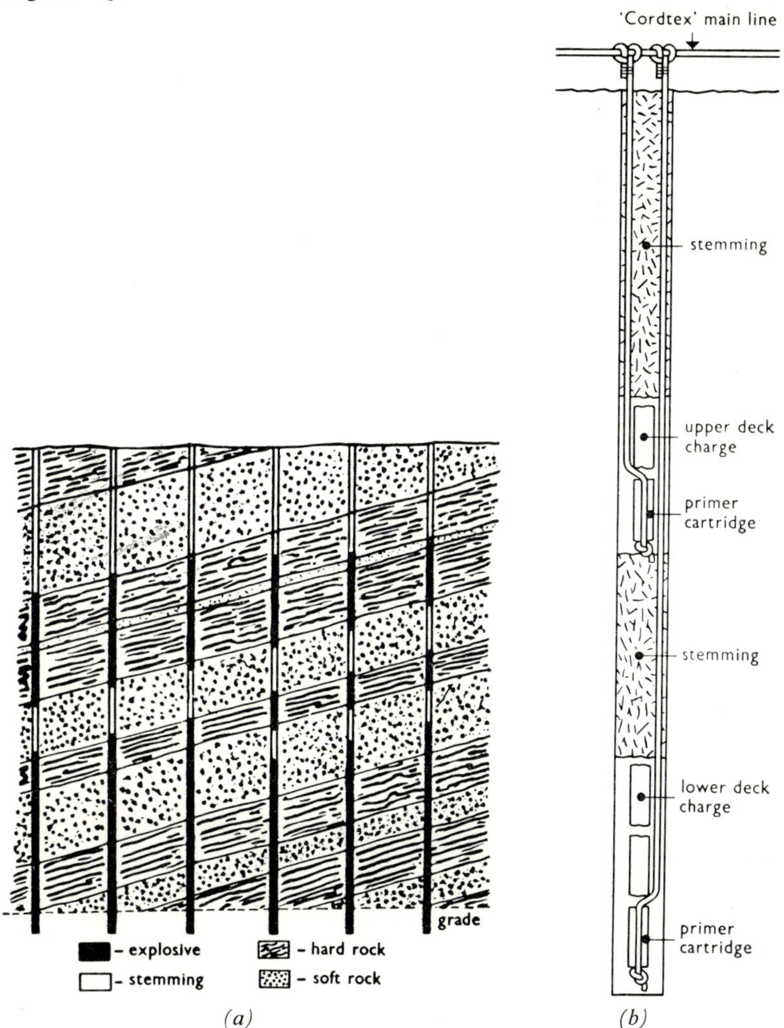

Fig. 35.   *(a) Deck loading of well-hole blast, (b) section through shothole showing deck loading.*

Charging ratios of 4 to 5 tons per pound of explosive are typical and Table VI gives typical charges and burdens for primary blasting together with blasting ratios and tons of rock per foot of drilling.

For certain conditions where vibration considerations do not restrict the quantities that may be fired, higher concentrations of explosives may be used. Primary blasting with heavier charges leads to higher efficiency in the working of mechanized quarries since better fragmentation is achieved, loading is easier, spillage is less and secondary blasting reduced.

Where hard bands occur and to obtain better fragmentation the technique of deck loading is used (Fig. 35) that is, separating the charges into sections by inserting lengths of stemming between groups of cartridges. It is sometimes preferable to use a separate primer cartridge and Cordtex line for each deck charge. The primer cartridge is the first cartridge placed in the hole, and as soon as it has reached the bottom of the hole, the Cordtex must be cut at the reel leaving sufficient slack for taking up in stemming and for making a joint in the main line. To prevent it from being lost during loading or stemming the Cordtex lead should be anchored. The hole is stemmed to the mouth with sand or quarry fines when all the cartridges have been inserted and after all the holes have been charged and stemmed the shots are coupled up for firing. Where detonating relays are being used the Cordtex lead or leads from each hole is connected to the main Cordtex line (Fig. 36 *a* and *b*) and a detonating relay inserted in this line between each shothole, the main line can then be initiated either by an electric detonator or by safety fuse and a plain detonator.

Where short-delay detonators are used the Cordtex lead or bunched leads from each hole is initiated by the appropriate short-delay detonator (Fig. 36 *c*, *d*, *e*, and *f*). Typical arrangements using detonating relays and short delay detonators for single-row, multi-row blasting and to give minimum vibration and/or improved fragmentation are shown in Fig. 36.

When charging, stemming and connecting up a blast, operators should stand on the side of the holes remote from the quarry face and no drills or metal equipment of any kind should be introduced into a hole once explosive has been lowered into it.

Where simultaneous firing is preferred Fig. 37 shows the method of connection for a single row of shots and Fig. 38 a suitable method of connection when several rows of holes are to be fired simultaneously. A ring main is used to ensure the initiation of all holes without risk of cut-offs and only one detonator is required for this arrangement.

Horizontal holes

Blasts using horizontal holes can be fired using detonating relays and Cordtex delay and short-delay detonators in the same manner as vertical holes. The shots can be fired either with a detonating relay between each shot or between groups, according to the spacing between the holes. To avoid cut-offs no detonating relay should be inserted unless the spacing is at least 6 ft, preferably more.

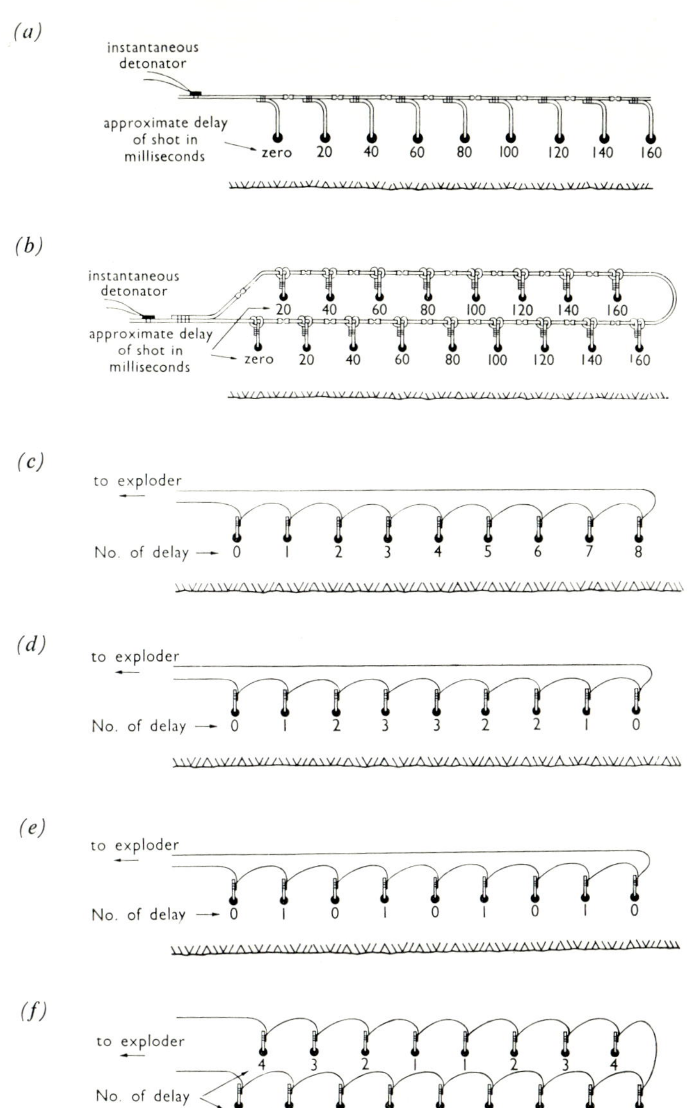

Fig. 36. *Arrangement of detonating relays in (a) single-row blasting, (b) multi-row blasting (c) arrangement of short-delay detonators to give minimum vibration, (d) to give improved fragmentation and reduced vibration, (e) to give maximum fragmentation, (f) in multi-row blasting (ICI).*

Blasts may also be arranged using both vertical and horizontal holes. Occasions arise, particularly with small-diameter vertical holes, when the toe burden is excessive and to assist in breaking out this heavy toe, horizontal holes are drilled as near floor level as possible (Fig. 39). The vertical and horizontal shots are often fired simultaneously, but sometimes better results are obtained with a short delay between the vertical and horizontal

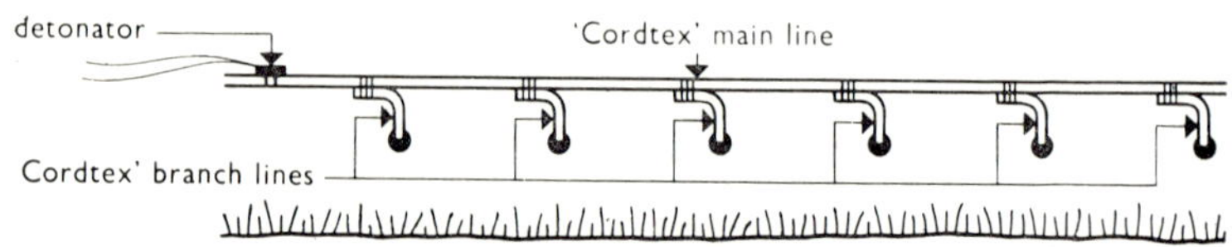

Fig. 37.   Connecting one row of holes for simultaneous firing (ICI).

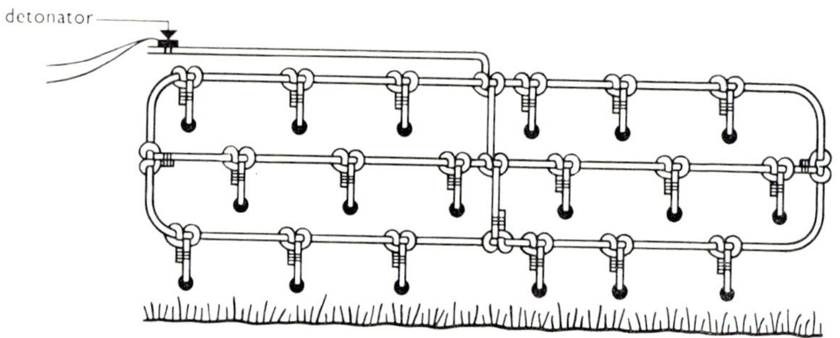

Fig. 38.   Connecting several rows of holes with ring main (ICI).

holes by introducing a short-delay detonator or a detonating relay in the appropriate Cordtex line between the detonator and the first shot, the two main lines being initiated by instantaneous electric detonators connected in series.

Small-diameter holes

Hand-held, jackhammer, pneumatic and wagon-mounted drills are applied to three methods of blasting, short-hole blasting, narrow-diameter long-hole blasting and springing or chambering.

Although most managements prefer full-face blasting, sometimes, particularly where the structure of the rock is disturbed and irregular, the faces are worked in steps or benches by blasting with short vertical holes 3 to 16 ft deep, drilled in a line parallel to the face. The burden, the horizontal distance from the bottom of the hole to the free face, should always be less than the depth of the hole. Average yields are $3\frac{1}{2}$ to 5 tons of rock per pound of explosive such as polar ammon gelignite (Fig. 40).

High yields of rock under favourable conditions can be obtained by drilling angled horizontal 'breast' holes near the quarry floor particularly in gneous rocks with well-marked columnar or hexagonal jointing. Horizontal holes, 12 to 16 ft in length, are drilled at intervals into the face at angles of 45° to 60° to the toe line. The toe is blasted out when the charges are fired

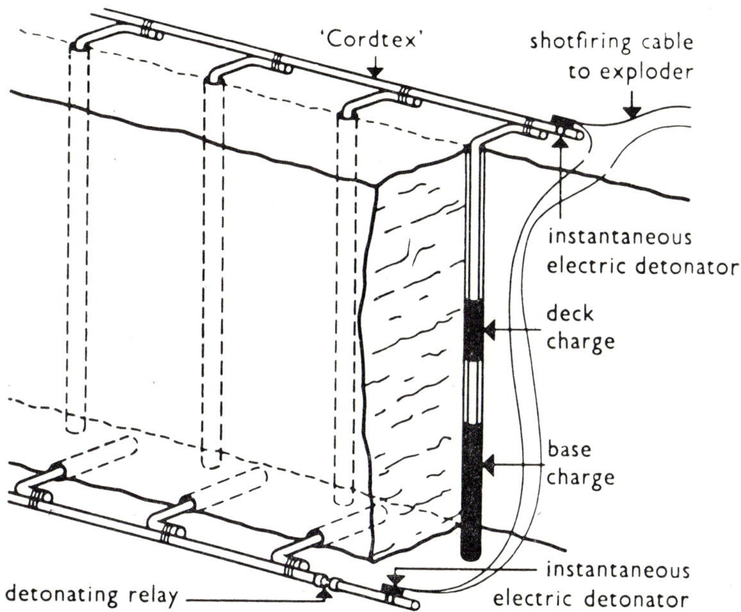

Fig. 39.   *Arrangement of vertical and horizontal holes in the same blast (ICI).*

and the unsupported rock above slips down along the well-defined columnar joints and collapses on the floor; yields as high as 15 tons per pound of explosive have been obtained.

When charging short shotholes they should first be cleaned by blowing out with compressed air. The cartridges are then inserted one at a time and each is pressed home to make good contact by means of a wooden rod. The primer cartridge is then pushed or dropped into the hole until it rests against the main charge, the primer cartridge should be inserted so that the detonator points to the full length of the charge.

Tamping or stemming of the charge follows and its importance cannot be over-emphasized since it increases the effectiveness of the charge by ensuring that the maximum work is done. Dry sand or quarry fines provide efficient stemming for vertical holes but for horizontal holes it is easier to use a cohesive material such as sand-clay stemming in the proportions of 3 to 1 or to use sand or quarry fines filled into suitable paper bags approximately the diameter of the borehole. By firing a number of shots in one operation initiated either by electric detonators connected in series or by Cordtex, either simultaneous or delayed firing may be adopted depending on conditions.

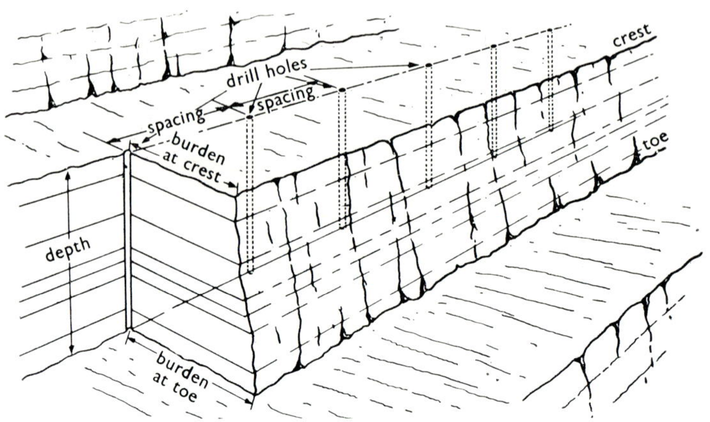

Fig. 40.   *Short-hole blasting or benching (ICI).*

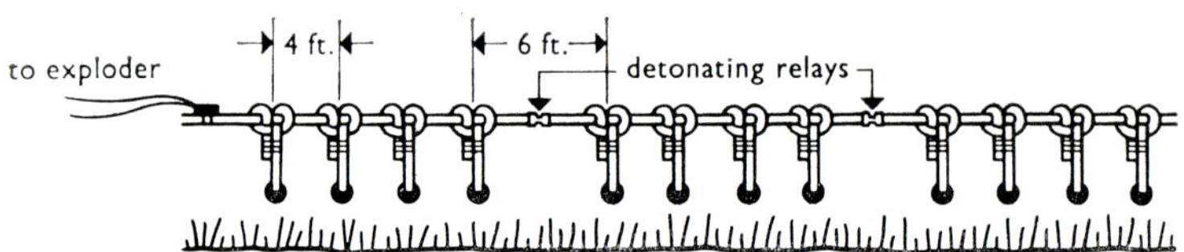

Fig. 41.   *Arrangement for small-diameter holes with spacing of less than 6 ft (ICI).*

Small-diameter long-hole blasting is applicable to a wide vairety of surface mining work. Well-fragmented rock piles of a few thousand tons can be obtained suitable for efficient easy excavator loading, the yield obtained often reaches 5 to 8 tons of rock per pound of explosive. The total charge in each blast is generally much smaller than in large-diameter holes or in heading blasting and, consequently, there is less risk of excessive ground vibrations which could cause damage to nearby property. Where the spacing of shotholes is less than 6 ft a delay of 20 milliseconds may result in cut-offs. In order to keep vibrations to a minimum, holes may be drilled in groups leaving a space of 6 ft and inserting detonating relays between groups. Figure 41 shows the arrangement using groups of four shots.

Tungsten carbide drilling bits are commonly used for small-diameter holes and are fitted with extension steels so that the full height of the quarry face can be drilled.

Blasting in small-diameter long holes is normally applied to vertical faces, although with diamond drilling equipment long holes can be drilled at any angle to follow the profile of the face. The holes are drilled to floor level in normal procedure at intervals of 6 to 9 ft and theoretically the burden to the free face should be about the same; high faces, 50 ft and more, are rarely quite vertical and toe burdens of 10 to 12 ft are common, so that to assist in breaking out a heavy toe, horizontal holes are as near floor level as possible. If each vertical hole is charged to within 9 to 12 ft of the mouth good fragmentation will be achieved. To assist efficient breakage of the toe it is often advantageous to charge the bottom 15 to 20 ft of the hole with polar ammon gelatine dynamite or a high-strength gelatinous explosive such as polar ammon gelignite primed with Cordtex may be used throughout or, in suitable conditions, the lower-density explosives such as Belex or the TNT–Ammonium nitrate powders may be used, particularly for the upper portion of the charge where lower energy concentration is often adequate.

Horizontal holes are drilled to give increased concentration of energy in the toe and should be charged to within 3 to 4 ft of the mouth with a high-strength gelatinous explosive such as polar ammon gelignite or polar ammon gelatine dynamite primed with Cordtex and stemmed efficiently.

Heading blasting

A specialized blasting method known as heading blasting uses small tunnels 2 ft to 3 ft × 4 ft to 4 ft 6 in (Fig. 42*a*), which are driven into the face of the rock to a depth that depends on the height of the face but is usually about 50 ft and from this side tunnels are driven parallel to the outside face of the rock and chambers are made at intervals of 25 to 35 ft along these side or back drives to hold the explosive charges. Figure 42*b* shows the method of initiating a heavy blast with Cordtex. The detonating fuse is slung from pegs driven into shallow holes 6 ft apart one or two inches from the roof of the heading and during stemming it should be covered with turves to protect it from damage. The Quarries (Explosives) Regulations 1959, require that a double-core detonating fuse be used for firing heading blasts. Taped double Cordtex consisting of two lines of standard Cordtex totally enclosed in a strong reinforcing tape, may be used. The Cordtex connexions shown in Fig. 42*b* are arranged to provide alternative paths for the detonation wave and to minimize the risk of a misfire if by chance a line of fuse should be cut during stemming. Each chamber should be initiated with a priming charge of about 5 lb of a gelatinous explosive with the Cordtex line to or threaded through it. This primer should be placed in the centre of the main explosive charge so that the satisfactory detonation of the lower sensitivity explosives used for heading blasts can be assured.

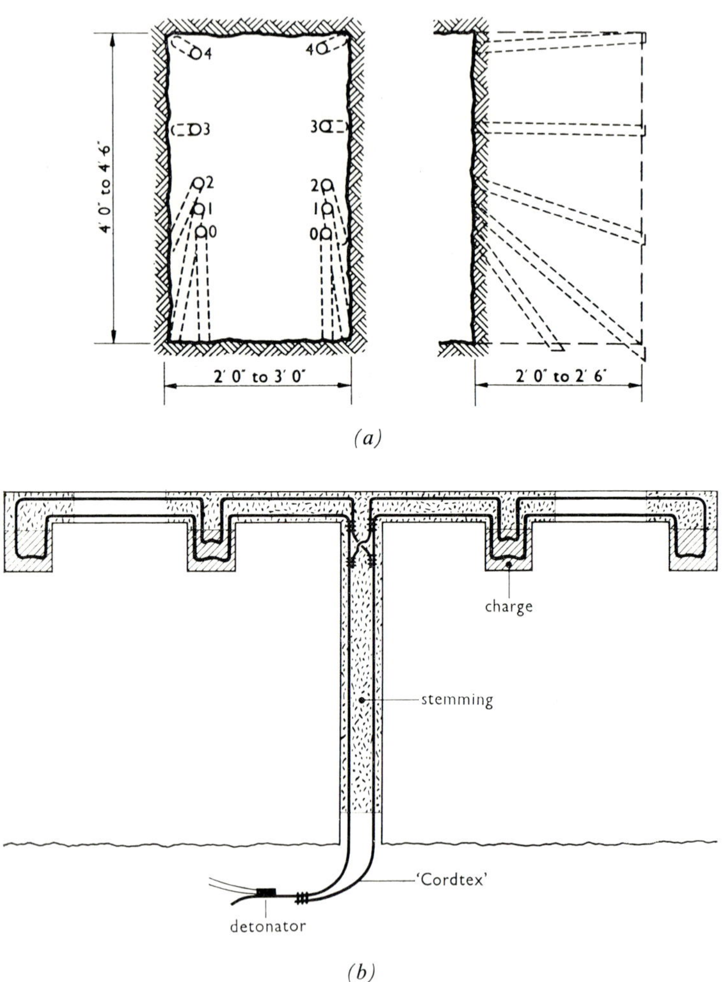

Fig. 42. Heading blasting; (a) drag cut for tunnel in heading blasts. Advance for round 2 ft to 2 ft 6 in. Explosive: polar ammon gelignite; charging details—two cut holes, No. 0 delay 6–8 oz, No. 1 delay 10–12 oz; two easers, No. 2 delay 8–10 oz, No. 3 delay 8–10 oz; two trimmers, No. 4 delay 6–8 oz. (b) Charges and stemming in heading blast, showing Cordex connexions (ICI).

SPRINGING AND CHAMBERING

Economy in drilling can be achieved in some quarries by 'springing' using blackpowder or 'chambering' using high explosives. In the former gradually increasing charges of blackpowder are fired in each hole to open up and enlarge cracks in the bedding and the joint planes, so that a large charge can be used for the final blast. The method is used mainly in quarries producing stone for monumental purposes. The results depend entirely upon the judgement of the shotfirer since the exact position of the blackpowder is often difficult to estimate and there are also certain hazards peculiar to the springing method such as smouldering material or hot spots left from previous springing which cause premature ignition and for this reason the hole must not be recharged until an interval of at least two hours has elapsed, after the firing of the previous shot. Steps must also be taken to ensure that the hole is cool and that any remnants of smouldering material are quenched and blackpowder must not be charged into any hole in which a shot has previously been fired by means of safety fuse.

It is difficult to estimate the weight of the final charge and overcharging can take place resulting in flying rock and a risk of double ignition since portions of the blackpowder can become separated in fissures in the rock.

Where there are few cracks or other planes of weakness as in igneous rocks the springing method using blackpowder cannot be applied successfully but similar results can be obtained by chambering using high explosives. Polar ammon gelignite or a free-flowing explosive such as Trimonite No 1 is suitable for this work. Fewer chambering shots are required with high explosive than with springing and the estimation of the final charge is easier because the position of the explosive is known more definitely. The final charges are normally from a half to two-thirds the weight of those of blackpowder. Similar safety precautions should be taken with high explosive except that the waiting period between shots should be a minimum of 30 minutes.

EXPLOSIVES IN OPENCAST MINING

Short-delay blasting using detonating relays is used in opencast mining to reduce ground vibration, improve fragmentation and reduce overbreak. It is usual, where it is important to reduce ground vibration, to initiate the charges in each hole on a separate delay but where ground vibration is not a major problem, initiation of rows of holes simultaneously but with a delay between individual rows along the overburden is often practised.

If the delays are so arranged as to 'peel off' from the next succeeding highwall, the result is less overbreak on the highwall side, leaving a more solid highwall and safer working conditions in the cut. Figure 43 shows the use of detonating relays in opencast mining and Fig. 44 the use of short-delay detonators and detonating relays in an opencast site.

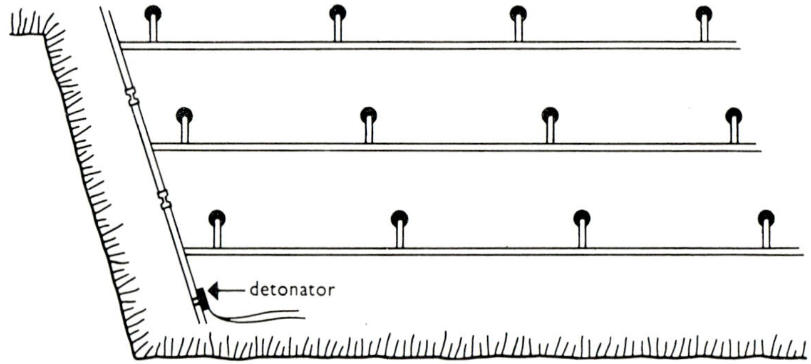

Fig. 43. *Typical delay pattern for opencast mining using detonating relays. For clarity details of the joints have been omitted.*

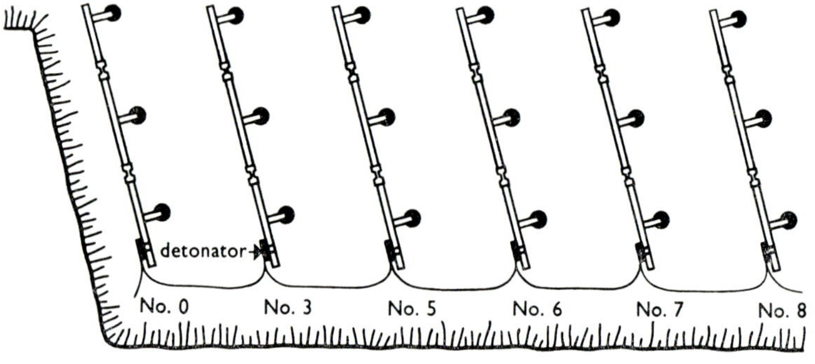

Fig. 44. *Arrangement for opencast mining using both detonating relays and short-delay detonators.*

SECONDARY BLASTING

Although secondary blasting is reduced as much as possible to save time and to improve safety, some further blasting is often necessary whatever method of primary blasting is adopted to break large pieces of stone into a size suitable for loading. Previously with hand loading of stone pop-shooting was the usual method adopted; today, when greater productivity is required, plaster shooting is more popular and since no shothole is needed it is quicker, more convenient and more economical. A charge of one or more cartridges is primed with a detonator and safety fuse or an electric detonator and laid on the surface of the boulder to be broken. It is then covered with a shovelful of plastic clay, which is pressed into position by hand, the surface of the stone being wetted before plastering and the clay is well pressed down to make good contact with the stone round the explosive. The most suitable explosive should have a high velocity and be of the high-strength gelatinous type and a special explosive with these characteristics has been developed for this work, plaster gelatine. Plastic igniter cord can be used to ignite the safety fuse of a group of shots.

In plaster shooting the weight of explosive required for boulder thicknesses will vary with the type of rock and best results are obtained with rocks which are hard and brittle. A guide to the charge for different thicknesses of rock is the following:

Stone thickness	Charge (oz)
1 ft to 1 ft 6 in	4
1 ft 6 in to 2 ft	6
2 ft to 2 ft 6 in	8
2 ft 6 in to 3 ft	10–12
3 ft to 3 ft 6 in	12–16

Plaster shooting has the following advantages:

1. No drilling is required thus saving labour and compressed air which is a consideration with hard and abrasive rocks difficult to drill.
2. A group of plaster shots can be prepared much more quickly than the same number of pop-shots.
3. The stone is broken where it stands and stones are not scattered over a wide area.
4. There is less likelihood of damage from flying debris which is advantageous where mechanized quarrying is in use since equipment need not be moved very far to a place of safety and less loading time is lost.

For pop-shooting a hole is generally drilled just beyond the centre of the boulder to be broken but with some rocks shotholes 12 in deep are sufficient to break large boulders.

The charge required varies with the size of the stone but for average work a boulder 3 ft × 3 ft × 2 ft would require a charge of $1\frac{1}{2}$ oz of polar ammon gelignite. The shots can be fired either by safety fuse and plain detonators or by electric shotfiring. The safety fuse can be ignited by fuse igniters or plastic igniter cord where a large number of shots can be fired with only one application of flame to the circuit. If electric shotfiring is practised the electric detonators are connected up in series and the shots fired simultaneously.

Pop-shooting may seem at first sight to be a very inexpensive and effective way of reducing stone to the necessary size, but it entails full-time drillers and the cost of running and maintaining a compressor and piping. It is also necessary to withdraw mechanical equipment to a safe distance since there is considerable scatter of rock when firing pop-shots. The one serious accident with explosives in 1965 occured to a shotfirer who sustained facial injuries and a fractured leg when attempting to remove stemming from a misfired pop-shot.

To reduce secondary blasting an impact pile hammer has been used to break boulders. Another alternative is the drop-ball technique, the ball generally weighing 35 cwt to 2 tons. The impact hammer gives better control, can be used to move boulders into position and does not produce flying splinters of rock.

To reduce danger from flying rock when blasting, the Columbus blast mat has been tried out at Allington Quarry, Kent.

The Columbus blast mat is blown into the air in a convex form when the explosive charge is fired due to its construction, at the same time restricting the scatter of debris. Each mat weighs 108 lb and is constructed of 20 in lengths of $2\frac{1}{2}$ in outer diameter $\frac{3}{4}$ in wall, 1 in bore special polythene tubing, formerly of Swedish manufacture but now made in the United Kingdom, and is in three sections, each containing 24 lengths, held together by a network of wires which are first drawn through holes disposed transversely through the tubes and secondly through the insides of the tubes themselves. The three-section mat is 5 ft^2 weighing just under 1 cwt, can be carried by one man and placed over the charge to be detonated. The first tests were carried out at Allington Quarry, on an 18 ft face, with little or no damage to the mat which has survived 300 blasts and in fact the prototype mat, a year old, is still in use.

DEALING WITH MISFIRES

The greatest care is requisite in dealing with a misfire, no person should be allowed to approach until it has exploded or until an interval of not less than one hour in the case of firing by fuse, or ten minutes in the case of firing by electricity has elapsed. With electric shotfiring, the cable must be disconnected and the handle removed from the exploder. After the prescribed time, the cable and connexions should be thoroughly examined and any faults remedied. The circuit is then retested from the firing point and, if in order, the shot is fired but, if on retesting, the circuit is found to be incomplete, or if the shot still does not fire, the fault must be somewhere in the borehole.

If the misfire is in a short small-diameter hole the normal method of dealing with it is to drill a relieving hole in line with the misfired hole and at a safe distance of not less, at any point, than 12 in from it. The broken material from the firing of the hole should be carefully searched for any undetonated explosives or detonators.

An alternative procedure, which may often be found preferable to the above, is to remove the stemming by washing out with water or blowing ou twith compressed air. Another primer is then inserted, and the charge restemmed and fired. The air or water must be fed from a non-ferrous nozzle or pipe or through a rubber hose. It is not safe to use an iron or steel instrument to dig or scrape out the stemming. Washing out with water is the preferred method but where blackpowder has been used, the whole charge should be washed out. This method is suitable where sand or coarse dust has been used for stemming, but is not possible for solidly tamped clay.

Where the burden on the misfired shot has been removed by other shots to such an extent that there is risk of excessive rock scatter if it is fired by another primer, the hole should be dealt with in stages. Relieving holes 3 to 4 ft deep should be drilled at the top and small charges fired. Further shotholes can be drilled as the resulting loose rock is worked off, but the bottom of the relieving holes must not come within 3 ft of the main charge.

The usual search for undetonated explosive must be made when the main charge is dislodged. This procedure also applies to misfires in deep holes where drilling a long relieving hole would involve the risk of drilling into the charged hole.

Misfired plaster shots

With a misfired plaster shot, the mud-cap or covering can be removed and another detonator inserted. The charge is then covered again and the second detonator fired. No metal tools should be used in removing the mud-cap.

THE STORAGE OF EXPLOSIVES

The Explosive Acts of 1875 and 1923 and various Orders made under these Acts govern the storage and transport of explosives. The regulations stipulate that a suitable licensed or authorized place must be allocated for storing explosives. The needs of the quarry will determine the type of storage accommodation required which may be in 'Registered Premises' or in a 'Store' or 'Magazine'.

Registered premises provide suitable storage for explosives where the consumption is small. Two forms are available, Model A storage requires a substantially constructed building and allows the storage of up to 60 lb of mixed explosives and detonators (or up to 200 lb of gunpowder if this is stored alone). Model B storage covers quantities up to 15 lb of mixed explosives and detonators (or up to 50 lb of gunpowder if this is stored alone) and may consist of a substantial receptacle such as a cupboard, box or drawer. It is permissible to keep detonators and high explosives in the same premises provided the detonators are in separate substantial locked receptacles.

These premises must be registered with the local authority and a police certificate or licence must be held by the person responsible and both registration and police certificate or licence must be renewed annually. .

'Stores for Explosives' are used for larger quantities of explosives which may be kept in buildings licensed by the local authorities as 'Stores for Explosives'. The occupier must hold a police certificate or licence as with Registered Premises which must be renewed annually. Detonators must be kept in a separate licensed annexe or building with this form of storage. Under the Stores for Explosives Order, 1951, the quantities allowed are shown in Table VII.

TABLE VII

STORAGE OF EXPLOSIVES

Division	Quantities of high or mixed explosives, lb	Distance from road, path, etc., ft	Distance from house, workshop, railway, etc., ft
A	150	75	85
B	300	75	130
C	1000	146	292
D	2000	230	460
E	4000	352	704

Greater quantities of explosives than those given in Table VII find storage accommodation in magazines, the licensing authority for which is the Home Office through HM Inspectors of Explosives, Whitehall, London, SW1. The distances from protected works are greater for magazines than for licensed stores.

Lightning conductors must be provided on all explosive storage buildings licensed for 1000 lb or upwards of explosives. All buildings must be secured against illegal entry and a warning notice to trespassers must be displayed.

Magazines or stores should be lined with wood or other suitable non-metallic materials. The floor, benches and shelves inside the building should be frequently swept clean and all brushings removed and destroyed. A pair of overshoes should always be kept near the door so that people entering will step from the dirty outside floor into the overshoes and, when leaving, step out of the shoes on to the outside floor leaving the shoes on the clean floor inside.

REFERENCES

'Effect of Weather on Sound Transmission from Explosive Shots', R. L. Grant, J. N. Murphy and M. L. Bowser, US Bureau of Mines, *R.I.* 6921. 1967.

Blasting in Quarries, 4th edition. I.C.I. 1965.

'Developments in Blasting Techniques in Opencast Mining and Quarrying', B. J. Kochanowsky, *Opencast Mining, Quarrying and Alluvial Mining,* Institution of Mining and Metallurgy. 1965, p. 227.

'Symposium on Ammonium Nitrate Blasting Agents', *Journal of the South African Institute of Mining and Metallurgy.* July 1964, Vol. 64, Part II.

'Primary Blasting in Quarries', R. Westwater, *Mine and Quarry Engineering,* July 1965, p. 285.

'Slurry Blasting Agents', V. O. and M. J. Cook and W. O. Ursenbach, *Mining Magazine,* August, 1967, p. 80.

'Rock Fracture by Water Jet Impact', I. W. Farmer, *Colliery Engineering,* January 1967, p. 23.

'The Blasting Action of High Explosives', B. J. Greenland, *Quarry Managers' Journal,* September 1967, p. 333.

'Down-the-Hole Drilling using Elevated Air Pressures', M. G. Adamson *Quarry Managers' Journal,* August 1967, p. 295.

'Slurry Explosives and Mixer Pump Trucks. Their Role in Rock Blasting', Brian J. Greenland and J. D. Knowles, *Quarry Managers' Journal,* May 1968, p. 187.

CHAPTER 6

QUARRYING HARD ROCKS

Every kind of fully or partly cemented or consolidated rock, igneous, metamorphic or sedimentary, has, in some locality, been used for building but certain properties render some varieties particularly suitable for the purpose. Essential properties, apart from considerations of divisional planes, jointing, bedding or cleavage, are those associated with strength and durability. The material must not only bear the stresses the engineer or architect wishes to subject it to but also it must continue to do so after it has been in place many years or even generations.

Durability means resistance to the point of immunity from the erosive action of an industrial or city atmosphere, or perhaps sea-water and also the mechanical effects of temperature changes and of frost action. However, some slight chemical weathering to 'mellow' the surface is aesthetically desirable.

The hardness of stone depends not so much on the mineral hardness of its major constituents as on the fit or state of aggregation of the minor particles between them.

A loosely cemented sandstone may be friable, that is, may be crumbled between the finger and thumb, while a compact limestone may be accounted a hard and durable stone, though the hardness of calcite is much less than that of quartz. Medium-grained rock is generally preferred to coarse-grained rock.

Jointing in an igneous rock or jointing and bedding in a sedimentary rock determine the size and regularity of the natural blocks obtainable and a rock may be valueless as a building stone if the joints are too close together.

In igneous rocks, and related to jointing, are the partings known as rift and grain which are due to some orientation of constituent mineral crystals having cleavage, or to microscopic flaws and may be related to strain suffered by the rock during or after consolidation. The quarryman and stone-dresser locate and use these planes of weakness though they may not be visible to most observers. In sedimentary rocks joints and bedding planes are the partings which should control both quarrying and working of the stones. Even the most reliable freestones only give good service in buildings if they are laid on their natural bed, that is when placed with the bedding planes horizontal.

Much construction work carried out in masonry or quarried stone is now carried out in artificial stone or concrete, often reinforced.

Many of the properties required of a good road metal are in common with what is required in a building stone with greater emphasis on compactness and resistance to abrasion and to weathering, but rock which is closely and irregularly jointed may also be satisfactory since the stone must be crushed and screened to size and for this reason many fine-grained igneous rocks and limestones that are unsuitable for building are acceptable as road metal.

At the beginning of the century a certain amount of incipient weathering was desirable so that the aggregate might 'bind' with water and fine-grained basic igneous rocks where they occurred in sufficiently large masses were greatly esteemed. With the increase in motor traffic, however, a new factor was introduced in macadamized road construction and ability to bind with water ceased to be of importance but behaviour with tar and oil-residue binders became of prime importance. Diorites and gabbros where available in large quantities and, particularly dolerites, are among the most reliable rock types for road purposes. Acid igneous rocks do not hold tar as well but their screened chippings make a non-slip surface dressing. Among sedimentary rocks some Lower Carboniferous limestones make sufficiently strong aggregates and compact Cambrian and Ordovician quartzites are useful for top dressing but most sedimentary rocks crumble or break too easily along planes of structural weakness such as bedding and lamination.

The physical properties of rock for aggregate, the crushing values, impact value and abrasion value are all of importance and will also influence the choice of crusher to be installed. If the figures are high, for example, a typical limestone would have an aggregate crushing value of 19, impact value 18 and aggregate abrasion value of 10, then impact breakers and hammer mills should be considered but if the figures are low, such as aggregate crushing value 12, impact value 12 and aggregate abrasion value 2, indicating a hard abrasive stone, then jaw or gyratory crushers will probably be required.

Of the various tests that are used to define the properties of roadstone that for measuring their susceptibility to polishing by traffic is one which is of considerable concern to the quarry since its value may be the main factor determining the suitability of the stone from that quarry for use in the wearing course of a bituminous road surfacing. The original test was standardized in 1960 and revised in March 1965 as an amendment of BS 812: 1960. The importance of the problem was first realized in the late 1940s when attention was drawn to the fact that road sites in S.E. England with a high incidence of skidding accidents frequently had surfacing made with a roadstone which, although excellent in other respects, become highly polished by the action of traffic.

There is correlation between skid resistance and accidents and between skid resistance and polished stone value but it is difficult to relate accidents directly to polished stone value. For adequate skid resistance a polished stone value of 60 on the 1960 scale should be specified and recent work has shown the figure should be 65 while in terms of 1965 values these figures should be 55 and 60 but it is better to have as high a polished stone value as possible.

GRANITE

Many of the breakwaters, docks and other major engineering works were constructed of Cornish granite and in 1946 J. C. Annear & Co. Ltd acquired two old granite quarries near Penryn in Cornwall and moved their concrete works to an area between the two quarries. In 1964 orders were placed for a new crushing and screening plant and a concrete batching plant alongside the old plant, with which it was fully integrated. An old quarry about a mile from the works, Rosemenowes Quarry, was included in a parcel of granite reserves purchased in 1962 and quarrying began again with the clearing up of the old waste banks obscuring the quarry face and further reserves were purchased immediately opposite the works.

Well-defined irregular bedding planes and jointing often characterize the granites of Cornwall, giving them a blocky appearance. The stone is a coarse-grained granite in which orthoclase felspar is dominant, generally as large white crystals in association with one of the plagioclase felspars. The other minerals, embedded in a mosaic of transparent or bluish-white quartz granules, are brown and white micas and black tourmaline.

Great difficulty was experienced when the old face was reopened in preparing it for deep-hole blasting as the old method of working had been to develop a number of small benches from which individual blocks of stone were blasted out. A clean open face was, however, obtained 150 ft long and 70 to 80 ft high by careful working to reduce the irregular profile.

Shotholes are drilled at 6 ft spacing with 6 ft burden using a Holman Vole rig mounted on a tractor, the compressed air power being supplied by a compressor also mounted on the tractor, the drill being equipped with a $3\frac{1}{4}$ in diameter cruciform tungsten carbide insert bit attached to the down-the-hole hammer. Each hole is drilled to 2 ft below the quarry floor in $1\frac{1}{2}$ to 2 shifts at a penetration rate of 8 to 10 ft per hour, half the face being blasted at a time. Shotholes are charged with $2\frac{3}{4}$ in diameter cartridges of opencast gelignite and detonated by Cordtex lines connected through short-delay detonators, the average yield being $3\frac{1}{2}$ tons of granite per lb of explosive. Fragmentation is normally good but when the top bed of rock is more than 10 ft thick or when an undetected jointing plane runs across the face near the line of drilling, difficulty may be experienced and excessive quantities of large blocks may be blown down in the rock pile. All secondary breaking is done by blasting, hand-held Holman F type drills with either $1\frac{1}{8}$ in chisel or $1\frac{7}{16}$ in cruciform bits are powered from an old Holman TA 13 compressor mounted on the back of a 12-year-old Muir-Hill wheeled loader with an air receiver and supply take-off fitted at the front. Particular care has to be exercised in secondary blasting to avoid damage to neighbouring houses. Generally $\frac{7}{8}$ in by $1\frac{1}{4}$ in polar ammon gelignite cartridges are used but plaster type explosive is occasionally used, the yield of secondary blasting being $13\frac{1}{2}$ tons per lb.

A Chaseside SLM 1000 shovel with a $1\frac{1}{2}$ yd^3 capacity scoop is used for loading at the face and is also used for stripping the 4 to 10 ft of over-burden and for general work in the quarry.

It is important that the stone produced is kept clean as it is ultimately used for concrete manufacture, so precautions are taken during loading operations to ensure that as little dirt as possible goes to the primary crusher which is a 36 in by 24 in Kue-Ken jaw crusher driven by a 50 hp AEI motor.

At the Croft Granite Division in Leicestershire of England China Clays Quarries Ltd of the group of the same name, granite is won from the quarry face by blasting, each blast yielding 12,000 to 15,000 tons with no secondary blasting, further reduction being obtained by a 35 cwt drop ball.

At present $2\frac{1}{2}$ yd^3 Ruston-Bucyrus 54 and $1\frac{1}{2}$ yd^3 Ruston-Bucyrus 18 face shovels are employed but it is intended to replace them with a 71 Ruston-Bucyrus with a $3\frac{1}{2}$ yd^3 capacity bucket.

The stone is loaded into three SN 35 Aveling-Barford 35 ton dump trucks working 19 hours a day for 5 days per week which transport the stone 200 to 600 yards from face to the primary crusher but includes a 50 ft ramp with a 14% gradient and future operations will entail a laden run down a $12\frac{1}{2}$% gradient. The trucks are powered by 450 bhp Rolls-Royce diesel engines and full advantage is taken of their 30 ft turning circle at the 48 in by 60 in double-toggle primary crusher. Secondary crushing is by a 20B gyratory and the stone is then elevated to the tertiary crushers and screens at the top of the quarry at ground level.

A large proportion of the quarry's output is taken by Croft Granite's own concrete works where all products are manufactured from granite aggregate, which occupies a site of roughly 80 acres adjacent to the quarry. Other uses for the stone produced include road sub-base, sewage filter course and stone for coating plants.

The maintenance of trucks is a very important consideration in the organization of Croft Granite Division. Full servicing facilities are pro-vided in a modern, well-equipped workshop and operating records of trucks are accurately maintained. Each SN 35 truck is withdrawn from service for four hours when it receives lubrication and routine servicing by its operator. Each Friday one dump truck is given a full day checkover by maintenance fitters. This schedule is rigidly observed and has resulted in the virtual elimination of down-time through mechanical failure.

Tyre cost can prove a major item in transport expenditure but regular attention to tyre treads and pressures has resulted in tyre life averaging 12 months for rear wheels and 18 months for front wheels.

Except for dense night fogs quarrying is carried on in all types of weather, a system of flood-lighting provides for safe working at night and safety organization is excellent, safety helmets being issued to everyone on first entering the quarry.

Planning permission at present granted for the area ensures production for a further quarter of a century.

GNEISS

Dalby is situated at the southern extremity of Sweden and lies on the edge of a ridge, Romoleasen, on its western slope; this ridge consists mainly of gneiss and has become a natural centre for quarries for supplying aggregate for concrete, asphalt work in coated materials and road construction.

The present quarry began operations in 1892 and after vicissitudes of fortune, mostly beneficial, the crushing plant was rebuilt in 1947 and 1948 and transformed into a fully up-to-date processing unit with an annual production of some 200,000 tons. Increasing mechanization and improved tonnages resulted and in 1963 a new jaw crusher was started and productive capacity reached 800,000 tons per year. Further increases in crushing capacity in hand will permit production to achieve one million tons of finished roadstone per year.

The firm Dalby StenBross AB, is situated within the north-western part of the Romole ridge, the area belonging geologically to the South Sweden gneiss region,

The older primary rock has been up-thrust and consolidated in a W.-SW. to E.-NE. direction, the strike is N.-NW. to NW. and the dip nearly vertical and after this up-thrust the bedrock has been subject to several slips and faults in different directions and penetrated by slide planes and fissures. Nowhere in the quarry is any rock which has retained its original deposition structure. The bedrock is composed of leptite gneiss, 'split' gneiss, chlorite slate and amphibolite, uralite diabase, younger diabase and callainite.

The rock is quarried by ordinary open pit mining using bench breaking and horizontal haulage, the rock being worked in vertical slices. Shotholes are drilled in rows vertically to bottom below bench floor level.

After firing, the rock is loaded at the face and transported to the processing plant.

The quarrying face at bench height is quite extensive and quarrying is not selective, the different kinds of rock enumerated are not separated but are all quarried, crushed and stored together. The first operation is the removal of overburden which consists of quaternary layers having a thickness of 6 to 30 ft and for stripping this an A-Kerman 1400 diesel-engined excavator with a $1\frac{1}{2}$ yd^3 bucket is used, the earthy waste being transported by two self-tipping Rockerman dumpers of 10 tons payload capacity, to the worked out western open pit which is being filled up. The dumpers make an average of six trips per hour and stripping proceeds at the rate of about 120 tons per hour.

The drilling of holes in the useful deposit is by two crawler tracked rigs mounting Atlas Copco medium-weight drilling machines on hydraulic booms, drilling 3 in diameter shotholes to a drilling pattern of 10 ft spacing and 10 ft burden, the effective drilling rate is 35 to 38 ft per hour.

Two mobile Atlas Copco PR 600 screw compressors are used to supply power to the drilling rigs, being installed in the quarry and the compressed air delivered to the drilling machines by flexible piping. The compressors are electrically driven, rated at 200 hp and have an output of 550 ft^3 per minute at 100 lb/in^2.

Nobel Extras dynamite B in 1·14 in (29 mm) cartridges is used for primary blasting while for secondary blasting Nitrolite is employed. The shotholes are charged by compressed-air-powered semi-automatic charging equipment made by Nitro-Novel AB, the rammer comprising a plastic hose. The cartridge is passed automatically into the loading hose by maintaining excess air pressure and enabling charging to continue without interruption, the cartridge moving steadily into the plastic hose with no delay, enabling the charging operation to proceed smoothly. The unit is handled by two men, one filling and one loading at the drill hole, the effective loading rate is some 14 lb per minute. Short-delay detonators are used to initiate the charge.

Shovels are used for loading the rock, both electrically driven and diesel driven, and trucks are used to transport the quarried rock to the primary crusher which is a 71 in by 55 in double-toggle jaw crusher manufactured by AB Abjorn Anderson, Svedala. It is driven by a 270 hp 990 rpm motor. The crusher weighs 130 tons and has an hourly capacity of 500 to 700 tons. One of the loading shovels is a Ruston–Bucyrus 54 RB with a dipper of 2·1 yd^3 capacity which has a Ward–Leonard electric drive, the other is a North West 80D with a 2·2 yd^3 capacity dipper also electrically driven.

In reserve there are an Aberman 1400 and a Ruston–Bucyrus 38 RB, both diesel powered. A Caterpillar type 955E Traxcavator tracked loader is used for cleaning up and maintenance of the quarry roads which are of a temporary nature and prepared as necessary with the help of the Traxcavator.

Quarrying is proceeding on both benches and the length of road from the upper bench to the primary crusher station is some 450 yd, from the lower bench it is about 300 yd, transport being by two Kockum–Landsverk type KL-420 dump trucks with a load capacity of 20 tons and two older Volvo–Titan semi-trailers which have a 22 to 35 ton payload.

During 1965 the average hourly production was 400 tons with a labour force of 46 divided so: quarry 13, stripping overburden 5, machine hands 7, workshops 13 and loading 8 persons. The finished products are delivered mainly to Malmo, a city 14 miles south-west of Dalby, distribution being by rail and by lorry. A coating plant is connected to the roadstone plant where most of the fine size material is used for asphalt products.

LIMESTONE

A typical limestone quarry is shown in Fig. 45.

The five limestone quarries of the Buxton Lime Group of Imperial Chemical Industries produce large tonnages of crushed and graded limestone for use in chemical processes and for conversion into lime in a variety of limestone products; Tunstead quarry is the largest.

Limestone of various kinds is found in almost all of the geological systems in Great Britain but it reaches its maximum development in the Carboniferous, where almost half the national output is obtained. The Pennines are mainly Carboniferous and here the principal lime-producing centres of the Midlands and North are to be found.

The greater part of Britain in early Carboniferous times was covered by sea except for a belt running from mid-Wales across England to the Wash and is known in stratigraphy as St. George's Land. The Carboniferous system, which includes in ascending order the Carboniferous Limestone, the Millstone Grit Series and the Coal Measures, although the succession is rather different in Scotland, had a duration in time of some 55 million years. *i.e.* from 275 to 220 million years ago. Crustal movement and erosion subsequent to deposition resulted in the limestone outcrops being exposed in their present positions. The area sixty miles round Buxton developed as a geosyncline, a great collecting dish for strata. In 1947 an exploratory borehole was put down by Imperial Chemical Industries at Woodale near Buxton on the axis of the limestone anticline and encountered the base of the limestone at a depth of 892 ft resting on what was assumed to be the oldest rock, the Precambrian. In North Derbyshire a limestone succession of at least 1500 ft is exposed and it would appear that the limestone thickness to its base is at least 2400 ft. At Settle and at Colwyn Bay there is also evidence of ancient rocks unconformably below the base of the limestone. Quarrying methods are influenced by local conditions of deposition. For example in the Buxton area individual beds are thick, sometimes 30 ft, while in North Wales beds average only a few feet in thickness and fragmentation is relatively easy. Thin bands of clay of volcanic origin, ranging in thickness from a fraction of an inch to several inches and in exceptional cases, nearly a foot, separate the bedding planes which are

Fig. 45.   *Marion type 151–M 8 yd³ shovel operating in a limestone quarry.*

nearly continuous; the clays are often colour-banded. In addition to the horizontal bedding planes there are two sets of vertical joints at right angles, those running parallel to the working face are known as 'backs' and it is these clay-filled joints which dictated in the first place the necessity for developing an efficient clay eliminating cleansing process, before mechanization of the quarries could proceed. Fortunately there is no clay problem in the Settle area.

The main face at Tunstead is 1¼ miles in length and the main continuity of the strata is broken by occasional faults but these seldom have a displacement of over 100 ft so that the succession is practically continuous throughout the quarry but, occasionally, local complex faulting occurs which, for reasons of safety, requires special quarrying methods. It is these complexes of shattered rocks that have deposited barytes and galena.

Limestone from Tunstead has a crushing strength of 24,000 lb/in^2 and that at Raynes quarry 22,000.

After extensive and searching tests on the various types of drill available, including two large rotary drills and after the decision had been taken to adopt angle-hole drilling at about 25° to the vertical and AN–FO as the blasting agent, it was decided that the difficulty of drilling downwards to a straight line off vertical with the two high-speed rotary drills, when detailed surveys showed a deviation of as much as 12 ft in a 180 ft high face, even at modest penetration rates, could not be tolerated. Tests with down-the-hole hammer drills, on the other hand, demonstrated clearly that a straight line could be drilled with a maximum deviation of only 7 ft, but as important developments in drills and drilling in the next quinquennium are expected it was stipulated that new drills should be written off over a period of five years. In addition, experience had indicated that tractor-mounted drills operating on a very rough quarry top had shown them to be put out of action for relatively long periods because of track maintenance. This will be of less importance in the future since the decision to leave, when stripping overburden, a cushion of a few inches of topsoil between the actual rock surface and the drill. On the results of the multitude of tests a specification was drawn up of requirements and eventually a drill rig was found which, when modified, measured up approximately to these requirements. This drilled a $4\frac{3}{4}$ in diameter hole at an average penetration of 16 ft per hour throughout the working day and a penetration rate of 25 ft per hour when actually drilling, which corresponded to a production rate of about 425 tons per hour. Bits have lives which average 5000 ft. Trouble was experienced with the outer sleeve of the hammer which was liable to cracking under conditions at Tunstead. This has been explained as due to the use of an hydraulic motor for the rotary motion instead of compressed air but it would appear that with some hammers shock is transmitted through the sleeve instead of the rock when a cavity is encountered.

The choice between rubber-tyred or track-mounted, depends mainly on local quarry top conditions, the rubber-tyred drills used were capable of travelling themselves.

AN–FO is used for primary blasting in all the ICI quarries. In some cases the bottom of the face is not fragmentated by the primary blast.

The only advantage claimed from firing angled holes is that a cleaner face results in which less precariously balanced pieces of rock are left behind as a result of the scouring action of the blasted rock on the face. Where the off-vertical angle exceeds a certain figure this advantage can be offset by the tendency of small pieces of rock to roll down the slope and be projected some distance horizontally on reaching the quarry floor.

In addition many quarries have closely spaced vertical and horizontal joints and after blasting the face while approaching a straight line in section actually consists of a series of steps and above a certain limiting angle individual pieces of rock can be left balancing dangerously on these steps as the main rock heap slides down the quarry face. This potential danger has had to be taken into account in sections of the quarry when experimenting with holes more than 20° to the vertical.

Toes or bottoms are dealt with when they occur at Tunstead by drilling $2\frac{1}{2}$ in diameter holes using a console which enables the drill to be hydraulically controlled from a point 60 ft away. Conventional high explosives were used but AN–FO blasting agent blown into the shotholes by compressed air pressure is projected.

It is intended to work within 100 yd of property at one of the smaller quarries and attention has been directed to the problem of ground vibration using a vibrograph which has enabled vibration limits to be forecast in any section of the quarry for a given design of blast. Measurement of noise produced by blasting in different weather is also carried out.

In 1945 a scrubbing process for dealing with dirt and clay enabled mechanized all-in loading of stone by $2\frac{1}{2}$ yd^3 excavators loading into 10-ton end-tipping dump trucks delivering to a 72 in by 45 in jaw crusher. Subsequently 3 yd^3, 5 yd^3 and now 6 yd^3 shovels have been adopted, the 5 yd^3 being replaced because of improved loading rates and by a radical change in design reducing maintenance costs although the 5 yd^3 shovels are retained as spares and to cope with long traverses of modern machines necessitated by mineralized faults which can be eliminated by the strategic placing of the displaced machines. A number of minor modifications has been made to the original machines, some of which are now available as standard equipment or optional extras including improvements to the lubricating systems which are operated in conjunction with a mobile unit designed at the quarry. This has reduced down-time and reduced the operating manpower required on excavators.

Detailed statistics of loading rates are maintained but these are affected by external factors such as quarry transport, rock fragmentation and local variation in digging conditions but actual loading rates on the 6 yd^3 machines averaged 432 tons per hour but for normal operations including all stoppages the figure of 290 tons per hour is used which has been confirmed in practice. The 6 yd^3 shovels have a rope crowd necessitating rope renewals every 116,000 tons and for the hoist ropes 203,000 tons.

For cleaning up and 'maids of all work' two large rubber-tyred shovels are used as auxiliary to the main loading operations including pushing up edges of newly blasted heaps of rock towards the shovels which would otherwise be operating inefficiently.

Transport from the quarry face to the various crushing plants is by means of 50-ton capacity side-tip vehicles at Tunstead, 15-ton end-tippers at Raynes and 20-ton end-tippers at Hindlow. The vehicles in the Buxton area have been designed as an integral part of a unit comprising shovel, crusher and quarry layout which gives level hauls from the face to the crushing plant, work study being widely used in determining the optimum sizes of the equipment. Initially at Tunstead 20-ton capacity side-tipping trucks were used in conjunction with $3\frac{1}{2}$ yd^3 shovels, the haulage unit being powered by different makes of engine, each developing 90 hp at 1000 rpm manufactured to ICI specification, the trailer portion being made by various contractors. The sides of the trailers were raised to increase capacity to 30 tons when 5 yd^3 shovels were used.

Work study investigations were inaugurated to determine the maximum size of vehicle which could be geometrically accommodated by the 60 in gyratory crusher which would allow full advantage to be taken of larger capacity shovels. This was found to be 50 tons maximum capacity or an average running capacity of 45 tons. With the trailer with a robust, all-welded body designed and fabricated by ICI and the specified power end with a tractor using the largest British commercial bogie available, with an engine and torque transmission system, the resulting vehicle would be considerably cheaper than an equivalent unit available from manufacturers. Method study indicated that with provision for a spare and with three 6 yd^3 shovels a fleet of eight vehicles would be required and this is the fleet at Tunstead which deals with the 4 million tons per annum working a conventional eleven shift week.

Tyre maintenance is an expensive item with such transport units, which costs about a farthing a ton. Trials with different tyre makes and treads have been instituted and a vulcanizing team is employed, equipped with large platens for tyre and conveyor belt repairs.

Careful attention to operating conditions in and around the shovel, in particular, pays excellent dividends. The quarry floor is built up to near level before the shovel moves forward and a small rubber-tyred bulldozer travels from one shovel to another to keep the area free of large stones.

In the Tunstead area a system of arterial concrete roads is adopted while in the smaller quarries the technique of road-making and maintaining roads with quarry material without disastrous potholing liable to occur during continuous heavy rain has been worked out. The roads are made by laying down a base course foundation of coarse rock and binding with $\frac{3}{16}$ in + 0 in material in damp condition and then running on the road. It is important that the vehicles should not run on a set course indefinitely and that the road should be well drained. There has been a progressive decrease in transport costs; improved fragmentation, maintaining a level floor and particularly improvement in vehicle design are the main causes.

The primary crusher at Tunstead is a 60 in Traylor gyratory installed in 1951 which is normally fed at the rate of 1200 tons per hour. There is no installed spare crusher but vital castings are kept in stock and a three months' emergency stock of stone is maintained.

Accidents directly associated with quarrying, generally occurred at the quarry face itself, before mechanization formed a major part but this is no longer the case and an analysis shows they are no longer of a type peculiar to quarries and the quarries are now no more unsafe than factories. In fact the rate on the British Standard scale is 0·38 so that only one man in three is liable to be off work once in his working lifetime. The improvement of the mental attitude of the individual worker to his job is the most important factor in accident prevention. In addition to other safety courses, every member on the payroll has had a full day's course to stimulate his safety consciousness.

'Job safety breakdowns' analyse in detail all jobs in the quarries with the specific object of focussing attention on any safety hazards. Visibility on locomotives has been improved so that persons can be seen down to a distance of 10 ft from the front buffers. With regard to the use of explosives only one accident has occurred in 20 years, no electric detonators are allowed down any shotholes, Cordtex being used.

The quarry manager and his assistants are trained mechanical engineers and the policy is the appointment of men trained in engineering for supervisory jobs down the line. Comprehensive basic training courses are available for foremen and specialist courses are held as required.

An incentive system applies to practically all the quarry personnel and a work study staff of six operates as an integral part of management.

To supply the cement works of Tunnel Portland Cement Co. Ltd at Padeswood, Flintshire, limestone is worked $7\frac{1}{2}$ miles away at the Cefn Mawr quarry and transported as a minus $3\frac{1}{2}$ in product by articulated bottom-discharge hopper vehicles. The Padeswood plant produces about 285,000 tons of cement annually but new extensions under construction will double production.

Cefn Mawr quarry was bought in 1949 and has since produced some five million tons at the rate of 300,000 to 325,000 tons annually.

Carboniferous limestone of the Corallian series averaging 98% $CaCO_3$ is the deposit worked which is free from large faults but some shale bands of considerable thickness have been met which have influenced the quarry's development as has also the dip of the limestone beds which is 20° to 22° to the east. A Caterpillar DC7 bulldozer was used initially to remove the overburden ranging from a few inches to 15 ft over the working.

A number of slides occurred when the faces were worked at a height of 110 ft so such high faces were changed in favour of 60 ft benches which are drilled at 5° from the vertical to a depth of 63 ft to give 3 ft of hole beneath the toe. Angle-hole drilling at 15° to 20° from the vertical was tried but the number of lost holes led to a decision to return to near-vertical drilling. A considerable number of oversize boulders needing secondary blasting resulted when 27 T well drills were used to drill to 24 ft burden and 24 ft spacing so 4 in Halco–Stenuick down-the-hole drilling machines were substituted by two units powered by a Holman Rotair 600 ft^3 per minute portable compressor, each achieving 16 ft per hour and producing 11 tons per foot of drilling.

These in turn have been replaced by a Holman HP Vole tracked rig powered by a 370 ft^3 per minute Rotair HP compressor for all the primary blasting drilling 35 to 38 ft per hour or three holes per shift of 6 hours actual drilling time, the machine drilling a 4 in diameter hole to a 12 ft burden and 12 ft spacing programme. Hammer life is 11,480 ft, bit footage is 500 to 2000 ft with no sharpening necessary, the earlier down-the-hole machines are now used to drill toe holes on face stubs when necessary through the dip of the strata. Each 63 ft deep hole is loaded with an average of 175 lb of explosive yielding some 700 tons of rock. The charge consists at the base of 50 lb of ICI $3\frac{1}{2}$ in diameter opencast gelignite, followed by 75 lb of Quarrex by Explosives and Chemical Co., followed in turn by 50 lb of ICI quarry dynamite with one stick of gelignite as primer. Cordtex runs the length of the hole and short-delay 25 millisecond detonators are employed for initiation. Seven to ten thousand tons per blast of ten holes take place three or four times a month. Opencast gelignite and quarry dynamite is used for the toe holes while for secondary blasting polar ammon gelignite is employed.

A relatively small amount of secondary blasting is required, only about 7% of primary blasting is oversize since the primary crusher is of large size. Drilling of boulders is by hand-held percussive machines using $1\frac{1}{4}$ in diameter bits. Blasting efficiency is 4 tons of stone per pound of explosive with 11 tons of stone per foot drilled. From several stations round the periphery of the quarry each blast is monitored on a Cambridge vibrograph, recording vibration frequency in cycles per second, maximum amplitude in thousandths of an inch, the square root of $1\frac{1}{2}$ times the maximum quantity of explosive in one hole, the site coefficient factor which is governed by the strata, wind force and direction and the prevailing weather conditions.

Two Ruston–Bucyrus 100 RB face shovels equipped with $3\frac{1}{2}$ yd^3 dippers, only one of which is normally in use at any one time, and a smaller 33 RB are utilized for jobs around the quarry that do not warrant moving the larger shovels.

Four Euclid dump trucks, supplied by Blackwood Hodge, comprise the transport fleet with three of the four vehicles in use at any one time. The body capacity is 8 to 9 yd^3 with an average payload of 15 tons. Engine exhaust brakes have been fitted to the trucks since the 600 to 700 yd haul loaded from the face to the primary crusher is mainly downhill. The surface of the haulage roads is made up to and maintained at a standard suitable for cars, a vintage steam-roller being made use of for the purpose. Before this standard was decided upon one Euclid tyre per week was being rendered useless, now they are worn down and retreaded as necessary at about a quarter the cost of a new tyre.

The trucks deliver to the primary crusher which is a Hadfield 60 in by 48 in jaw breaker set at 9 in and driven by a 250 hp Crompton Parkinson 440 V motor *via* a David Brown gearbox and Smith laminated-steel shafting $2\frac{1}{2}$ in in diameter.

Until 1946 the Adam Lythgoe organization, founded in 1913, was basically a local, but vigorous, agricultural marketing business, selling lime, manures and farm produce in the South Lancashire and Cheshire area. Then at the end of the war the advent of mechanized contract lime-spreading combined with the Government lime subsidy, made it possible for the company to expand rapidly and extend its trading area from Hereford to Aberdeen.

Five limestone quarries in various parts of the north of England were acquired to ensure adequate supplies of ground limestone and among these was Minera lime works at Wrexham in North Wales with an annual output of 70,000 tons of agricultural and 300,000 tons of industrial limestone. Quarrying at Minera has been going on for at least 200 years and the Roman occupation forces mined lead in the area. The works were purchased by Adam Lythgoe in 1954 when the output was 25,000 tons. In 1964 it had risen to 300,000 tons of which 75 to 80% was sold to engineering and industrial consumers. In 1961 the first steps towards increased output were taken and the practice of employing a quarry-master abandoned in favour of direct employment by the company of staff and labour.

Previously the quarry face had been worked in one bench to a height of 200 to 220 ft and it was deemed advisable to split this into two benches each 110 ft high to give more flexibility and safer working. These have now been established but the intended configuration although nearing completion has not been entirely achieved. Figure 46 shows present and future transport routes. On the left will be noted a clay-filled seam which creates difficulties in working and an outside contractor was employed to remove this and a rock pile of 40,000 tons of clay-contaminated rock pile resulted.

With the quarry redevelopment additions to the processing plant have continued and in 1960 a contract was entered into with John Summers steelworks at Shotton, Cheshire, for the supply of crushed limestone suitable for their sintering plant.

The area of Carboniferous limestone in which Minera quarry lies, runs in an arc from south of Prestatyn, on the coast of North Wales to a point north of Llangollen. The present working area presents a clean face of well-bedded stone with little weathered rock in the upper strata. Present reserves have been estimated at 16 to 20 million tons. The crushing strength is 24,000 lb/in² and the $CaCO_3$ content is over 98%. The overburden is an 18 in thickness of soil on ground rising gently in the direction of face advance. It is removed at yearly intervals by an outside contractor using a Caterpillar DC7 tractor and scraper equipment.

The drilling pattern was established in 1964 with burden and hole spacing at 11 to 12 ft and the holes angled 15° to the vertical as it was

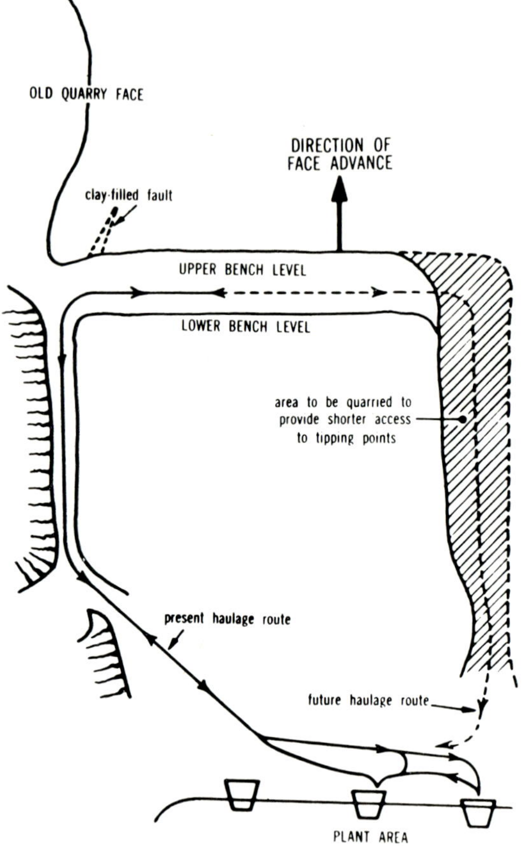

Fig. 46. Present and intended future haulage routes at Minera.

considered that a greater inclination would increase the formation of a step-like profile on the quarry face thus increasing the risk of loose blocks coming to rest on ledges formed by bedding planes.

All quarrying has been confined to the upper bench for two years so advancing it well ahead of the bottom bench so that rock blasted from the top bench will not interfere with operations in the lower working area.

A Halco–Stenuick 4 in down-the-hole drill is used to drill holes to the full height of the bench, powered by a mobile Holman Rotair compressor at the drilling site although a Broomwade stationary 320 ft^3 per minute compressor is installed. A Holman Vole drill has been purchased to increase drilling capacity.

Only small quantities of blasting agents are stored at the quarry, the requirements for face blasts are delivered on the day that blasting is programmed. Each hole is charged with 50 lb of Gelamex at the bottom followed by 310 lb of AN–FO blasting agent in loose powder form supplied by Explosives and Chemical Co. Ltd. An average blast produces four to five thousand tons of stone from four to six holes at 800 to 900 tons per hole and a ratio of $2\frac{1}{2}$ to 3 tons per lb of explosive. Fragmentation is generally good but some secondary blasting is necessary because the top few feet of the face tend to be dislodged and not broken by the blast. The blocks of stone thus produced are drilled with hand-held Atlas–Copco BD11P machines powered by an Ingersoll–Rand 125 ft^3 per minute compressor.

For increased loading capacity a 54 RB shovel with a $2\frac{3}{4}$ yd^3 dipper is used to supplement the 37 RB shovel with $1\frac{1}{2}$ yd^3 dipper and the 24 RB shovel with $\frac{7}{8}$ yd^3 dipper at present used for loading.

For transport three 15-ton capacity Euclid, one Foden 10-ton capacity and four AEC 12-ton capacity dump trucks comprise the internal quarry fleet, a total capacity of over 100 tons, which gives ample capacity and flexibility to operate both transport from the face loading points to the crusher hoppers, a distance of 350 yd and other necessary secondary haulage between stone bins, stockpiles and the rail-truck loading points. Maintenance is carried out on site in a covered repair shop.

The carboniferous limestones, Keuper marls and sandstone beds which comprise the Mendip Hills have been considerably disturbed by Hercynian earth movements to form anticlinal folds. The southern extremity of the Lower Oolites lies to the east, limestone beds extend from the coast of Dorset north-eastwards to the border of Lincolnshire and Yorkshire. The oolites are worked mainly for building stone, the famous Cotswold stone, but some areas also yield stone of sufficient hardness to be used as aggregate and for road surfacing. These two limestone strata, apart from the Kentish ragstone, are the nearest extensive hard stone resources to London and SE. England.

In 1957 the Amey Group purchased the Crookswood quarry at Holcombe 14 miles south of Bath and a new quarry plant was installed during 1958 for an annual production of 150,000 tons. Since then, additional plant enables a production of quarter of a million tons per year to be achieved.

The area of the original quarry is still being worked for present production, but a new face is being developed to the east which will open up considerable reserves estimated at 10 million tons. A public road presents a difficulty in the present layout as it separates the old quarrying area and plant from the new area and at present, stone from the new face has to be transported over the road to the plant which is some 100 ft below the level of the new face. At a later stage it is intended to drive a tunnel deep under the road to provide more convenient access and in the meantime a diamond-drill hole through the intervening strata will carry a compressed air supply line.

The new face is being developed initially by driving an access at a width of 30 ft inclined downwards to gain face height and then turning at right angles with a cut approximately 70 ft wide to advance through three wide vertical clay joints after which the working face will be established. By advancing the face in 12 ft cuts the development of the access has been achieved, vertical holes being drilled at 12 ft burden and spacing to 2 to 4 feet below the bottom of the face. A 4-foot wall of limestone was left to support the unconsolidated strata as the face approached the 6 ft wide clay joints and the shotholes are drilled at a lesser burden to ensure that the seam is completely collapsed. A 30 RB shovel with a 1 yd³ heavy-duty dipper loads out the rock pile in four days into two Foden dump trucks. The material is transported to the plant when working through stone but when the clay seams are blasted, the stone which cannot easily be separated from the clay is tipped on a waste area.

The original face, from which the majority of the stone passing to the processing plant comes, amounting to some 575 to 600 tons per day, is a single bench 75 to 82 ft high and holes are drilled at 15° to the vertical although drilling at 20° has been tested extensively and it may be possible to increase to 30°, the nature of the near-vertical bedding planes seeming to eliminate the danger of large blocks being held on the face after the blast, although a difficulty may arise in the loading of loose explosive into shotholes inclined at 30°.

Charging of shotholes is complex since two types of explosives are used—polar ammon gelatine dynamite and Nobelite H, a prepared AN–FO mixture by ICI. These are carefully placed in the hole with a total of 43 ft of drill dust stemming (Fig. 47). Each hole is threaded with Cordtex which is connected to a main Cordtex line through a short-delay (17 milliseconds) detonator. Six to ten holes comprise a blast with a ratio of 4·8 to 5·1 tons of rock per lb of explosive or about 900 tons per shothole.

A smaller 22 RB shovel is used at the producing face and the transport fleet consists of the two 9 yd³ Foden and two older 7 yd³ ERF dump trucks. Loaded material is dumped into the 30-ton capacity main hopper of the plant with a gravity feed to the primary crusher which is a 40 in by 32 in jaw crusher.

About three miles south of Buxton in Derbyshire on the A6 Matlock road is the Topley quarry of Derbyshire Stone Quarries Ltd at the edge of an almost continuous exposure of Lower Carboniferous limestone extending to a depth of 1500 ft. The area available for quarrying is some 165 acres giving reserves for a life of over 25 years.

The installation of a new processing plant and the redevelopment of the quarry represented a further stage in the Derbyshire Stone Group's long-term programme to meet the anticipated growth of demand for quarried products arising from future extensions of the M1 motorway, general road construction and improvement schemes and the implications of the Buchanan report on 'Traffic in Towns'.

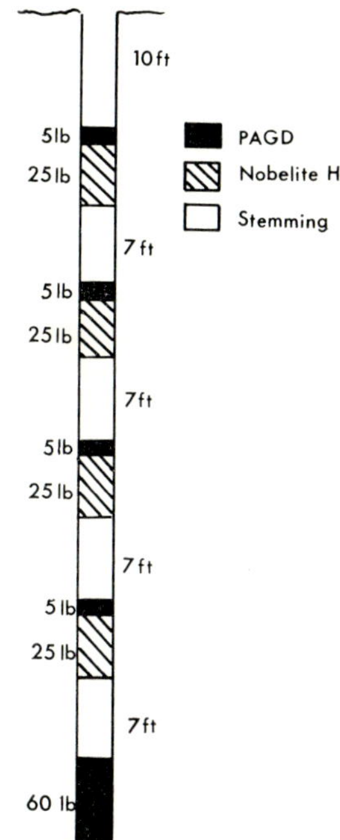

Fig. 47. Shothole charging at Crookswood limestone quarry.

The quarry was acquired from Newton Chambers & Co. Ltd in 1956 and the working face is still in course of reorientation, the two upper benches of the previous working having been combined to give a 30° inclined face of 140 ft vertical height; the present face length is some 500 ft, and is being extended to 300 yd.

The present production, that delivered to the primary screens, is 13,000 tons per week of which from 10 to 17% goes for lime-burning. It is anticipated the output will rise to 16,000 tons per week. The limestone has a crushing strength 25,000 to 28,000 lb/in^2.

The limestone is overlain by from 1 to 2 ft of soil which is removed periodically by scrapers and bulldozers to give a clean and even rock surface.

The limestone is thinly bedded and well jointed and free from clay or dirt seams and breaks freely giving good fragmentation with an economical use of explosives. Shotholes 6 in in diameter are drilled by a Halcotrack rig with a down-the-hole percussive hammer. The spacing of the holes is 27 ft and they are drilled to a depth of 165 to 167 ft at an average hourly rate of 14 ft. An Ingersoll–Rand portable compressor of 600 ft^3 per minute capacity supplies power to the drill. A short 175 ft face at right angles is lengthening the main face and is in a direction at right angles to the intended development direction. Seven holes are drilled on this section in each blast which provides 45,000 to 50,000 tons of stone.

Two 25 lb cartridges of opencast gelignite are placed at the bottom of each hole for ensuring complete detonation in the case of water accumulating in the hole, followed by AN–FO blasting agent to within 25 ft of the collar of the hole. A further two 25 lb cartridges threaded with Cordtex are placed above the main charge to initiate the detonation and the hole is filled with loose material as tamping. There is a series of 17 millisecond delays in each Cordtex lead which is connected to a main trunk line to which a two minute safety fuse is attached and lit by hand when the shot is fired.

The requirement of the blasting programme is to produce sufficient run-of-quarry material immediately suitable for the lime-kilns in a size range 8 in to 5 in. It is important that the kiln stone be given priority since excess material can be reduced, by secondary crushing to provide road-stone, a deficit can only be made up from quarry-run or crusher-run stone. Large blocks are broken by a 30 cwt drop-ball operated by an old 24 RB shovel.

A 71 RB face shovel with a $3\frac{1}{2}$ yd^3 dipper capable of an hourly output of 500 tons and an NCK machine with a $1\frac{1}{4}$ yd^3 dipper providing 200 tons per hour are available for loading but only the 71 RB is required at any one time to provide sufficient stone for the crusher and the other machine provides a spare or is employed for cleaning up.

Regular maintenance is carried out on these primary items of equipment, one hour at the end of each shift being devoted to greasing, cleaning and minor repair work. For cleaning up the quarry floor immediately after a blast a Bray front-end wheel-loader is used.

To transport the limestone two 20-ton capacity Foden dump trucks are deployed which deliver to the crusher hopper, a distance of only 100 yd. As the main face is developed it will advance away from the processing plant and after a period of say 5 years conveyor transport may have to be considered.

The primary crusher is a Sheepbridge 4850 double-rotor unit powered by two 150 hp electric motors.

On the European continent the crusher is regarded as the most important item of plant at a limestone quarry and the other operations of winning, loading and transport are all designed to fit in with crusher capacity particularly as multiple units are employed in loading and transport so that spares are easily provided but in the great majority of cases only one primary crusher is installed.

The operating costs of a typical European limestone quarry, inclusive of capital charges, may be divided as follows:

Winning	28%
Loading	24%
Transport	26%
Crushing	22%

SLATE

With the increasing use of concrete tiles and other substitutes for slates the industry was a declining one. However, the tonnage produced in 1964 was 106,325 tons and employed 1465 men and the output has increased lately from 78,000 tons in 1960.

In the past the Welsh slate industry achieved such a reputation that some of the quarries extended over several miles. One of the largest was the Penrhyn quarry in North Wales with its multiple benches each about 70 ft in height. The work on all these benches proceeded simultaneously both on the surface and underground by means of adits. A large number of quarries are now abandoned and buried beneath spoil banks because in quarrying roof slate about nine-tenths of the slate won is dumped as waste.

Besides Wales, slate quarrying is being carried on in the Lake District, Cornwall and Scotland in the Ballachulish area of Loch Leven.

Slate veins were worked to produce slate of a certain quality, texture and colour. The most productive area is the belt of country between Snowdon and the North Wales coast including Bethesda, Llanberis and Nantlle. Perfect cleavage was also impressed upon certain Lower Ordovician sediments south of Snowdon and these were mined at Blaenau Ffestiniog. A less perfect cleavage is developed in the overlying Upper Ordovician and Silurean rocks throughout North Wales.

Slate is used for roofing, wall-lining panels, external cladding, flag-stones, steps, sills and mantels. Granulated slate is used for artificial stone, surfacing, aggregates and expanded slate for concrete aggregates. Pulverized slate is used as a filler for asphalt roofing felt, damp-course, paint, plastics, fertilizers and insecticide. It is also used for rock-wool, bricks and tiles. Roofing slates were dug by the local inhabitants from outcrops of Cambrian rock in the Llanberis Pass and the area has been exploited by Dinorwic Quarries, the largest slate working in the world, once with 4500 employees, but at present only some 700 men are employed.

The Cambrian slate belt extends south of Nantlle to Bethesda, outcropping as the Elirdir Mountain rising to over 2000 ft OD, capped with gritstone. The quarries occupy 800 acres, the 'veins' or beds of slate striking northwest–southeast and the main groups in the Nantlle area of Caernarvonshire are shown in Table VIII.

TABLE VIII

STRATIGRAPHICAL SUCCESSION IN
THE NANTLLE SLATE QUARRYING
AREA

Slate veins	*Thickness, ft*
Ordovician	
Dark Slate Group	
Cambrian	
Cymffyrch Grit Group	600
Green Slate Group	300
Mottled Blue Slate Group	600
Pen-y-Bryn Grit Group	200
Striped Blue Slate Group	600
Dorothea Grit Group	100
Purple Slate Group	2000+
Clog Grit Group	2000+
Cilgwn Conglomerate	500
Pre-Cambrian	
Tryfau Grit Group	900
Clogwyn Volcanic Group	1000+

The veins worked in the Dinorwic area are known locally as the New Quarry (hard blue rock), G & W (soft purple), Mottled (mottled blue), and the Red, separated by grit bands totalling 250 yd.

Two quarries are being worked, a mile apart, the Dinorwic, nearing the end of its working life, and the Marchlyn is a new operation to phase in with the gradual run-down of Dinorwic with its narrow galleries, railway lines and inclines of traditional slate quarrying methods, and replace them with the latest techniques and ultimately, output is expected to rise.

At the Dinorwic quarry working follows the various veins up the slopes of the Elidir Mountain for 1500 ft from Lake Padoon. A minimum of fragmentation and as much whole rock as possible is required in slate quarrying, so blackpowder is used exclusively for production blasting. Full use is made of the experience of trained quarrymen in recognizing rock that will cleave easily and making use of the natural joints and bedding planes to bring it down carefully, and various techniques are adopted including springing, levering (Fig. 48) and the use of small-diameter shotholes. Obvious waste rock is blasted with Burrowite explosive firing at definite shift times. Seven active 60 ft galleries are in production and the maximum height of the working in the New Quarry vein is 540 ft.

Fig. 48. Prising slate from the bottom workings of a North Wales slate quarry.

'Sinks' or conical holes in which slate is worked down to considerable depths for the remainder of the workings and the normal system of benching is being used, so that the main workings may be regarded as one large sink with a side missing.

In this old traditional method a team consisted of two rockmen and two dressing-shed men, the latter receiving slabs from their partner rockmen and the team dividing the money received for finished slates monthly. A small number of men in the upper galleries work under this system but the bulk are paid on individual piece-work but still finding their own tools and explosives which encourages maximum economy

Rock drills in use are mostly Flottmann machines and drill steels are forged in 12 ft, 8 ft, 4 ft and 2 ft lengths at the quarry shops with re-sharpening equipment throughout the works, the diameter of the plain steel bits used are $\frac{7}{8}$ in and $1\frac{1}{4}$ in. In addition a $2\frac{5}{8}$ in diameter down-the-hole Halco–Stenuick machine is available and is transported around the quarry by truck. The slabs produced are often too large and need reducing by splitting along the cleavage with wedges or splitting at right angles to the cleavage by drill-holes, plugs and feathers. Whilst drilling and working on the faces, men use safety ropes attached to the gallery above (Fig. 48).

The slate slabs are chained into open trucks running on 1 ft 11 in gauge throughout the quarry, waste rock also being hand-loaded into three-sided trucks and most galleries are served by steam or diesel locomotives for truck handling and, alternatively, the different galleries are connected by self-acting inclines at 1 in 3 to 1 in 6 or 'Blondin' ropeways by Clarke Chapman. The eighteen diesel locomotives in use are two or three cylinder Ruston machines, some 20 years old, and three Hunslet 0–4–0 saddle tank steam locos are also still operating.

Compressed air requirement is 2500 ft^3 per min at 100 lb/in^2 and is supplied by two compressors through a 6 in diameter main, one a Belliss and Morcom 1550 ft^3/min and a Tilghman 1200 ft^3/min twin cylinder and four standby compressors.

Six tons of blackpowder, over a ton of Burrowite explosive, 7000 detonators and 17,000 yd of safety fuse comprise the explosives used yearly.

The neighbouring Marchlyn quarry, a mile away, is situated on the western slopes of the Elidir Mountain at 1500 ft OD.

The slate veins lie under 60 ft of weathered slate and boulder clay amounting to $1\frac{1}{2}$ million tons of waste over two years. Four Caterpillar 977 Traxcavators with $2\frac{1}{2}$ yd^3 buckets were used to load seven 11 yd^3 Foden dump trucks with Cummins engines running a shuttle service to the spoil heaps. The weathered rock and clay were pushed over the top of each partially developed gallery by a Caterpillar D8 tractor with a bull-dozing blade and ripping tine, any hard rock being blasted with $3\frac{1}{4}$ in cartridges of opencast gelignite in 4 in holes drilled by two Halco–Stenuick drilling rigs. An average waste blast comprises four horizontal or vertical holes charged with 200 lb of explosive bringing down 1500 tons with a 12 ft burden and spacing pattern.

Four 60 ft slate galleries have been developed with a fifth well advanced. The two lowest are producing excellent slate blocks produced by a blast comprising two 28 ft 'horizontal' holes having an upward inclination each charged with 100 lb of blackpowder which brings down a thousand tons with similar burden and spacing pattern, the drills achieving 10 to 14 ft per hour, each drill compressed air driven from a Broomwade WR 210 rotary portable compressor with a spare compressor in reserve. The short life of front tyres on the Foden trucks and wear and tear on the Traxcavators is fairly severe but the use of Michelin Metallics has overcome the tyre difficulty and repairs are dealt with in a new fitting shop at Marchlyn where a Murex arc-welder is used to repair and build up worn Caterpillar buckets and blades.

Twenty-one men are employed at Marchlyn quarry including five experienced rock men whose job is to do light blasting of good rock and to split the slates brought down by heavier blasts. A normal working day is 7.30 am to 6.30 pm and 3.30 pm on Saturdays, flood-lighting being used during winter months and payment is by day wage.

When production has been built up to target, some production machinery will be transferred to Dinorwic to remove the gritstone and weathered slate forming the summit of Elidir Mountain to attempt to provide the old quarry with further reserves of slate in the north-west corner.

A medium wheelbase Thames Trader 75 is used to carry split slabs of slate, placed in special slings which hold $1\frac{1}{2}$ tons and hoisted into position by one of the Caterpillar 977 Traxcavators to the dressing sheds at Muriau where they are off-loaded by a Jones KL44 mobile crane on to flat trucks for wheeling inside.

Fresh waste slate is used to make bricks and tiles at Dinorwic which seem to find a ready market. A variety of hip and ridge tiles are also manufactured by hand since only a small quantity are involved.

A team of 65 men do all repairs and this figure includes engineers, carpenters, builders and electricians, extensive workshops being provided near the quarry offices.

To the south of Dinorwic quarry the Dorothea Slate Co. Ltd realized that their 440 ft deep quarry at Talysarn was becoming uneconomic and machinery obsolete so they decided to spend £25,000 on new equipment and recent market trends indicate they came to the right decision.

The slate veins at Dorothea are the Mottled Blue, the Striped Blue and the Purple (*see* Table VIII) which dip south-east at 80° being separated by the Pen-y-Bryn and Dorothea grits respectively.

The quarry consists of an irregularly shaped hole 440 ft deep and 500 ft across at its widest part, a dolerite dyke forming a rough division halfway but current working is in the northern half, working in the other half at 500 ft deep being interfered with by water problems (in one inrush eight men were drowned) and now forms a useful waste dump. Water drains into the working and into the waste dump and a 500 ft vertical shaft and is pumped by a Berisford 8 in centrifugal pump with a second similar reserve pump for the wet season.

Work at present is proceeding in a 60 ft high face on the other side of the dolerite dyke and here ten rock men are employed and work very much on their own initiative, rock being brought down by straight drilling and blasting, by springing or by the use of bars, making full use of natural joints and bedding planes whenever possible (Fig. 48).

Drills of Ingersoll–Rand and Holman SL9 types are employed using hand-forged steels and tungsten carbide bits. Blackpowder is used for blasting good slate rock to achieve the least possible fragmentation, but for waste, where good fragmentation is required, polar ammon gelatine dynamite explosive is used. A recent innovation has been a Halco–Stenuick drill, drilling 4 in diameter vertical shotholes for bulk blasting. A new face is expected to be needed in 12 to 18 months and is being developed along the north-eastern side of the pit where the best slate lies beneath 60 to 120 ft of near-useless material consisting of 30 ft of superficial sand and gravel, dolerite dykes and slate rock with poor cleavage. Bulk blasting of a 40 ft face is the first step, the intention being to have ultimately two 40 ft or one 80 ft face. When drilling vertical holes the entire 40 ft of the face is utilized with a 3 ft spacing and a 10 ft burden. The penetration of horizontal holes varies but the 3 ft spacing is retained and any combination of vertical and horizontal holes may be used to suit conditions. Initiation is by safety fuse and Cordtex.

A small Blondin ropeway lifts slate out of the quarry and removal of blasted waste rock is by a Chaseside loader with a 2 yd^3 bucket working with two SWB Dodge dump trucks with Telehoist 7 yd^3 bodies, twin-ram tippers and Firestone All-Traction tyres. Sand and gravel will be moved prior to drilling and may be sold as ballast for subsequent washing and screening, present stripping of overburden only extends 50 ft laterally but future operations will necessitate the removal of trees, an old house and possibly some old tips. Handling of waste in all slate quarries is a problem particularly when the waste rock often constitutes 99% of the total. Waste 18 in maximum is loaded by an Atlas-Copco rocker shovel into two Thwaites Sprite dumpers operating a shuttle service to the tips a few hundred yards away through a level in the dolerite dyke. They are fitted with Petter single-cylinder diesel engines and carry about 25 cwt.

Air at 100 lb/in^2 is supplied by two Holman compressors delivering 305 ft^3/min. A further Holman TH 185 supplies an additional 175 ft^3/min to supplement existing capacity.

Orders are sufficient for six months' production and some have had to be refused. Production of materials at Dorothea Quarry per month are

Waste rock to tip	1000 tons
Slab slate to mill	275 tons
Roofing and damp-course slates	100 tons

In the Furness district of north Lancashire, some nine miles from Barrow, are the Kirby slate quarries of Burlington Slate Quarries Ltd on Kirby Moor where slates have been worked for over 200 years and there is ample evidence that sufficient reserves exist for a further one or two centuries of production. The workable area in lease is 508 acres which will allow the quarries to be extended on strike for 1¾ miles. The beds of slate occur with gritstones and flagstones with a series of upper grits followed by 100 to 200 ft of upper flagstones, then 50 ft of lower grits followed by a considerable thickness of lower flagstones from which the Burlington blue-grey slate, known as Westmorland blue, is worked (Fig. 49). The beds are almost vertical and erosion since the Silurian period has exposed the succession at surface.

Overburden comprising 30 to 40 ft of degraded material produced by weathering and minor sedimentation is above the solid slate so that only at a depth of 50 ft and below is good slate found, and the main quarries

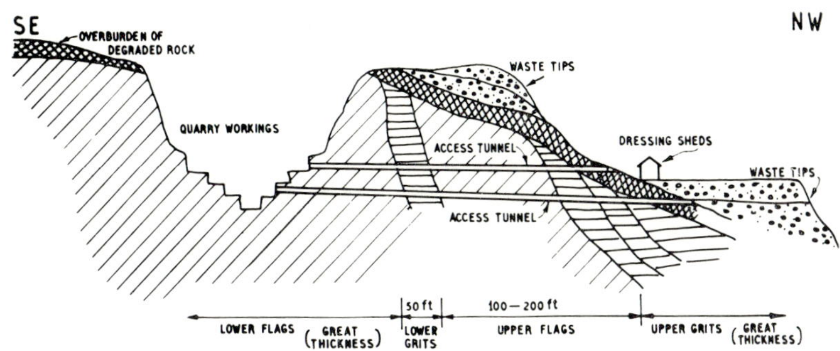

Fig. 49. *Diagrammatic section through Burlington Slate Quarries Ltd showing the geological succession and access tunnels.*

follow the strike line over a width of 150 yd over a length of ¾ mile at present but with extension towards the north-east projected to 1¾ miles.

The plant at the quarry is shielded from blasting (Fig. 49) and comprises dressing and sawing sheds and tunnels were driven from the bottom of each quarry through the sheltering rise to its associated dressing shed and, as working depth increased, new tunnels were driven at lower levels, so that there are now several access roads, some still in use, the gradient being 1 in 400 away from the quarries for drainage and giving a gradient in favour of loaded carts and other transport.

New working methods were introduced after the halt in production caused by World War II, this being the introduction of a Puckering electric hoist of 3½ tons capacity as an additional means of access to the quarry floor now 350 ft deep, the majority of the hoist shaft is through made ground and is supported by a steel framework lined with corrugated iron, the lowest 20 ft being in solid ground. Battery locomotives hauled to and from the shaft on rail track have been replaced by dump trucks. In 1949 wire-sawing replaced the traditional 'sumping' using blackpowder in blast holes. Slate is won from benches (Fig. 49) and by reducing the size of the barriers which separate the former individual quarries, particularly by taking out the middle leaving the sides for support of the quarry sides.

The wire-sawing adopted in the quarry is a scaled-up version (Fig. 50) of the method used in the dressing shops for cutting the rough blocks to

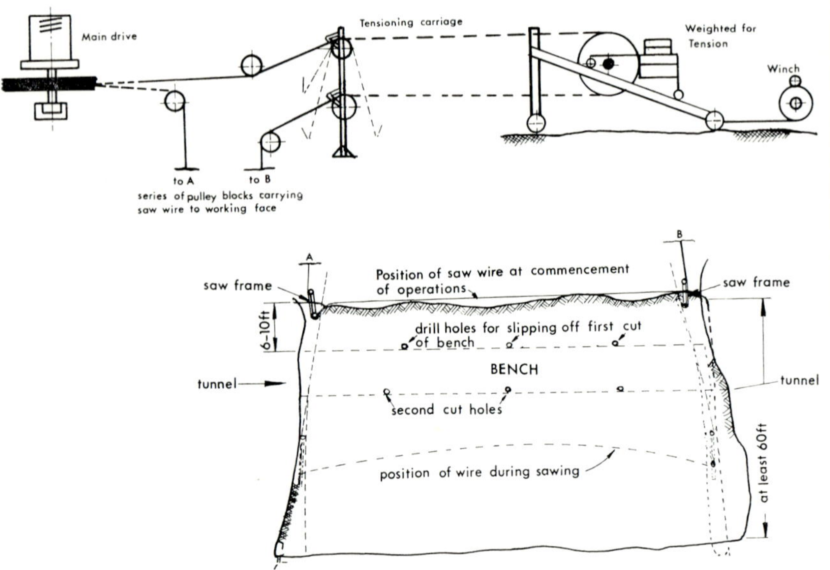

Fig. 50. *The quarry wire-saw system showing to an enlarged scale the application of the wire at the bench.*

size and it has entirely replaced the 'sumping' method while achieving the same result with the advantage that no good material is spoilt by fragmentation, which occurs as a risk with explosives.

The wire used is a 0·08 in diameter 3-ply cutting wire made by British Ropes, spliced to form a continuous length, two saws being used with a reserve, each unit comprising a main drive of a friction pulley driven by a $12\frac{1}{2}$ hp motor, a tension head and a series of pulley blocks carrying the wire to the working site. The area sawn in one operation averages 400 to 500 yd^2 over a length of 70 to 80 ft and a width of 60 ft or more, a bench being prepared for sawing by driving two tunnels, one on each side of the bench at quarry floor level (Fig. 50). These are extended a short distance beyond the back of the bench and are generally 4 ft high and 5 ft wide and do not normally require support, one man is kept tunnelling, averaging 20 ft to 30 ft per week, stress cracks from previous benches produce relatively easy conditions for tunnel driving; a round of 10 to 15 ft long holes, the exact pattern of the round depending on conditions, each hole is charged with $\frac{1}{2}$ lb of dynamite and fired in sequence by safety fuses.

The block is then undercut which forms the bench by sawing horizontally through the area between the two tunnels. Two pulleys, mounted on a fabricated beam manufactured by British Laundry Machinery Co. of Barrow-in-Furness, constitute the saw frame and they are set up at the entrance to each tunnel, a head frame with a tensioning device at one end and a tail frame at the other end, the saw drive passes through the first pulley and is guided towards the second, tensioning pulley.

Figure 50 shows the position of the wire saw before cutting has commenced and at a stage when the operation is nearing completion. The frames are placed so that the wire is bowed, thus ensuring that pressure is maintained along the whole of the cutting face. The frames are held in place by anchor ropes and timber blocks and as cutting proceeds the tensioning pulley is advanced by turning the fixed threaded shaft on which the pulley bearing is mounted and when the pulley reaches the end of the shaft, sawing is stopped and the frame is repositioned further into the tunnel and repeated until the bench is completely undercut. The wire is fed with sand-water slurry which acts as an abrasive and also keeps the wire cool, the wire travelling at 30 mph, the tension being adjusted to give a cutting speed of about 1 in per hour over a length of 70 to 80 ft, the wire which is about $\frac{1}{4}$ mile long is only used for one week and is then replaced. Breakage of the wire is not common but if a break occurs within the cut it is practically impossible to thread a new wire into the saw slot and the operation has to be recommenced.

When sawing is complete it is left until required for quarrying and the sawing process is kept well ahead of production requirements. The winning of the slate from a bench is carried out in stages in which a first cut is taken by drilling three or four holes along a line some 6 to 10 ft in from the edge of the bench, which are drilled down from the bench top 12 to 20 ft vertically by a Halco–Stenuick 4 in diameter down-the-hole drill. When the slate from this blast amounting to 900 to 1400 tons has been used, a second cut is taken and so on until the whole bench has been removed.

Removing the slate from the centre of the barriers and reducing the width of the remaining buttresses is by undercutting the centre portion by wire-sawing from two tunnels, one to be used for transport is driven 6 ft by 4 ft or larger (Fig. 51). From the buttress section slate is won by driving a 4 ft by 5 ft tunnel and developing a blasting chamber at quarry floor level in the centre of the back of the block to be removed, the area again being undercut by wire-sawing between the tunnel and the adjacent tunnel through the barrier, the burden above the chamber being from 60 to 90 ft, a typical blast producing six to eight thousand tons of slate of which 5% will already be waste because of fragmentation, a charge of 1500 to 2000 lb of blackpowder in 28 lb boxes is packed into the blasting chamber, each box being threaded with Prima cord detonating fuse, the

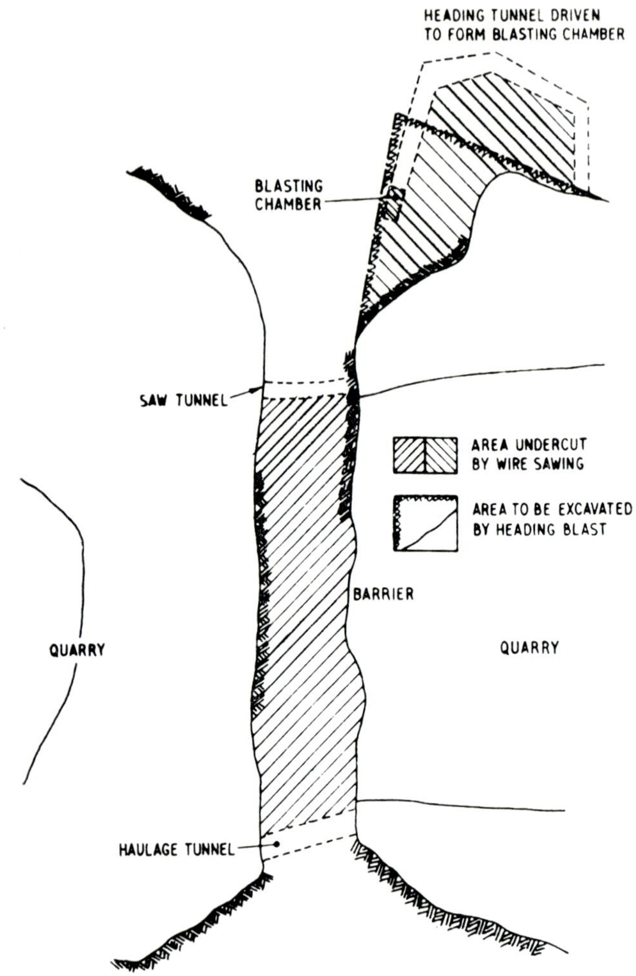

Fig. 51.   Arrangement for quarrying the barriers by removing the centre portions and reducing the buttress width.

blast being fired electrically from a refuge in one of the main access tunnels of the quarry, remote from the barrier being worked.

The large blocks of slate resulting from the blasting operations are split into 'clogs' or slabs of 5 tons in weight by plugs and feathers or by drilling short 'pop' holes for blackpowder. The clogs are loaded into dumpers by a Chaseside SL 400 front-end loader or by Smith 10 or 12 face shovels or when won above or below by two Anderson–Grice jib cranes. The dumpers used are three Northfield F7 $7\frac{1}{2}$ yd^3 with Perkins diesel engines and they transport both slate and waste out of the quarry via the Cavendish level, a 440 yd tunnel some 350 ft below the top of the quarry and now 16 ft by 16 ft in size with a 5 in thick concrete road surface laid throughout the tunnel with a drainage channel cut on one side.

Compressed air is supplied for drilling from a 4 in ring main from a Tileman Boiler Company stationary compressor and three mobile compressors, two Ingersoll–Rand 250 ft^3/min and a Holman 240 ft^3/min, these being used for compressed air tools in areas remote from the ring main.

The total labour force is 120 men, all from the locality, and an active apprenticeship scheme is in operation.

REFERENCES

'Granite Quarrying Developments at Penryn', W. J. Welsh and B. E. Wood, *Quarry Managers' Journal*, February, 1966, p. 41.

'A Notable Granite Quarrying Operation in North Wales', *Quarry Managers' Journal,* July and August 1962, pp. 265–272 and pp. 305–310.

'Shap Granite', E. G. Holland, *Mine and Quarry Engineering*, August 1959, pp. 345–351, and September 1959, pp. 396–401.

'Quarrying at Dalby', B. Bojession and H. Jacobson, *Mining and Minerals Engineering*, October 1966, p. 384.

'The Limestone Quarrying Operations of Imperial Chemical Industries Ltd, Great Britain', R. G. Jackson, *Opencast Mining, Quarrying and Alluvial Mining*. Institution of Mining and Metallurgy, 1965, p. 260.

'Cefn Mawr Quarry', *Mining and Mineral Engineering*, December 1966, p. 465.

'Quarrying Limestone at Minera', *Quarry Managers' Journal*, May 1965, p. 173.

'Crookswood Quarry', *Quarry Managers' Journal*, October, 1965 p. 387.

'Limestone Products from Topley Pike', *Quarry Managers' Journal*, February 1965, p. 43.

'Hillside Quarry', *Mine and Quarry Engineering*, December 1952, p. 514.

'Factors influencing Aggregate Impact Value in Rock Aggregate', D. M. Ramsey, *Quarry Managers' Journal*, April 1965, p. 129.

'Trends in the Use of Aggregates for Concrete in Building Construction', A. Short, *Quarry Managers' Journal*, April 1965, p. 151.

CHAPTER 7

WORKING IRON AND COPPER DEPOSITS BY OPEN PITS

IRON

Four minerals are the main source of the iron of commerce. They are magnetite, Fe_3O_4, with a theoretical iron content of 72% and sometimes known as 'black' or 'magnetic ore'; haematite, Fe_2O_3, is a red ore with 70% iron; limonite, $2Fe_2O_3 3H_2O$, is a yellowish-brown ore with varying amounts of water in its composition and contains 60% iron; siderite, $FeCO_3$, known also as chalybite or spathic iron ore has a theoretical iron content of 48%. In addition other important iron deposits such as those worked in the Midlands and in France are partly composed of greenish hydrous iron silicates such as chamosite and greenalite with siderite.

Although the theoretical iron content is never attained in deposits, rich deposits of magnetite may contain up to 68% iron such as those of Kiruna and Gellirare in northern Sweden.

Hematite occurs in different forms; when in good crystals with an adamantine lustre it is called 'specularite' or 'specular iron ore' (looking-glass ore); when in thin plates or scales it is known as 'micaceous haematite' and in kidney form it is called 'kidney ore'. In Cumberland and the Furness district of North Lancashire haematite ore-bodies occur in large irregular masses or 'flats' related to the stratification of the enclosing rocks and in pockets and hollows all within limestone beds of Carboniferous age. The ore-bodies were formed mainly by replacement of the limestone by hot ascending mineralizing solutions but some are thought to be derived from the red ferruginous sandstone of the Trias above which the limestone occurs, as at Llanharry in South Wales.

The famous haematite deposits of the 'iron ranges' near Lake Superior in the USA contain 50 to 60% iron.

The common impurities in iron ore are silica, alumina, calcium, magnesium, manganese, titanium, sulphur, phosphorus and arsenic. Manganese and titanium may enhance the value of the ore but sulphur, phosphorus and arsenic are undesirable impurities but removable by metallurgical treatment. The Kiruna deposit was not worked for many years after its discovery because of high phosphorus content until the 'Thomas process' removed the phosphorus successfully.

The Northampton ironstone is the most extensive and accessible British iron ore and contains a compact group of beds, all ironstone, averaging 8 ft in thickness but varying from 6 to 15 ft, lying under a 50 mile strip of dissected plateau 5 to 10 miles wide extending east of the Middle Jurassic (oolite) escarpment from Grantham in Lincolnshire across Rutland to Northampton. These reserves have been estimated at a thousand million tons of ore carrying over 30% iron, the ore being a concentrate of chamosite ooliths passing upwards and downwards into Northampton Sands and is therefore a siliceous ore. The most favourable overburden is a soft sandstone of the Lower Estuarine Series up to 30 ft in thickness; elsewhere massive Lincolnshire limestone up to 100 ft in thickness has to be blasted and loaded out before the iron ore is reached. Even less desirable is overriding boulder clay filling old hollows.

The Frodingham or north Lincolnshire ironstone was a narrower ore bank in the Lower Lias but, because of its accessibility and self-fluxing properties, much of its reserves have been worked. The ore deposit consists of a pack of lenses 30 ft thick in the central area, reducing to 20 ft in three miles on either side. It consists of rusty chamosite ooliths, now limonite set in precipitated siderite, and chamosite mixed with one-third shells and shell fragments reducing the percentage of iron to 20% and making it a calcareous or limey ore.

Other native ores are the Cleveland ironstone of North Yorkshire of the Lower Oolite and Upper and Middle Lias consisting of siderite mudstone, magnetite-siderite mudstone and chamosite-siderite mudstone. The ore is mined by underground workings and is relatively expensive and has decreased in production.

The blackband and clay ironstones of the Coal Measures contain some 34 million tons of ore with more than 30% iron. Again these are too expensive to work in the face of competition from cheaper or richer ores.

Midland ironstone opencast workings

The iron ore deposits of the Jurassic system, composed of sands, clays and limestones which dip generally to the east at less then 1°, are worked by the United Steel Co. at Colsterworth North and at Exton Park. The deposits are in the lower part of the Inferior Oolite and are known as the Northampton Sands ironstone (Fig. 52). The ore horizon overlays Lias clay and the ironstone varies between 8 and 12 ft thick. The overburden consists of sands and clays, a bed of limestone and the surface is of boulder clay and soil and at Colsterworth North the overburden is from 35 to 50 ft and at Exton Park from 50 ft to 90 ft thick.

For stripping the overburden a very large shovel in one case and a very large dragline in the other are used; at Colsterworth North a Marion 5325 shovel fitted with a 17 yd^3 dipper dumps the overburden across the mine cut, while at Exton Park where the overburden is thicker, a Ransome and Rapier W1400 walking dragline with a 20 yd^3 bucket is used.

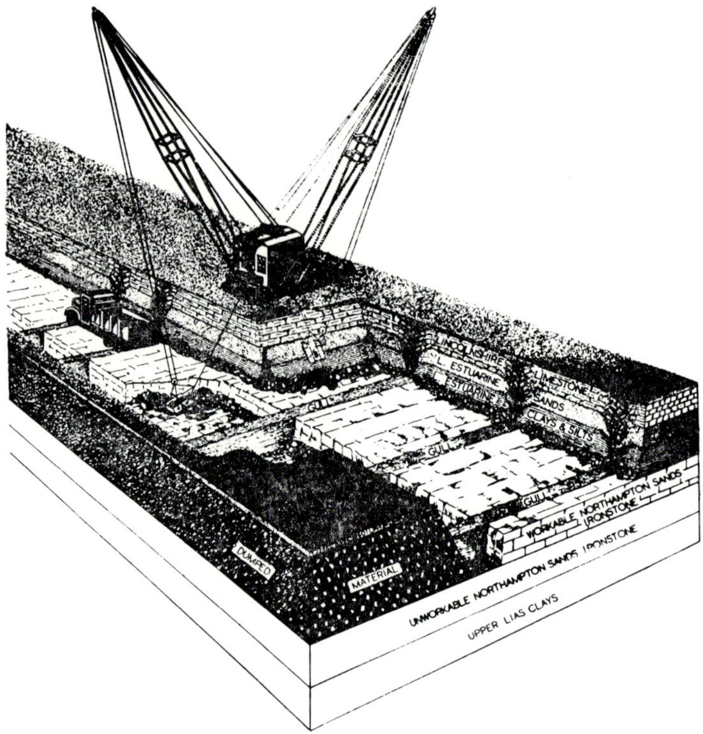

Fig. 52.　Northampton Sands ironstone opencast site, dragline removing overburden and ironstone.

The topsoil is removed at both sites by a D8 bulldozer and three Caterpillar 619 scrapers ahead of the overburden strip and spread to restore the surface of the worked out areas by travelling round the end of the cut when the face shovel is used or leaving it in piles on the cut side for the dragline to transfer across the cut as required, in both cases it is respread on top of the worked out area where overburden other than topsoil has already been dumped and levelled so that the reclaimed area can return to agricultural production. Land restoration receives particular attention as required by the Town and Country Planning Act 1947 and the Mineral Workings Act 1951 as an essential part of the mining cycle, the land returning to normal farming after some three years of special care. Further examples of land reclamation after opencast mining are given in Chapter 14.

Although the main proportion of the overburden in both cases is dealt with by the large shovel or the large walking dragline respectively, the clearing of the benches is the work of bulldozers of the Caterpillar D8 or D7 type. Drilling and blasting of the ironstone then follows where necessary. In the previous removal of the overburden, however, its hardness and depth differ at the two sites. At Exton Park the upper 15 ft can be dug without previous blasting by the dragline, the bottom 35 ft then being drilled and blasted before excavation. At Colsterworth North the full 35 ft requires drilling and blasting. In both cases drilling is performed by one Bucyrus–Erie 30 RC rotary drill, which is transported between the two pits by low-loader as required; a smaller pit is also serviced by the same drill, Colsworth No. 2. The blastholes are $6\frac{1}{4}$ in diameter, tricone bits being used and giving a penetration rate of 80 ft per hour. Drilling takes place on a three shifts per day continuous programme with four crews. The rig weighs 54,000 lb, has a 40 ft 4 in derrick and is powered by a General Motors Co. 6–71 two-stroke cycle diesel engine developing 165 bhp at 1800 rpm but the bit speed revolution is variable up to 100 rpm and the compressed air for flushing at 40 to 50 lb/in^2 is provided by a Gardner–Denver WCH diesel-driven single-stage compressor.

The spacing of the blastholes is 15 ft on a square grid, the charge per hole at Colsterworth North is 28 lb of AN–FO and 5 lb of quarry dynamite initiated by Cordtex and electric detonators. An average of 20 holes, but up to 60 holes, are fired as a primary blast with $\frac{1}{2}$ sec delays between every six. A 20 yd bench for the Marion 5325 shovel is produced by four rows at Colsterworth North and the same rows produce a 17 yd wide bench for the W1400 Ransome & Rapier walking dragline at Exton Park where the blasthole charge averages 18 lb because the limestone is softer.

The ironstone bed, after the removal of the overburden is dug and loaded by 100 RB Ruston–Bucyrus face shovels, into standard gauge railway wagons (steel hopper wagons) running along the bottom of the cut, the 25 tons per wagon load taking an average of 3 minutes. The ironstone bed is drilled and blasted occasionally and an electric rotary drill of United Steel's own design is used to drill $2\frac{3}{8}$ in blastholes which are charged as required with $1\frac{3}{4}$ in diameter cartridges of 60% opencast gelignite.

The Marion type 5325 crawler-mounted electric face shovel, used at Colsterworth North weighs 1200 tons and is electrically driven, powered by a single Westinghouse Ward–Leonard motor generator set driven by a 900 hp synchronous motor driving six dc generators the motor being supplied through a 3·3 kV $3\frac{1}{4}$ in diameter trailing cable. Two generators rated at 225 kW each supply the two hoist motors each of 250 hp and three of 68 kW supply the three 94 hp swing motors, the sixth rated at 68 kW supplies the single crowd motor of 94 hp. The hoist motor, by a system of shafts and gearing, drives the eight crawler tracks. Four self-levelling hydraulic jacks support the main frame and these are in turn supported on twin track bogies, each crawler track having a length of 14 ft 9 in between sprocket centres or 15 ft 8 in overall and a width of 3 ft. The track base overall is 39 ft by 44 ft 9 in, giving a total bearing area of all eight tracks of 295 ft^2 and a bearing pressure of 60 lb/in^2.

The superstructure proper revolves on a circular ring roller path, containing 78 hard steel rollers of 10 in diameter, the ring being supplied with centralized lubrication operated from a single point.

A small cab slung below the ring provides a control centre from which the crawler tracks can be steered when it is required to move the machine, better visibility and easier control is obtained with this positioning rather than would be obtained by operation from the main cabin.

The superstructure of the machine is designed as a cantilever with the machinery house 28 ft high and the A-frame structure rising to a height of 99 ft from ground level. The shovel at Colsterworth has a slightly longer boom than usual being 145 ft from pin to the sheave wheel while the dipper stick, as on all large Marion shovels, has a patent knee action which is claimed to increase positive break-out at the bucket lip and improve reliability. The principal advantage of front-end knee action is that the weight and reactive load of the crowd mechanism is removed from the boom to the point near the centre of rotation which allows a larger bucket capacity, which is 17 yd^3. The six-tooth all-welded dipper or bucket is welded to the dipper stick, the length from knee to bucket lip being 88 ft 6 in and with the main boom angle at 45° this geometry allows a maximum dumping height of 104 ft 3 in at 133 ft radius or 72 ft 9 in at a maximum radius of 148 ft 6 in. The clean-up is 84 ft 3 in and normal cycle times are between 45 seconds and one minute depending on cutting height, swing and dump height. The rear-end clearance of the superstructure is 33 ft radius at 18 ft 1 in above ground level or 37 ft maximum.

Apart from the steering of the crawler tracks the whole machine is controlled from a single cabin on the left-hand side of the front end of the machine. No air filtration or pressurization, as on the dragline at Exton Park, is provided and this has considerably simplified the design of the structure and reduced capital costs particularly as under the conditions existing in the Midland ironstone fields there is little need for air filtration.

The motor generator set and the hoist and swing mechanism are housed within the machine room, the two 250 hp hoist motors are coupled to the 52 in hoist drum which is grooved to accept the 5½ in circumference single hoist rope. A bail pull of 185,000 lb is obtained with the triple-reeved rope. High above the machine deck at the apex of the A-frame is the 94 hp crowd motor supplied with dc from the main generator.

The maximum output in yd^3 per hour is 700, average 600, the capital cost is £550,000 and the power consumption per yd^3 is 0·51 kWh and the overall cost per yd^3 is 10d.

The shovel is stabilized with 340 tons of ballast. The required output normally is obtained on two shifts, boom floodlights are provided for operation after dusk. Two men are employed on each shift as drivers to reduce operator fatigue but acting as oilers when not operating the shovel.

The W1200 Ransome and Rapier W1400 walking dragline was built in 1956 and was then the largest in the world, weight 1675 tons. NCK–Rapier Ltd manufacture a larger model, the W1800, one of which is installed to remove ironstone overburden at another quarry of United Steel at Scunthorpe in Lincolnshire.

The W1400 is also electrically driven, power at 66 kV through a 3·7 in diameter trailing cable is fed to twin Ward–Leonard motor-generator sets by BTH, each driven by a 1500 bhp 1200 KVA synchronous motor driving three dc generators of which one of 400 kW supplies two of the four 225 hp hoist drum motors; the second generator, also of 400 kW, supplies two of the four motors for either the drag or the walking motion and the third generator of 200 kW supplies one of the two slewing motors.

A 48 ft diameter tub supports the revolving superstructure and carries the main roller path which consists of a live ring of 120 alloy steel tapered rollers each 10 in in diameter. The revolving unit which carries the jib is designed as a cantilever structure, and carries also the A-frame structure and machinery house. It has a tail radius of 68 ft 6 in and a width of 49 ft, with the roof some 45 ft above ground level. Two shoes, each 48 ft long and 9 ft 6 in wide and carried on vertical legs driven by eccentrics, constitute the patented walking gear, each step of 6 ft 10$\frac{1}{4}$ in takes 30 seconds which gives a travelling speed of 0·1 mph when the dragline requires to be moved.

The drag and hoist machinery consists of two complete winding units, practically identical, with 5 ft diameter drums driven by four 225 hp motors. The twin drag ropes are of 7 in circumference and the hoist ropes 6$\frac{1}{2}$ in and the maximum pull on each drum is 100 tons, rope speeds ranging from 290 ft/min at full load to 513 ft/min at no load. Two motors drive the rotate motion at 1$\frac{1}{4}$ rpm giving a jib-head speed of 2000 ft/min and a torque at the centre post of 7,032,000 lb/ft.

The all-tubular welded jib was designed and welded by Tubewrights Ltd and is normally set at 30° but can be used at 35° if an increased dumping height is necessary. The drag bucket is of all-welded construction with renewable teeth and a capacity of 20 yd^3, an Esso bucket of the same capacity is also used. They weigh 22 tons and can move 27 tons of over-burden in a cycle of 1 minute giving an hourly output of 700 yd^3, while with a dump radius of 260 ft, cycle times involving a 90° to 110° swing are 49 seconds.

To enable optimum visibility and control of operation two cabins are provided and one man operates with a spare driver-oiler on each shift and an extra cleaner on the day shift only. Floodlights are provided on the boom for after dusk operation and the output required normally entails working two shifts per day. The whole superstructure is pressurized by filtered air to give clean air working conditions and keep the machinery free of dust.

The output averages 700 yd³ per hour and has achieved 800 yd³ per hour, the capital cost was £850,000, the power consumption per yd³ is 0·95 kWh and the overall cost per yd³ overburden is 1s.

In comparing the two excavators the shovel requires less maintenance and is cheaper in capital cost for the same duty and there is a more positive break-out force at the bucket lip which gives a quicker cycle and enables larger stones to be handled. In the same conditions more blasting is required for dragline operation since the lump size must be less and this additional blasting may affect the stability of the bench on which the excavator is working.

On the other hand the dragline is better suited for work in soft conditions as the bearing pressure on the bench is 19 lb/in² against 60 lb/in² for the shovel, so that it can stand on the top of the bench rather than in the pit or cut, and the shovel has a more limited radius of operation. The dragline can also deal with a greater thickness of overburden than the shovel and is more adaptable to varying topography which may mean variation in the depth of overburden and can level off its own working place more effectively than the shovel, and subsequent restoration of the land surface is expedited as the greater reach enables greater spread to be obtained when dumping.

The choice of excavator depends to a large extent on the type and thickness of overburden. Other things being equal a shovel which is cheaper would be selected if the thickness of overburden does not exceed 50 ft with auxiliary machines for restoration work.

Opencast mining of taconites in the Ukraine, USSR

There has been widespread growth in the opencast mining of ferruginous quartzites in the Krivoi Rog iron ore, followed by concentration. The iron ore basin at Krivoi Rog is one of the most extensive in the USSR and two types of ore are particularly suited to exploitation by surface mining.

These are:

(1) Ferruginous quartzite deposits in the form of intensely compressed synclinal folds. These deposits have the advantages of requiring comparatively little capital investment for their exploitation and a low stripping ratio of 0·125 to 0·25 yd³ per long ton.

(2) Thick, 270 to 330 yd beds of quartzites of moderate and steep dip. Their development entails much heavy cost in development work and a higher stripping ratio of 0·6 to 0·7 yd^3 per long ton of ore.

The ferruginous quartzites are of two main types; namely magnetite (non-oxidized) ores easily concentrated by magnetic separation and haematite (oxidized) which occur in a contact zone and are mined as ancillary supply but are difficult to concentrate by magnetic methods.

In planning and developing the site the first item to be decided upon is the position of the railway for transport to and from the opencast site and then the position of the processing and concentration plant, including the crusher.

Outer cuts initiate stripping operations which, within the topography of the site, lead to the entrance cuts baring one or more mining levels with the aid of working cuts. Inclines of outer and entrance cuts have gradients of $2\frac{1}{2}$ or 3% depending on the quarry dimensions and the advantages of reducing the number of fast-end rounds.

The benches on the upper levels consisting of loose and mixed rocks have a height of 33 ft, the height of the hard rock benches below is 49 ft. The angle of slope of the upper benches is 45° and that of the lower slopes is 70°, the general angle of the slope of the quarry sides varies from 25° to 40°.

The opencast pit, the crushing and concentrating sections, all work three shifts a day, the number of days worked per year being about 300.

Heavy churn or well drills are used for drilling blastholes, the string of drill rods weighing $2\frac{1}{2}$ to $2\frac{3}{4}$ tons. Chisel-shaped bits 8 to $10\frac{1}{2}$ in diameter are used, the drilling rate in medium hard rocks varies from 18 to 25 ft per shift of seven hours. Improvements in drilling performance have been achieved by the adoption of jet-flame cutting and rotary drilling using tricone roller bits. At one quarry in 1962 87,000 ft of blastholes were drilled with the use of jet-flame drills which increased the efficiency of blasting.

Seven jet-flame drills at present are used at this opencast site for routine production, their productivity increases the harder the rock.

Rotary drills are used successfully for drilling the softer rocks and jet-flame drills for the harder rocks.

For mining ferruginous quartzites the blasthole programme was introduced in the period 1954–57. Shotholes were drilled 8·6 to 9·4 in in diameter in one line and fired in groups of five to forty holes. The toe burden on the bench foot was 28 ft to 37 ft depending on the hardness of the rock, blastholes being 14 ft apart but because of the too wide spacing, fragmentation was bad and the amount of rock blasted did not exceed 16 yd^3 per ft of blasthole, the specific consumption: ore blasting ratio being 0·66 lb of explosive per yd^3 of rock. Adoption of a $10\frac{1}{2}$ in diameter drilling bit enabled the distance between holes to be increased which was also assisted by multiple row delay firing and as a result in 1962 0·35 yd^3 per ft of blasthole was achieved. Since then delayed-action blasting has been further improved with the results shown in Table IX.

TABLE IX

Quarry No.	1	2	3	4	5
Rock hardness according to ProtodyaKonov scale	6 to 20	6 to 18	6 to 16	6 to 14	8 to 16
Bench height (ft)	31 to 46	31 to 46	37 to 124	31 to 87	15 to 46
Hole spacing (ft)	22 to 31	22 to 40	25 to 26	25 to 31	22 to 25
Toe burden (ft)	23 to 37	34 to 37	28 to 29½	25 to 28	22 to 28
Charge concentration factor	1·0 to 0·9	1·0 to 0·9	1·0 to 0·9	1·1 to 0·9	1·0 to 0·9
Specific consumption of explosive lb per yd^3	0·13 to 0·29	0·21 to 0·20	0·19 to 0·20	0·12 to 0·15	0·19
Rock blasted per foot of blasthole, yd^3	25	23	24	37	22½
Oversize rock %	up to 1·0	0·5 to 1·2	up to 1·0	0	0·5 to 1·0

The working of high benches has been developed recently to improve the quality of blasts. The high benches are blasted in rocks with thin layers of medium hardness. From four to ten rows of shotholes have to be blasted before the rock broken by the previous blast is removed; this favours fuller use of the explosives' energy, permitting an increase in spacing of shotholes and reducing the specific consumption of explosives. The shothole distribution for blasting high benches is the same as for ordinary benches (Fig. 53).

Fragmentation of overburden and ore is good. In the period 1960–63 about 15 million yd^3 of overburden was blasted using the high bench method, uniformity of fragmentation with absence of large boulders was achieved and this may allow simultaneous loading from multiple benches to be applied in the surface mining of hard rocks.

The loading of soft overburden and blasted rock for transport is by face shovels with dipper capacities of 4, 5 and 7 yd^3, the 4 yd^3 is generally used for loading ore and the 5 and 7 yd^3 machines for overburden; the utilization of shovels reaches 88% and only 12% are kept in reserve or under repair and the average annual output handled per shovel is 1½ million tons, the mineral weighing 2 tons per yd^3, 77% of the material consisting of hard overburden or ore.

Both rail and road transport are utilized, some 61% by road. The rail rolling stock consists of 80 to 60 ton dump cars drawn by electric locomotives of 100 to 150 tons and heavy two-section diesel locomotives weighing 252 tons. Each electrically hauled train deals with 1·3 to 1·5 million tons per annum, the average haulage distance being 3⅜ miles up an incline of 3½ to 4%.

The net train load of a hauled train is 480 tons and it deals with 900 thousand tons or $5\frac{1}{2}$ million ton miles per annum.

Locally built 10 to 25-ton capacity dump trucks are used for transport within the opencast pits and when a pit is being developed motor transport is used for the first 90 yd after which a combination of motor and conveyor transport is used.

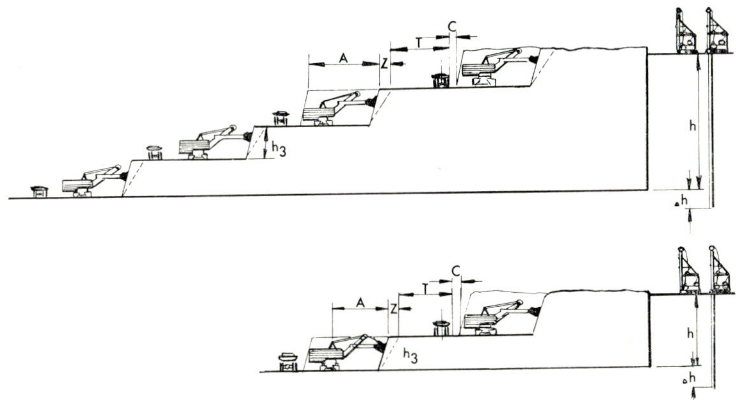

Fig. 53.   High bench mining method: (A) excavator cut radius, m; (h3) height of single benching, m; (T) width of transportation strip, determined according to transport method used; (Z) width of safety zone, determined according to possible prism of sliding, m; (C) gap between low edge and transportation strip, m; (h) height of multiple bench, m; (Δh) depth, m.

Surface mining of iron ore at Mount Goldsworthy

The first shipment of ore to Japan from Mount Goldsworthy, in the rich Pilbara iron fields of Western Australia, took place in May 1966.

Before 1950 the Commonwealth Government at Canberra were concerned at the supposedly meagre reserves of iron ore in Australia and prohibited export and only the Broken Hill Proprietary Co. Ltd with its domestic outlets had any incentive to prospect for potential iron ore fields. It was realized, however, that reserves were considerably more than anticipated and in 1960 the export embargo was withdrawn. To exploit the reserves a consortium of 5 interested mining firms was formed (*see* Chapter 2, p. 39), one of these was Mount Goldsworthy Mining Associates (MGMA), a company operating the open pit of the same name 70 miles east of Port Hedland and 900 miles NNE. of Perth, Western Australia. MGMA is owned in three equal parts by Consolidated Goldfields Australia Ltd, a subsidiary of the London-based international mining group of Consolidated Goldfields, Cyprus Mines Corporation of Los Angeles and the Utah Construction and Mining Co. of San Francisco; Goldsworthy

Mining Ltd, an Australian company wholly owned by MGMA, operates the surface mine under contract to MGMA and provides transportation, port handling and administrative services. The agreement between MGMA and the State Government ratified by the Iron Ore (Mount Goldsworthy) Agreement Act 1964, gave MGMA the right to prospect Mount Goldsworthy at a specified minimum cost and MGMA undertook to spend a minimum of £16 million on development within ten years, and rates of royalty to the State of $7\frac{1}{2}\%$ of fob revenue per ton for lump ore, 4 in to $\frac{1}{2}$ in, and $3\frac{3}{4}\%$ fob revenue per ton for fine ore less than $\frac{1}{2}$ in were also agreed.

The iron deposit occurs in an area of Pre-Cambrian of Archean and Proterozoic age in which thick and extensive banded iron formations (jaspilites) occur, similar to the taconites of the Lake Superior region they are of sedimentary origin and consist of thin alternating layers of chert and siliceous iron oxides. The concentration may have been affected by surface weathering from Pre-Cambrian times. Three main grades of commercial iron ore deposits resulted. The first are lode-type and are conformable lenses of massive haematite in a banded iron formation and Mount Goldsworthy No. 1 ore-body is a typical example and gives a direct shipping highest quality ore of up to $69\frac{1}{2}\%$ iron. They are found in a vertical or steeply dipping Archean banded iron formation possibly of early Pre-Cambrian age.

The second are outcrop replacement or crust ore deposits which are shallow cappings of haematite with some goethite and are surface remnants of banded iron formations of both Archean and Proterozoic age.

The third are detrital and bog iron ore deposits which originated through erosion of pre-existing iron ore deposits and transport of the iron as detritus, or in solution and redisposition as conglomerates or as bog iron ore deposits precipitated in shallow lakes and swamps and these may prove to be the most important as there are massive reserves of beneficiable goethite ore consisting of pistolitic and fragmental goethite of average grade between 50% and 60% iron.

There are five ore-bodies at Mount Goldsworthy and the programme by MGMA included geological mapping, surface sampling and some geophysical prospecting, diamond and percussion drilling and the driving of adits. Four thousand feet, and three Government holes totalling 3500 ft, of diamond drilling in twelve holes were positioned to provide a series of parallel cross-sections of the ore-bodies at right angles to the strike of No 1 ore-body.

Consideration was given to the problem of shipping the ore to world markets concurrently with the mining investigations and the negotiation of a sales contract for $16\frac{1}{2}$ million tons of ore to be delivered over seven years at a rate of $1\frac{1}{2}$ million tons in the first year, 1966, and $2\frac{1}{2}$ million tons in the remaining six years. The main contract specifications were:

iron 64%, minimum 61% silica and alumina 9% maximum, phosphorus 0·07 maximum, sulphur 0·05 maximum, copper and other metals 0·20% maximum, free moisture 2·0% maximum.

Construction of the plant and development commenced in January 1965 and in 15 months shipment of ore took place.

Extraction at first took place from No 1 and part of No 3 of the five ore-bodies, the No 1 ore-body extends for some 440 ft above datum to 320 ft below, while most of the No 3 ore-body intended for initial mining was more than 320 ft above datum. The pit was started in January 1966 with the cutting of the main access road up the north flank of the outcrop and the opening of the first two benches with a minimum of stripping.

Tentative pit specifications are the following:

Bench height 40 ft, bench width 100 ft, transport road maximum down grade 8%. The main crushing plant and its dumping point has been sited on the northern side of the ore-bodies at an elevation of 200 ft which will give a down-grade haul for the greater part of the 16 million tons of the initial contract. From the crusher a main haulage road has been established to the NW. corner of the proposed pit and then runs along the north edge of the eastern pit, which will be worked from the northern side throughout its life. The main pit is initially to be opened from the south with access through a central valley and this system will be maintained for the first two 40 ft benches. The third 320 ft bench will be opened directly from the north side and this will remain the pattern in depth, the only modification being that when mining commences below the crusher level the cuts will be established from the NW. corner rather than centrally.

Initially, practically no stripping of overburden was required except for the access road, the material from which was utilized to make the road down to the crusher dumping station. However, the stripping ratio will increase as the pit becomes deeper and in the first year to March 1967 the ratio of tons of overburden to tons of ore produced should be 0·46 to 1, increasing to an average of 1·09 to 1 for the next three years.

A Bucyrus–Erie rotary drill type 40 R is in use and a second is to be obtained and these are expected to use a 9 in diameter tricone bit with tungsten carbide inserts and $6\frac{1}{2}$ in diameter down-the-hole percussion bits, the holes in all cases being drilled vertically and carried 4 ft below grade giving a total depth of 44 ft in normal conditions, the spacing being 20 ft between holes and the burden 15 ft. AN–FO blasting agent is used and for secondary drilling and for sampling an Ingersoll-Rand CMS–IR crawler-mounted drill is used.

The blasted ore is loaded by a $4\frac{1}{2}$ yd^3 dipper P & H diesel-electric shovel and a further P & H $2\frac{1}{2}$ yd^3 dipper Type 955A diesel-electric shovel is also used. Le Tourneau–Westinghouse rear dump trucks of 65 ton capacity are used for transportation to the crusher; eight trucks are in use and of are Australian manufacture.

The rear dump trucks deliver the ore direct to an Allis Chalmers 42 in primary crusher.

The labour requirements of such a modern open pit and for ore handling are relatively small and amount to 307 for a production rate of $2\frac{1}{2}$ million tons per annum.

The initial capital expenditure for an annual output of $2\frac{1}{2}$ to $3\frac{1}{2}$ million tons amounts to £5·4 to £7·3 per ton of annual output. The expenditure is much lower than the figure often assumed for the development of a new iron ore open pit at a remote site which is from £10 to £13·3 per ton of annual capacity.

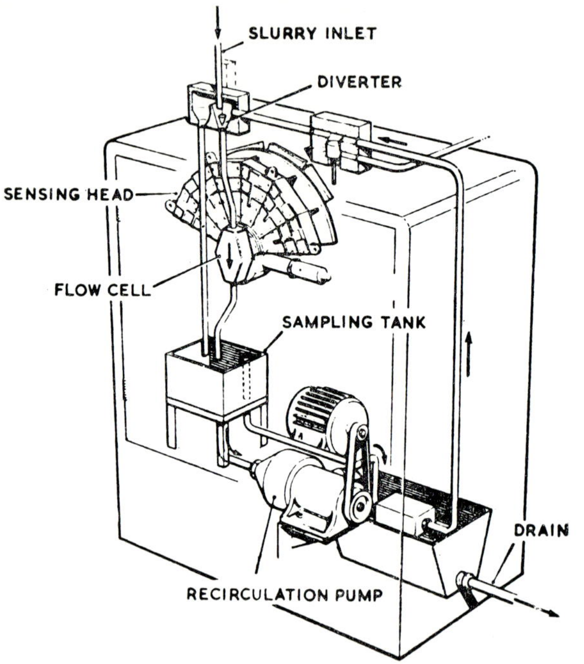

Fig. 54.   Elliott on-stream analyser.

Iron ore on-stream analyser

At a Western Australian iron ore shipping port an Elliott-Automation on-stream analyser (Fig. 54) together with an Elliott Arch 1000 computer has been installed to control accurately the grade of iron ore shipments. It analyses continuously slurries of ground ore made from dry samples delivered to the stockpiling and loading point from the mine. The analyser supplies the average content of phosphorus, alumina, silica and iron. The computer converts the scintillation and proportional counts into a per-centage analysis of the elements in the ore, performs structure analysis in terms of particle size and automatically works out the correct procedure for stockpiling. The results are accurate in the case of phosphorus to 0·02%, alumina to 0·4% and silica to 0·23%.

Mobile plant for iron ore concentration

The Utah Construction and Mining Co, one of the three partners in MGMA working at Mount Goldsworthy, Western Australia, when working an alluvial deposit at Iron Springs, Utah, are able to make an economic

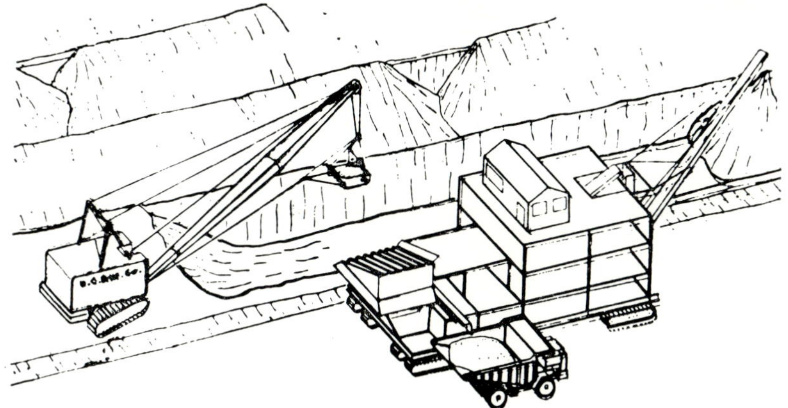

Fig. 55.   Arrangement of dragline and mobile iron ore benefication plant.

recovery of iron ore at a cut-off grade of 6% by utilizing a mobile plant which concentrates the ore and dumps the waste at the site of excavation.

Vegetation having been cleared from the mining area, the ground is terraced into two levels (Fig. 55). A dragline standing on the higher terrace digs a cut 40 to 50 ft wide and up to 60 ft deep, loading the material into the mobile plant feed hopper. When the cut is finished the dragline moves back about 30 ft followed by the mobile plant and mining continues.

The benefication plant comprises a grizzly and feed surge bin, screening section, electromagnetic separator drums and concentrate and waste delivery conveyors. The raw feed is delivered on to the grizzly and all minus 8 in material passes into the surge bin and the quantity of rejected plus 8 in material is small.

From the bin the ore gravitates to an apron feeder and conveyor for delivery to two single-deck screens working in parallel to provide plus $1\frac{1}{4}$ in and minus $1\frac{1}{4}$ in products. The coarse fraction goes to an electro-magnetic belt cobber which sorts an open-hearth concentrate with little loss of recoverable iron in the waste product. The cobber, which rotates at 250 ft/min, has a maximum field strength of 2300 gauss at the surface and 450 gauss at a radius of 10 in. The fine screened product is treated with four 30 in by 60 in Stearne high-speed drums. An alternating polarity causes violent agitation of the material passing over the drums and results in a minimum of waste being trapped in a flow of concentrate. Drum speeds are maintained at between 52 and 60 rpm and field strengths vary from 1640 gauss at surface to 350 gauss at 1 in radius.

The drums are protected against abrasion and impact by rubber covers and during a year's trial wear has not been significant. One problem has been the build-up of damp material on the drums reducing their efficiency, but the fitting of heaters has been effective in effecting a cure.

Concentrates from the cobber and drums average 50% iron with the initial feed running at about 10% iron. These are then loaded into trucks for transport to an iron springs concentrator where further processing provides a 64% iron feed for blast furnaces and a plus $1\frac{1}{4}$ in feed for open-hearth furnaces. Waste from the mobile plant is deposited directly into the cut.

The plant weighs over 400 tons and is mounted on four 51B Bucyrus–Erie shovel bases with crawler tracks, and to counter the effects of uneven ground four hydraulic cylinders, one at each corner, enable the super-structure to be levelled. Electric power is supplied at 4160 V by trailing cable, the plant demand being about 300 hp.

Production with eight men working three shifts a day for seven days a week is approximately 300,000 tons of crude ore per month.

COPPER

Copper occurs in a multitude of forms: native, as sulphide carbonate or oxide and in combination with other elements, in all some 360 forms.

Copper minerals, particularly the sulphides, are unstable when weathering occurs and are very prone to oxidation and enrichment. The primary copper mineral is often chalcopyrite which, within the near-surface zone of weathering, becomes oxidized to form green and blue copper carbonates, malachite and azurite, but much or all of the copper content may be carried down as a soluble sulphate, leaving at the surface a zone from which the copper has been leached and where the iron, now limonite, remains as a reddish-brown gossan, or capping, referred to as 'iron hat'. The copper in solution may be redeposited some distance below surface as sulphides, chalcocite and bornite richer in copper than the original chalcopyrite containing 79·9% and 63·3% copper respectively against only 34·5% copper in chalcopyrite.

'Porphyry coppers' are one of the chief sources in the USA and are low-grade disseminated copper sulphide ores which have been enriched by secondary processes often to chalcopyrite, as at Bingham in Utah.

Large deposits of disseminated copper ores occur in Katanga and Zambia which are now among the chief copper fields in the world. Chile is also a large producer.

Opencast copper mining in Zambia

All the ore-bodies worked on the Copper Belt and in the Katanga Province of the Republic of the Congo are found in the lower section of the Katanga geological system which lies unconformably on a much

Fig. 56. N'changa open pit, Zambia; benches 36 ft one and a quarter miles long and expected to reach a depth of 1000 ft.

older and intensely metamorphosed complex of schists, gneisses and granites with irregular intrusions of younger granites. The whole system has been subject to considerable upheaval and contortion followed by major erosion so that there is now a 'peninsula' of basement rocks running NW. with the mines and claims all along the shoreline and the Katanga series dipping away from the other rocks in a NE.–SW. direction. The ore horizon can be traced along this shoreline but does not always contain copper in viable quantities.

The Copper Belt ore bodies are predominantly sulphides, usually bornite, chalcopyrite and chalcocite but certain mines, particularly N'changa, have significant quantities of oxides (normally acid soluble) as malachite, azurite, cuprite, tenorite and chrysocola. Cobalt is recovered at Rhokana in the form of carrolite.

Fig. 57.   *New 10 yd³ (16 tons) P and H shovel dumping load into 65-ton capacity Haulpak*
at N'changa open pit.

N'changa has two open pit copper mines which produce 270,000 tons of ore per month of which 170,000 tons from the N'changa open pit (Fig. 56) is dropped through ore-passes to underground workings and 100,000 tons are transported by rail from the Chingola open pit.

The volume of overburden removed averages over 10 times the volume of ore removed and the overburden stripping equipment consists of a bucket wheel excavator capable of digging over 2000 tons per hour and electric shovels of 10 and 6 yd³ capacity in conjunction with 65-ton capacity Haulpak diesel dump trucks (Fig. 57).

The overburden rarely needs drilling and blasting but as depth increases (and it is expected to reach 1000 ft) it is anticipated that increased blasting will be required.

The ore itself is drilled by Ingersoll-Rand drillmasters giving a 6 in diameter hole or GD Airtrace drilling a 3 in diameter hole and fired with ANBA or conventional explosives. Ore is loaded by $2\frac{1}{2}$ or $3\frac{1}{2}$ yd^3 shovels into 35 or 45 ton capacity diesel trucks.

Three new open pits, 8 miles from N'changa, are currently being developed and by 1970 it is anticipated that $\frac{1}{2}$ million tons of ore per month will come from the open pits.

Problems anticipated in the open pits are largely those associated with slope stability, economic working depths and ideal size of equipment. A soil mechanics laboratory has been established at N'changa.

The slopes of the larger and deeper open pits such as N'changa may extend to depths exceeding 1000 ft and present formidable and special problems with reference to stability, the aims of which should be to improve the mining economics and provide for safe operation.

The cost of mining depends upon a purely geometric relationship, the amount of waste removed compared with the volume of ore recovered, known as the stripping ratio, R, which is a direct function of the slope angle. The steeper the slope angle, the smaller the volume of overburden removed and the more favourable the economics of the operation.

However, the slope must also be secure against sliding and, generally, the less the slope angle the more secure against sliding and the safer it will be. A slope failure or slide may be disastrous to life and equipment and can affect the viability by altering considerably the volumes of waste to be removed. It is therefore necessary to strike a balance which renders the mining operation both safe and viable. However, slope stability plays but a minor part in the daily planning of open pit operation but it assumes a major role in the planning of the ultimate open pit, which introduces the complication of a time factor in slope stability studies which has to be correlated with the working life of the pit.

In large opencast pits special slope stability problems result from the following:

(a) Pore-water pressure as a transient condition may exist behind the slope while mining proceeds continuously. Pits are often worked below the level of the water-table and a flow net into the pit occurs.

(b) The regional tectonic compressive or tensile stresses are interfered with by large and deep pits with an uncertain effect on slope stability.

(c) Composite soil and rock slopes may be the result of changes in the nature of the ground with depth.

(d) Cleft-water pressures may affect the stability of rock slopes but are difficult to determine.

(e) Theories of slope stability current may not apply to large slopes.

(f) A time parameter may play a part.

For the calculation of slope stability essentially there exist two groups of theories:

(a) The plastic flow or zone failure theories. In these it is assumed that the material is in a state of failure at all points within the sliding mass.

(b) The line failure theories.

In these it is assumed that failure takes place by shear along a surface of particular shape which will be plane or cylindrical over the major portion of its length. The plane is of course a particular case of the cylinder where the radius is infinite. The upper mass of material slides as a unit intact block over the lower stationary mass and the failure is sometimes known as a 'block-over-block' failure.

Failure along a circular arc is commonly observed in soils that have some cohesion. On rock slopes failure is frequently observed mainly along a pre-existing plane of weakness. If the slope material has cohesion at the top of the slope, tension cracks develop and further complicate the pattern of the failure surface.

Line failure theories have been used at N'changa since zone failure theories require much advanced mathematics while the circular arc theory only employs the basic laws of statics and is much easier to apply. It is difficult if not impossible to apply zone failure theory to a slope cut in a number of different strata which also have planes of preferential failure such as joints and faults.

Pore-water pressures are not taken into account in zone failure analysis. The analysis must be carried out in terms of total stresses. Circular arc analysis can be carried out in terms of total or effective stresses. Line failure mechanism is supported by numerous field observations since, for example, roads or benches in the upper mass retain their relative positions and only when the displacement becomes large does the sliding mass become distorted, the distortions being principally confined to the upper and lower regions. Since Fellenius first introduced the circular arc analysis in 1929 it has been very successfully applied.

The essential information required to be produced in slope stability investigation must include:

(a) A detailed knowledge of geological structures such as fault and bedding planes, folds and joint structure.

(b) The shear strength of the materials through which the slope is cut.

(c) Pore- or cleft-water pressures.

(d) Any surface movements.

(e) The magnitude of regional stress fields.

Most of the required information is obtained from bore-hole cores spaced at intervals along section lines. Jointing systems may vary from point to point in a rock cutting. Details of the methods adopted at N'changa are given in a paper by Steffen and Klingman.

Open pit copper mining in the United States of America

The Mission open pit of the America Smelting and Refining Co. at Sahuarita, Arizona, is working a low-grade copper sulphide 'porphyry' type of ore-body in an extensive zone of altered sedimentary rocks, fractured by three fault systems, folded and with monzonite porphyry. The deposit is divisible into three areas: east with a softer grinding ore of higher than average grade; central area with a thin layer of ore at shallow depth over limestone; and the west area of less complex structure and copper mineralization is more continuous with lower than average grade. The material mined from the pit is divisible into four different rock types and the overlying alluvium and each of these is broken by methods tailored to its individual characteristics. The alluvium overburden can be removed successfully without blasting but it was discovered that the output of stripping shovels could be increased by as much as 40% when it was well broken. In the alluvium the drilling and blasting methods require 9 in diameter blastholes to a depth of 45 ft, which is 5 ft below the base of the alluvium to give a 40 ft bench height, the hole spacing being 30 ft with 24 ft between rows when multiple row shots are drilled. An explosive (lb) to material (ton) ratio of 0·7 gives good fragmentation.

A tough conglomerate underlies the alluvium which is hard to break and has caused fragmentation problems. It overlies the bedrock and is 20 to 50 ft in thickness, and dips gently to the north-east which causes it to occur at different levels in the pit and the bench grades are sometimes raised or lowered to get the formation in an advantageous position as best fragmentation occurs when the conglomerate is the central portion of the face. If fragmentation is dependent on explosive distribution, tests have shown that 9 in diameter blastholes closely spaced with three or four deck loads per hole could break adequately a full face of conglomerate and this led to the drilling of $6\frac{3}{4}$ in diameter holes on a 15 ft hole spacing and 10 ft between rows with a solid column of explosive loaded throughout the layer of conglomerate, all being drilled 10° off vertical and holes in two rows which minimized back-break and excessive toe burden.

The other rock formations, ore and waste, are argillite, tactite and hornfels and in these formations holes are 9 in diameter and 55 ft deep to provide 12 to 14 ft of hole below the base to give a 40 ft bench, the hole spacing depending on the formation type and characteristics. The argillite is a hard abrasive, highly fractured rock but breaks easily though difficult to drill, which is on a 24 ft hole spacing with 18 ft between rows pattern. Tactite is of moderate to extreme hardness, drills slowly and has few fractures. Hole spacing is reduced to 21 ft to give proper fragmentation with multiple rows with high deck charges in difficult areas.

Hornfels is soft, friable and easy to drill, breaks easily and is drilled on a 24 ft spacing by 18 ft between rows pattern with an explosives ratio of 0.7 to 0.8 giving satisfactory fragmentation.

The production targets for Mission open pit are

(1) To produce 100,000 or more tons per day of broken alluvium and rock material for shovel handling.
(2) To provide efficient digging of the blasted material by proper fragmentation and provide an acceptable crusher feed.
(3) Approved safety procedure in the conduct of drilling and blasting programmes.
(4) These objectives to be achieved at minimum cost.

Criteria for the long-term mining of the deposit were developed by geological mining and metallurgical teams. Plans to a scale of 1 in to 100 ft were made of the drilling grid, property boundaries, bench plans were drawn for each 40 ft level with details plotted on the median line. One set shows a geological compilation of rock characteristics and geological structures. A second set consisted of ore reserves on a geological-block method including mineral grade, ore type, overburden and other rock types, alteration and specific gravity. The geometric shape for the end of each annual planning period over the projected life of the pit, vertical sections and topographic maps were obtained by photogrammetric methods to a scale of 200 ft to an inch with contours at 5 ft intervals.

Dilution of ore was a problem because of the large area of contact between ore and waste, so a strip 10 ft in width on each side of a contact was assumed to become mixed during winning, reducing the recovered grade to the average of the blocks in contact, and, because the average inclination of the ore-waste contact is 30°, the 10 ft strips were plotted as 20 ft on the bench plans and both grade and specific gravity of each pair of strips were determined by averaging the opposite ore and waste polygons. When the average grade was lower than 0·4% copper the strip on the ore polygon remained waste and when the average grade was 0·4% or above, the ore polygon remained unchanged and the 20 ft strip was transferred from waste to dilution ore at the grade and specific gravity value of the waste polygon.

The mill capacity was determined at 5·4 million tons per annum or 15,000 tons per 24 hours. The topography over the pit from west to east slopes from 3300 to 3100 ft.

Benches were initially 50 ft but, as already indicated, this was changed soon after stripping begin to 40 ft height so that the shovels could move efficiently, while the height of the ore benches remained unchanged at 40 ft, the final slopes of the pit are designed for 37° ($1\frac{1}{3}$ to 1) in gravels and 45° in rock with 30 ft and 25 ft berms or cuts respectively between benches.

The pit is 5000 ft long, 2000 ft wide and 500 ft deep with a planned ultimate depth of 600 ft and Mission is a good example of a large open pit on a flat working slope because of its shape. The main transport road, of the two permanent roadways, is along the south side rising from the lowest bench in a counterclockwise direction on a 7% grade, the road surfaces on the east wall and leads to the crusher and the east dump where sub-marginal grade rock is piled.

The second exit is along the north side of the pit to the north dump or gravel disposal pile; these roads are 50 ft wide and temporary and are laid out on 10% maximum grade.

Planned development aimed at establishing benches of sufficient length to expose one foot of ore face for each 5 tons of daily production and expose one year's supply of ore in advance of mining and ore was mined at a daily rate of 16,600 tons and 66,400 tons of overburden were stripped and ultimately ore production has been maintained at 20,000 tons per day and the stripping has been increased to 86,000 tons per day.

Included in short-term planning are grade and tonnage forecasting of pit operation up to and including quarterly periods and are as consistent as possible with longer term plans. Grade is determined from blasthole drilling and the geology of the ore becomes more defined as exposures occur. Final ore reserves consisted of 1000 ore and waste polygons replacing the triangular method of calculating ore reserves.

Coast and Geodetic Survey monuments are used as major reference points for a major triangulation network and then expanded over the property for photogrammetric mapping methods. A daily routine of mapping working bench progress is done by two men using a plane table and a map of the geological features of each bench is recorded, both on a scale of 1 in = 100 ft. The geologists periodically examine the pit slopes for potential zones of weakness and an inspection for any shedding from the walls.

Diamond drilling is with wire core lines with core recovery averaging 92·4% recovery from NX and BX wire line core sizes, drilling rates are 62·4 ft for rotary and 21·2 ft for core-boring. The factors used for calculating tonnages are 17·0 ft^3 per ton for gravels and 11·5 ft^3 for waste rock and ore. The cores are logged, split and a specific gravity determination made on each run, portions are sent for assay, the short iodide method being used to determine the copper assay.

Blasthole samples enable shovel crews to separate ore from waste.

Fragmentation is essential for crusher feed which must be minus 54 in and it was found that with 9 in diameter blastholes better explosive distribution and the desired fragmentation was obtained with an increased drilling rate, the drilling being by two model 60 R Bucyrus–Erie drills, and two model C–850 Reich drills, both models using rotary roller cone hard formation rock bits and tungsten carbide button bits for extremely hard formations. Each machine drills with two joints of integral drill stem $27\frac{1}{2}$ ft long with a 5 ft long bar–stabilizer running behind the bit. Masts 75 ft long allow 55 ft of hole to be drilled in one continuous pass. One Reich drill is arranged to drill $6\frac{3}{4}$ in diameter holes more rapidly but only on the conglomerate benches. Penetration rates by the 60 RBE drill and 9 in bits are: alluvium $2\frac{1}{2}$ to $9\frac{1}{2}$, argillite 1 to 2, hornfels $2\frac{1}{2}$ to 3 and tactite $\frac{1}{2}$ to 2 ft per min. The 60 RBE drills use approximately 50,000 to 60,000 lb weight on the bit and rotate at 60 to 70 rpm. The Reich drills with 45,000 to 50,000 lb weight rotate at 80 rpm, at $6\frac{3}{4}$ in diameter the pull-down weight is 20,000 to 25,000 lb and 80 rpm rotation. Tungsten carbide button bits are used when the penetration rate is less than 1 ft per min and down pressure is reduced to 40,000 lb and rotation to 40 rpm. All drills are capable of angle drilling and detergent is added to the drilling water giving a better penetration rate and longer bit life.

The charging and blasting of primary and secondary holes is carried out by a team of ten men in the day shift, the blasting agent used for all long holes is prilled fertilizer grade AN mixed with fuel oil. Three-hundred thousand pounds of AN is stored and delivered to the blastholes in a 20,000 lb capacity mixing truck, the AN being mixed with fuel oil as they are dispensed into the blasthole. Wet holes are charged with nitro-carbo-nitrate slurry in 50 lb plastic bags. For special jobs and for secondary blasting 60% amogel cartridges are used. Holes charged with AN–FO are detonated by primacord and $\frac{3}{4}$ lb of booster explosive. Secondary drilling is by $1\frac{1}{2}$ in holes in boulders and a crawler-mounted drill is used for $3\frac{1}{2}$ in holes in areas of hard toe and lumps in the bench floor.

Since secondary blasting is potentially hazardous and very expensive, every effort is made to reduce or eliminate it by improving the effectiveness of primary blasting.

REFERENCES

'Overburden Stripping by Dragline and Face Shovel', *Mining and Minerals Engineering*, May 1965, p. 330.
'Opencast Mining of Taconites in Krivoi Rog, Ukraine', Novozhilov, *Opencast Mining, Quarrying and Alluvial Mining*, Institution of Mining and Metallurgy 1965, p. 344.
'Mount Goldsworthy Iron Ore', R. I. J. Agnew, *Mining Magazine*, December 1966, p. 432.
'Mobile Plant for Iron-Ore Concentration', C. K. McArthur and P. R. Porath, *Mining Congress Journal*, 51, No. 10, p. 28.
'Palabora', I. C. Herbert, *Mining Magazine*, Vol. 116, No. 1, January 1967, p. 4.
'Slope Stability at the Open Pit of N'changa Consolidated Copper Mines Ltd.', Steffen and Klingman, *Journal of the South African Institution of Mining and Metallurgy*, Vol. 67, No. 4 November, 1966, p. 140.
'Further Developments in Open Pit Mining at N'changa', R. V. C. Burls and W. Holt, *Journal of the South African Institution of Mining and Metallurgy*, April 1967, Vol. 67, No. 9, p. 421.
'Stresses and Displacements Surrounding an Open Pit in a Gravity Loaded Rock', W. Blake, *U.S. Bureau of Mines, Report of Investigations No. 7002*, 1967.

CHAPTER 8

OPENCAST COAL

The continued fall in the output of coal from deep, or underground, mining in a period of growing demand for war purposes, both internal and external, and the necessity for maintaining as far as possible the supply for domestic purposes during World War II led to the exploitation by opencast methods of those deposits of coal which occur at, or relatively near to, the surface.

Opencast coal working prior to the outbreak of the war had not been practised to any extent in Great Britain and was limited to small haphazard working in periods of cessation of output from collieries through national strikes. Systematic opencast working has however been carried on in the USA over a long period and provides a substantial portion of the national output and the production from some sites exceeds 10,000 tons per day.

In this country the control of the working of opencast coal was at first the function of the Ministry of Works in conjunction with the Ministry of Fuel and Power but after the nationalization of the British coal industry the National Coal Board set up an Opencast Executive which took over control of the bulk of opencast coal working, except a small amount (0·4 million tons) produced by private operators. In 1965–66 the total output of opencast coal was 7·1 million tons at a cost of 75s. 9d. per ton, and the National Coal Board made a profit on opencast coal of £4·8 million or 14s. 3d. per ton, against a loss of 1s. 0d. per ton on deep mined coal which cost 92s. 6d. per ton.

Opencast coal mining has proved an important and profitable source of supply since it provides a flexible means of meeting fluctuations in market demand and deep mined production.

Modern earth-moving and tree-planting equipment makes it possible to restore land on a scale and at a speed not possible ten years ago and at many opencast sites, restoration has resulted in substantial improvements to the agricultural and amenity value of the land used, particularly through improvement of drainage.

OPENCAST COAL WORKING IN THE
UNITED KINGDOM

Two large opencast coal sites, at Acorn Bank in Northumberland, which is a strip-mining site where the coal is obtained from a series of parallel cuts, and at Westfield, on the Fife–Kinross boundary in Scotland, which is an open pit comparable with a normal quarrying operation, are operating under contract to the National Coal Board Opencast Coal Executive.

The Acorn Bank pit was commenced in 1955 and three seams extend over the whole of the 440-acre site. The High Main seam with an average thickness of 3 ft 3 in is excavated in one leaf, the Top Grey seam is in two leaves, each about 15 in thick, separated by a parting 1 in to 4 in thick is worked in two separate leaves while the Bottom seam, with a total thickness of about 7 ft is in four leaves, with a parting of several feet in one area of the site between the top two leaves, while the other partings are only an inch or two thick.

Two other seams, the Top Yard and the Yard, whose recovery by deep mining has been prevented because they lie between upthrow faults, have also been worked in one area of the site.

All the seams are in a band of strata 30 ft in depth in the north but in the south-east the seams are as much as 60 ft apart.

The seams are deepest at the eastern end of the site and excavation was commenced at this end rather exceptionally because it was required that the site be restored progressively. The initial box-cut was 3200 ft long by 100 ft wide and 230 ft deep. This initial excavation of ten million cubic yards provided the waste room for the normal strip mining which followed.

Most of the site was covered by a layer of drift material to an average depth of 0 ft consisting of boulder clay overlain by yellow clay and with a bed of sand between the clays in many places. This was excavated by 120 RB 5 yd^3 capacity electric face-shovels and loaded into 22- and 27-ton capacity dump trucks. The yellow clay and the sand were dug separately when the latter was thick and stockpiled as subsoil for later restoration work, the remaining material being transported and dumped over the de-coaled area.

Below the drift the overburden consists of shales and sandstones which must be blasted. Joy 58 BH Champion drills with tricone bits are used to drill $6\frac{5}{8}$ in diameter holes in the overburden, generally on an 18 ft square grid. Primed with 10% gelignite the holes are charged with AN–FO, the charging ratio being $3\frac{1}{2}$ yd^3 per lb. Deck loading allows the bulk of the explosive to be placed in the hard, heavy sandstone beds. This material is loaded by three 120 RB face-shovels down to the dragline bench level, approximately 90 ft above the top coal seam. Where two 1150 B draglines, one with a 215 ft jib and 22 yd^3 bucket, the other with a 180 ft jib and a 25 yd^3 bucket are used the first operates from the centre to the northern end of the coaling cut where the coal is deeper, the other operates from the centre to the southern end of the coaling cut. The machines could be put out of step by mechanical breakdown and any phase difference of more than a week is corrected by adjusting the position along the length of the cut where the first machine to finish the old cut starts the new cut, the average width of the cut being 80 ft.

Two derricks follow the dragline while it is excavating the end of the previous coaling cut along the highwall, lifting out the coal as it is dug by the face-shovels in the cut below. A drill then follows the derricks along the highwall and the overburden of the next cut is drilled and blasted, firing occurring after all the coal has been removed from that section of the previous cut immediately below.

The dragline is moved to a position about half-way along the cut, digs down to the top seam of coal, and works along the half-length of the cut, the derricks follow, lifting out the coal.

The drill is then taken down to the de-coaled cut and drills down to the next seam of coal, the holes are then charged and fired. The dragline then moves on the spoil bank, travels back to a position opposite the centre of the cut and digs down to the next leaf of coal and at the same time the machine also digs the rehandle material from the upper cut.

The dragline is used in this system for removing the thicker parting because usually the parting between two of the seams is thin enough to be removed by face-shovel and dumper and this is carried out during the night shift using the face coaling excavators, the waste being run to the de-coaled end of the cut.

In operating a site, with three main leaves or seams of coal, accurate programming is necessary to avoid a machine having to stop while others complete their section of the work. The two draglines are the major items of plant and other work is planned to keep these two machines always employed.

The necessity for progressive restoration precludes the use of a permanent access ramp into the cuts for coal removal and is a major restriction in the operation of the site. A twin access road tunnel formed in pre-cast concrete units through the spoil banks to the bottom of each cut was one of the methods originally considered for providing a route to the coal but this method would have been difficult because the seams of coal were at different levels and might be hazardous because deep mine workings can cause subsidence. Use of 4 high-speed electric derricks was the method finally adopted. They travelled on rails along the top edge of the highwall and were of the double-drum type with jibs 152 ft long and capable of lifting 13 tons at a maximum radius of 140 ft. The driver's cab is situated near the top of the mast and the derrick has separate hoisting, luffing and slewing and can make all three motions simultaneously. Each derrick operates with two or more coal skips of 10 tons capacity. The derrick lowers an empty skip into the cut and this is swung by hand into a position near a face-shovel excavating coal and is unhooked. A full skip is then picked up with the main rope attached to a bar which lifts just forward of the centre of gravity of the skip and the second rope is attached to the tail of the skip and the full skip is lifted out of the cut and swung into position over a 40-ton capacity coal hauler. The main rope is lowered with the tail rope held and the coal is emptied directly into the coal hauler without the use of a banksman or a hopper. The derrick operators quickly became skilled and achieved outputs of up to 150 tons per hour. The derricks are fitted with special bogies with four 25-ton hydraulic jacks, each bogie is lifted in turn where it stands on a special crossing rail, the wheel housings are rotated through $90°$, the bogie is lowered back on to the cross-track and the derrick can then travel at right angles to its original track and be repositioned for the succeeding cut. The major difficulty with the derricks is the need to work them very close to the edge of the highwall in order to be able to reach into the cut and for such heavy derricks a stable highwall is therefore essential. Frequent faults occurring on the site, the highwall has cracks and these are extended by heavy blasting. The drilling pattern and the sequence of delay detonators have been adjusted in order to disturb the highwall as little as possible by blasting for the adjacent cut. The derricks have been operated for many years without an accident but it is questionable whether this is indeed the best method to move the coal.

The surface of each coal leaf is cleaned by bulldozer as it is exposed and final cleaning is by hand. Shovels of $1\frac{1}{2}$ and $\frac{3}{4}$ yd^3 capacity are used and the dippers have a flat lip designed for clean loading. The bottom leaf of the Bottom Grey seam is usually too hard to dig without blasting. Hand-held drills are used to bore holes at 4 ft centres, these are charged with 2 oz of gelignite and fired separately.

The 40-ton capacity coal-hauler units are modified Euclid B.1 chassis fitted with a fifth wheel and towing 60 yd^3 semi-trailers, two trailers to each unit, one being filled while the other is in transit. The 40-ton units can only operate 'off highway' and a private coal-haul road $1\frac{1}{2}$ miles long crossing a river and two public roads on three Bailey bridges links the site with the screening plant and the saving in cost by using the large units has justified the cost of the private road.

When the site is worked out some 16 million pounds of explosive, mostly AN–FO, will have been used. A central mixing shed using a pan-type mixer is used with a hard rubber roller replacing the iron one, and the scraper blades are of wood not steel. The driving motor is outside the central mixing shed and drives through a long shaft. The fuel oil is coloured so that the mixer operator can see whether each batch is sufficiently mixed when it is packed in polythene bags 6 in in diameter for transport to the blastholes.

The topsoil and 3 ft of subsoil were placed in dumps initially around the perimeter of the site. After some years, it has become possible to lift the top- and subsoil ahead of the coaling cuts and spread them directly on to the levelled spoil banks of the decoaled area. After these have been carefully graded to prearranged contours, the subsoil is spread in two layers 18 in thick and finally the topsoil is spread in a 12 in layer. Each layer is rooted and all stones large enough to turn a plough are removed. The work is then inspected before the next layer is begun.

At the Westfield opencast coal site the phase one contract undertaken by Costain Mining Ltd was commenced in January 1961 and is scheduled for completion by the end of 1969. It specifies all saleable coal to be recovered at a rate of 13,000 tons per week over an area 2500 ft by 1000 ft to a maximum depth of 450 ft leaving a worked out area, with a finished volume of 34 million yd^3 which will later be used as a disposal area for overburden and waste rock excavated during the second phase contract to

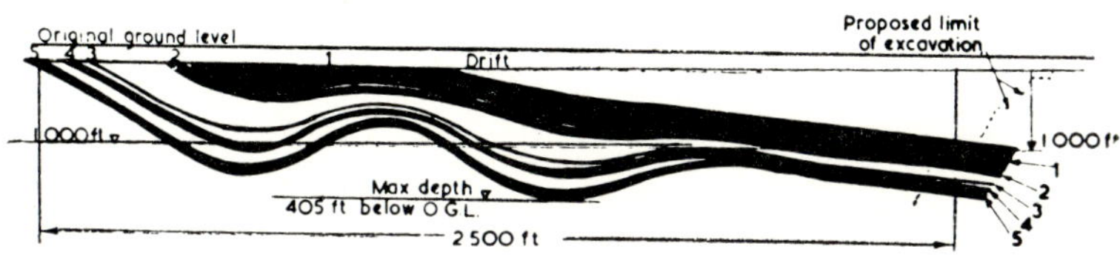

Fig. 58. Westfield opencast coal site: E.–W. section through centre-line of pit. Main coa seams to be won (1) Bogside Thick, (2) Bogside Main, (3) Un-named seams, (4) Westfield Thick, (5) Westfield Main.

win coal to the north of the present site. A third and final phase of the programme will eventually leave an excavation 900 ft deep which will be flooded and landscaped to form a Scottish loch.

The coal seams belong to two groups, the main one being the Boglochty series (Fig. 58) and the other, part of the Coal Measures in the north-eastern part of the site. The seams crop out under the surface drift to the south and the west, mining to the north being limited by the Ochil Fault which faults out the coal strata, and to the east by the tendency of the seams to deepen and thin out.

Three possible methods of winning the coal at Westfield were considered; underground or deep mining and conventional opencast strip mining were discarded because of acute folding and faulting of the strata and by neither method could the economic winning of the coal have been effected.

The third method, and that adopted, was conventional quarrying with 10 to 15 ft benches, the coal and waste rock between the seams being excavated separately as encountered.

The southern part of the site was partially covered by a peat bog which had to be removed before excavation of the underlying strata could begin and to remove the peat a dredging operation was chosen, a cutter-suction type dredge with a $7\frac{1}{2}$ ft diameter head mounted on a 71 ft long arm, was taken to the site in sections and built up in a specially prepared dry dock close to the bog area. The dock was then flooded and the dredge cut its own way into the peat bog. The dredged material was pumped into lagoons prepared by surrounding waste land away from the opencast site with earth banks having a maximum height of 65 ft and containing a total of $3\frac{3}{4}$ million yd^3 of drift material excavated from the northern part of the site, some 4·3 million yd^3 of peat, sand and silt were removed at an average rate of 90,000 yd^3 per week.

The provision of an area for dumping waste comprising overburden, the rock partings between coal seams and waste from the coal washing plant had been provided by the NCB Opencast Executive prior to the letting of the phase one contract and this area, some $1\frac{3}{4}$ miles from the site, is a shallow valley containing a bog and surrounded by poor land of no particular agricultural value. A conveyor system served by a crushing plant to deliver 4·3 million yd^3 per annum was decided on rather than transport by trucks which would have involved crossing a water course, a railway and two roads requiring substantial bridges and a suitable road.

The topsoil and drift material overlying the rock overburden is stored for relaying on the waste dump. The rock, which is a medium-soft sandstone, is quarried and delivered to the crushers for reduction to a size suitable for transport on the conveyor belt.

Two Joy 58 BH Champion rotary drills and one Reich 675 track-mounted machine drill $6\frac{1}{4}$ in diameter holes to a maximum depth of 95 ft at a rate of 65 ft per hour for blasting an 80 ft high face. The burden and spacing employed vary with face height. For benches over 30 ft high a 21 ft burden by 24 ft spacing pattern is used and for less high benches, such as the coaling benches, an 18 ft by 18 ft grid is usual.

A lorry-mounted Reich 600 drill and a CPTG 800 drill are available for drilling, the softer coal and strata in the partings between seams and benches are cut at a height convenient to the excavators, generally from 10 to 15 ft, and blasting is employed when the hardness of the strata requires, which applies mainly to the sandstone and shale partings and to some of the lower grade coals.

AN–FO blasting agent was used initially, but during the winter of 1962 increasingly difficult conditions due to groundwater were encountered because water penetrated the polythene bags containing the explosive and trials using rigid cardboard and polythene containers and 16-gallon metal cans were not entirely successful since blasting was not uniform. A further problem was to obtain enough fragmentation which would provide a good crusher feed without recourse to secondary blasting. More powerful explosive and one unaffected by excessively wet conditions was required.

It was decided to investigate the use of slurry explosives and the first group of these to be tested were sensitized but although these were water compatible their power was low and difficulty was experienced in initiation and transmission of the explosive wave. A second group sensitized with nitroglycerine had a high loading density and resistance to water. A primer with very high detonation velocity was used and a number of trial blasts were carried out on a 70 ft face and the results when compared with those obtained with AN–FO were encouraging and have since been confirmed. An order for nitroglycerine slurries was placed and it was decided to prepare the ammonium nitrate and other intrinsically non-explosive materials on site and transport the explosive base only from the factory thus reducing transport costs. The Opencast Executive allowed the building of a factory on the site and no objection was raised by the local planning authority or the Home Office. The slurry used has a dense sugary constitution and is loaded into polythene bags, each holding 10 lb; the cost of slurry explosive is about twice that of AN–FO but this is discounted by the increased strength, allowing increased spacing and burden of holes and less drilling. Hole spacing was increased from 12 ft to 18 ft by 18 ft and yield per lb of explosive from 0·87 to 2·5 yd^3.

The blasted rock is loaded by four 150 RB Ruston–Bucyrus face-shovels with 5 yd^3 capacity dippers, into a fleet of 18 yd^3 capacity rear-dump trucks consisting of 17 Aveling–Barford and 8 Euclid trucks for delivery to the waste crushers, costing some £18,000 per truck. The crushers should not be fed with any stones too large to enter the gap as any blockage will seriously disrupt the whole cycle, so the blasting is designed so that the fragmentation is sufficient to produce all stone less than this maximum size but shovel operators must take care not to load any oversize pieces and they are trained to this end.

For removing the coal and partings between seams, four 54 RB $2\frac{1}{2}$ yd^3 Ruston–Bucyrus electric face-shovels are used. It was thought at one time that it might be necessary to use draglines since, in parts of the site, the seams have an inclination of 45° but after trials it was decided to excavate seams and partings in benches 15 ft high. At this height the shovels can load without having to travel up the face. The ash and sulphur contents of the coal vary within a distance of 50 ft and it is necessary to strip a variety of coals so that a satisfactory blend of raw coal can go to the coal preparation plant to which it is taken in 36 yd^3 rear dump trucks or 20 yd^3 side dump trucks up a gradient mainly 1 in 12, at a rate of 20,000 tons per week of which 6500 to 7000 tons are discarded as working waste. Some 50,000 tons per week of overburden and partings are transported to the crusher and then conveyed on a 48 in wide steel-cored belt at 800 ft per minute to the waste pit. One double-roll and one single-roll Sheepbridge crusher each with a throughput of 1100 tons per hour and each driven by a 200 hp motor are used to reduce the overburden and partings to 10 in size while two hammer crushers with 450 hp motors are used to reduce whinstone boulders.

A computer-operated dumper-truck control unit has been installed to meet the problem of routing dump trucks so as to optimize turn-round by dispatching each one to the most appropriate loading point as it returns empty. There are two entrances to the open pit and eight or nine shovels working two main types of material to be handled and two types of dump truck—rock-bodied and coal-bodied—arriving at the rate of a hundred an hour. The unit, built by Lintott Engineering Ltd, consists of a device at each of the two entrances for identifying dump trucks as they enter the site, a computer for storing loading times and for calculating the best shovel to send a truck to, and to signal at each entrance to inform the truck driver of his selected destination.

The dump truck is identified as to type and required material by means of a system of electric lamps and photo-electric cells situated at the control office, the control unit consists essentially of a series of electronic clocks, one for each shovel, and each clock measures the time a truck would have to wait before loading at any shovel, and a scanning device selects the shovel with shortest waiting time. When the control unit has calculated the best destination and the information is displayed for the truck driver to see on an illuminated screen on the side of the road used by the trucks, the truck can continue through the entrance at its normal speed since the whole process of identification, calculation and direction takes such a short time.

Haulage roads within the pit are maintained in good condition using a shale forming one of the partings between the seams which makes a good, cheap surfacing material. The life of tyres is 5000 hours for front wheels and 2500 to 3000 for rear wheels and all roads are graded, kept rolled and made good as soon as wear occurs, particularly in wet weather.

The crushed rock is dumped directly on to the 48 in wide steel-cored conveyor belt travelling at 800 ft per minute with a rated capacity of 3440 metric tons per hour. The overall length of the conveyor system is 11,000 ft divided into three fixed sections leading to the tip on which is mounted a movable belt and spreader installation. The movable belt, which is laid across the width of the tip, rests on sleepers so that the complete assembly can be placed into a new position as the face of the tip advances. Also mounted on the sleepers, on either side of the conveyor and parallel to it, are two flat-bottomed rails. These serve primarily to carry a tipper unit which elevates the belt and transfers the rock to the connecting bridge leading to the spreader conveyor which is mounted on tracks and has a 92 ft long jib which can be swung through an arc of 300°. A high-speed conveyor mounted in the jib throws the stone out over the edge of the tip. The movements of the jib and the movement of the tipper along the transverse conveyor are controlled by an operator from a cabin attached to the spreader unit. The dump has a shallow convex profile which must be maintained by careful disposition of the waste by the spreader operator, any irregularities of the surface being smoothed out by bulldozers. As much overburden as possible is stripped from the valley floor and conveyed to a second spreader unit for distribution on top of the rock pile. The tip is being formed into a hill having a maximum slope of 1 in 8 and blended in with the local scenery. As each section of the tip is completed it is graded, the subsoil and topsoil replaced, and restoration completed as soon as possible by planting grass. It is intended to reinstate the site as good agricultural land and already sheep and cattle are grazing on that part of the site which has been restored.

Two main sumps, into which water collecting in the pit is delivered by small diesel pumps, are served by 6 in electric pumps to a settling pond from which the clarified water is pumped into a local stream. Since the water contains abrasive grit, stage pumping in 120 ft lifts is to be adopted as the pit is deepened instead of the use of high-lift pumps.

The power supply at both Acorn Bank and at Westfield is electric, to drive the larger machines. At Acorn Bank the supply is through overhead lines and since the two big draglines are American built and operate on 60 cycles current, two frequency changers have been installed for their supply. The total installed horsepower is 6000.

An 11 kV overhead ring main has been installed around the pit at Westfield and the excavators and pumps are supplied from this, through trailing cables through eleven 3·3 kV skid-mounted transformers and field switches. The total installed horsepower is 12,853.

On both sites a large hangar has been erected so that all but the largest machines can be repaired and maintained under cover and these shops are equipped to carry out major repairs and large stocks of spares are maintained. A water-brake dynamometer is installed in each site's machine shop, and test runs of rebuilt engines are carried out before re-installation.

Communications are vital to ensure proper control on these large sites without unnecessary delay and both sites have been equipped with radio and internal telephone systems. The radio controller at Westfield, where mining operations are concentrated in the pit, is sited overlooking the pit and has proved of great value, for an alert radio controller can sometimes anticipate a possible hold-up in the operation of the plant and take early action to counter it.

Each site has a fleet of Land-Rovers for speedy movement of personnel and at Westfield they are equipped with radio and operate as taxis under the radio controller. Radio communication to remote sites and off-shore drilling rigs is dealt with in a later chapter.

Some items of equipment are common to both sites, for example, Joy 58 BH drills, 27-ton dump trucks and Caterpillar D8 tractors, and although the organization and maintenance arrangements are similar, some cost items are so different even on the same basis. They include:

(1) Tyre costs on 27-ton dump trucks at Westfield are 22% greater than at Acorn Bank although road maintenance at both sites is of a high standard. Trucks at both are loaded by 5 yd³ shovels and maintenance organization is similar. The two differences which contribute to the inequality are that at Acorn Bank the shale for surfacing the roads is soft and bulldozers and graders used for forming and maintaining the roads can shear off any inequalities to produce a well-graded running surface.

(2) At Westfield the rock contains sandstone which is too hard for the grader and dozer blades to break so that the running surface is poorer and in wet weather this is very obvious.

At both sites the gradients are similar but at Westfield, owing to its greater depth, the length of haul at maximum adverse gradient is longer. Bit cost at Westfield is 73% higher though the bits are of similar size and manufacture and footages per hour are comparable. The higher proportion of abrasive materials in the strata penetrated at Westfield appears to be the only reason for the higher drilling cost.

Again at Westfield track costs are 35% higher than at Acorn Bank and again this would be due to the more abrasive strata at Westfield.

Both sites are operated under contracts to the NCB Opencast Executive but as the circumstances in each case differ so do the financial terms of the contracts.

At Acorn Bank no inspectors are employed by the Executive but their samplers take frequent samples of the coal and the coal floors which are analysed for ash content and the results communicated to the contractor. Payment is made per ton of clean coal produced and includes all operations from stripping off the topsoil at the beginning to replacing it after extraction of the coal and levelling of the site. The Executive can refuse coal with an ash content exceeding 10% and the contractor is required to use coal analysis figures to produce a blended product to meet this limit. The initial excavation of 10 million cubic yards represented a heavy capital lock-up and was met by an advance by the Executive repaid as coal production proceeded.

The normal system of payment was modified at Westfield because although the primary object of the contract is to recover all the clean coal in the basin the secondary object is to produce a pit to receive the overburden from the phase two operations. Two rates are therefore paid: one covers the cost of all excavation and disposal of overburden, the second covers the cost of washing and handling the coal after delivery to the coal washery. In all 22 seams or leaves of coal and over most of the area 128 ft of *in situ* coal was worked which, after deducting shale and clay partings, left 82 ft of recoverable raw coal.

OPENCAST COAL MINING IN THE USSR

In the USSR great importance is attached to opencast mining and this method is used for mining 22% of coal, 55% of iron ore, 50% of non-ferrous ore, 40% of manganese ore and almost 100% of non-metallic ores.

It is planned within the next ten years to increase opencast output to three-quarters of the total volume of the output of useful minerals and in particular 500 million tons of coal per annum by opencast mining is budgeted for. By 1980 the output of all kinds of fuel is to be increased fourfold and to satisfy the fuel requirements of the national economy coal output must be increased from 550 million tons in 1964 to 650 million tons in 1970 and 1100 to 1200 million tons by 1980. Technical and economic performance is expected also to improve the industry in the eastern regions which will receive priority since mining and geological conditions are favourable and costs are low.

The total published coal reserves of the USSR are very large, the proved and probable reserves are 1085 thousand million tons, with possible reserves of some 500 thousand million tons in addition.

A threefold increase in output per manshift is called for in the long-term plans of the industry as compared with 1964 and at the same time the working week is expected to be cut initially to between 30 and 35 hours compared with the present 36 to 41 hours. The increase is expected to be obtained by wide advances in coal-mining technology, the preferential development of opencast and hydraulic mining, the development of large and very large opencast and underground operations, the reconstruction of the working capital for mining generally and a radical improvement in production organization. Opencast coal production relative to 1964 is to be increased more than fivefold until it amounts to 53% of the country's total output of coal, compared with 23% in 1964.

Extensive rich coal deposits exist in the eastern parts of the USSR and also favourable conditions for opencast working. In the Kansk-Achinsk Basin at the Ekibastuz deposit and other eastern regions planning and development of large, 8 to 25 million tons per annum, very large, 30 to 60 million tons per annum, coal opencasts are in progress. Research and experience has proved that increasing their size to the optimum which the reserves and technical resources permit, reduces specific capital investment and improves technical and economic performance considerably. For example, the Irsha-Borodinsk opencast which originally had a planned capacity of a million tons per annum, now produces $5\frac{1}{2}$ million tons annually and at the same time cost has fallen by nearly 60% and the output per man-hour has increased to double the output originally planned. A further revision to increase the capacity to 40 million tons per annum budgets for a threefold reduction in capital investment, a sevenfold increase in output per man-hour and a fivefold decrease in the unit cost of the coal.

Bucket wheel excavators with capacities from 1000 to 3000 yd^3 per hour have been produced since 1959 together with belt conveyors and tip spreaders and, recently, excavators for opencast working with a bucket capacity of 17 yd^3 have been produced and plans are in hand for the assembly of a unit with a continuous output of 11,500 yd^3 per hour. The bucket wheel excavator has buckets of $3\frac{1}{2}$ yd^3 capacity together with a 730 ft boom for the transport dumping method of working and 360 ft for the auxiliary transport system. Delivery of walking draglines have been made with bucket capacities ranging from 13 to 34 yd^3 and boom lengths of 220 to 310 ft, while shovels with a bucket capacity of 47 yd^3 and a boom length of 200 ft have also been delivered. A dragline is being produced with a bucket capacity of 110 to 120 yd^3 and a boom length of 325 ft. It is expected that draglines will constitute not less than 30% of the total excavator fleet.

Dump trucks of 80 to 100 tons capacity and automatic self-dumpers of 27 to 40 tons are in production and dump trucks with capacities of 140 to 180 tons are contemplated.

The main consideration in planning new opencast mines is the use of fully automated, highly productive stripping equipment which when deployed in minimum numbers, will produce at a high unit rate with a corresponding increase in output per man-hour.

The choice of equipment depends on the mining geology so that at sites where the seam is flat and comparatively thin, up to 65 ft thick which in the British Isles would be considered exceptionally thick and comparable with the Thick Coal of Staffordshire and Warwickshire, about 60 ft but now worked out, a dragline system of working is adopted using draglines with bucket capacities of 30 to 120 yd^3 and boom lengths of 310 to 375 ft.

The normal methods adopted are strip mining or cut-and-fill when overburden material is handled by draglines on to spoil banks in mined-out areas. If there is insufficient room to place the waste rock in the spoil banks, rehandling is adopted and for this purpose draglines are stationed on the spoil banks.

Shovels are used to load the coal into road or rail transport, roads made or rails being laid on the bottom of a bench, one, two or more benches being used for loading.

If the general dip of the strata does not equal 10° and the thickness of overburden is considerable combined methods of mining are adopted. For example, at Vakhrushev in the Urals, in mining by opencast a seam $97\frac{1}{2}$ ft thick the upper benches of the rock overburden are excavated and transported to spoil banks by road or rail while the lower bench is excavated by the cut-and-fill method, in this case the overburden waste is dumped on to spoil banks in a mined-out area as at Westfield.

One system in which a dragline is used is known as 'excavator quarry' and employs the same dragline stationed on an intermediate overburden bench (Fig. 59) to remove overburden and load the coal and consequently there is no need to drive the usual entrance and working cuts for working these seams since the coal is transported along the floor of the seam. From

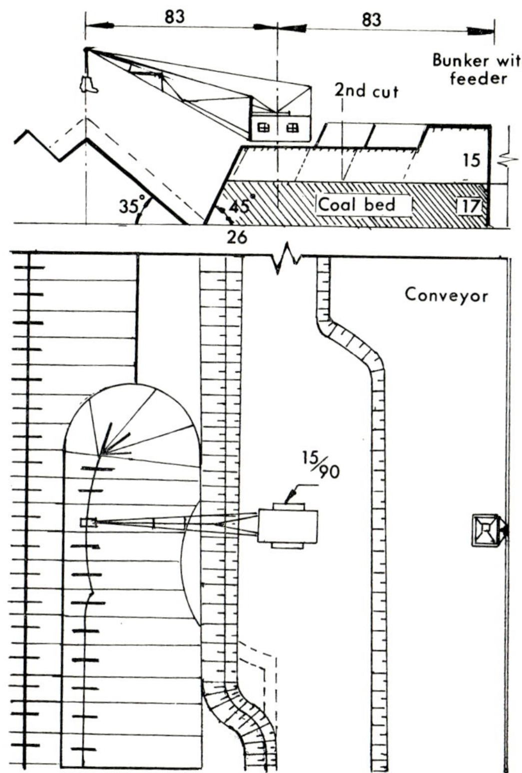

Fig. 59.   *Transportless method of mining using the cut-and-fill system with a single walking dragline for both stripping and loading.*

the same station the dragline first removes overburden, recasting it from the upper and lower parts of the bench to the spoil bank. Then after stripping the overburden, the dragline loads out the coal into a self-propelling bunker or stockpiles it on the surface. A shovel or other excavator reloads the coal on to a conveyor, dump truck or rail car. Draglines with buckets of 20 yd^3 or more capacity, such as the ESh-15/90A (Fig. 60 and Table X), are capable of excavating hard rock or ore if it is first prepared for excavation by blasting and wear on buckets and ropes does not appear to be unreasonable. The dragline excavates overburden from the next cut after mining out the coal or ore so that the dragline alternates the removal of overburden and the mining of coal or ore. If two or three draglines are employed in an open pit, mining goes on continuously round the clock. This 'excavator quarry' system is used in mining brown coal or iron ore in the Tula region, the Ackerman iron-ore

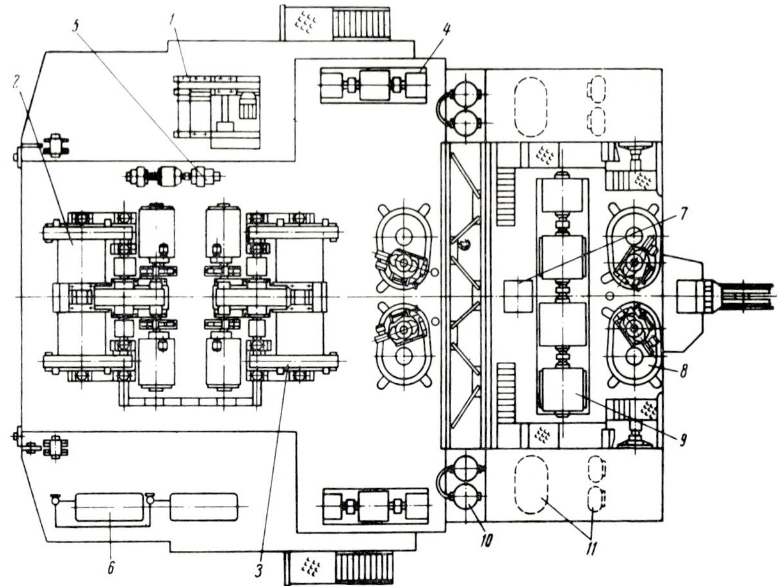

Fig. 60.   Location of the equipment on the slewing platform of the ESh-15/90A excavator: (1) jib hoist, (2) traction haulage, (3) lifting hoist, (4) pump, (5) exciters, (6) compressors, (7) high-voltage power supply, (8) slewing reduction gear, (9) converter set, (10) tanks for hydraulic liquid, (11) power transformer.

deposits in the Orenberg region and is projected for the coalfields in Nusarovo in Siberia, where the seam is 52 ft thick and is to be worked by draglines with a bucket capacity of 67 yd^3. It is expected that the cheapest coal will be won with an output of 80 tons per man-shift.

Where the rocks are friable and poorly cemented, the continuous technique is used with rotary excavators, belt conveyors and belt boom tip spreaders (Fig. 61).

The ESh-80/100 draglines at Nazarovsk and Abansk of the Kansk–Achinsk Basin, with seam thicknesses up to 65 ft, have an annual output of 8 to 12 million tons of coal. Rotary excavators ERShR-2600 at Irsha-Borodinsk and Itatsk, seam thickness 140 to 293 ft, produce 25 to 30 million

TABLE X

TECHNICAL DETAILS OF USSR ESh-15/90A WALKING DRAGLINE
by Uralmashzavod

Item	*ESh*–15/90*A*
Bucket volume cu m (yd³)	15 (20)
Boom length m (ft)	90 (293)
Maximum digging radius m (ft)	82 (260)
Maximum dumping radius m (ft)	82 (260)
Maximum digging depth, m (ft)	41 (130)
Maximum dumping height, m (ft)	37 (90)
Power of synchronous motor drive (kW)	1900
Power of hoist motors (kW)	650 × 2
Power of slewing motors (kW)	210 × 4
Diameter of slewing platform, m (ft)	14 (46)
Crawler track length, m (ft)	13 (41)
Width of crawler track, m (ft)	2·5 (8·1)
Average pressure on ground during movement (kg/cm²)	2·45
Average pressure on ground during work (kg/cm²)	1·0
Walking speed, metres ph (ft hr)	60 (195)
Working weight, tons	1600
Calculated duration of working cycle with 120° slewing angle and average digging depth (sec)	63

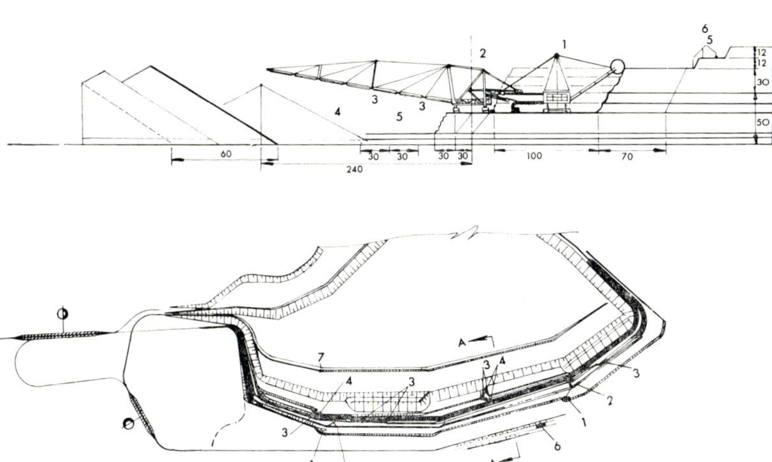

Fig. 61. Opencast system at Irsha-Borodinsk utilizing tip spreaders: (1) ERShr R–2600–50/5 excavator, (2) OShr R–11200/225 tip spreader, (3) ER–2000–2500 excavator, (4) OSR–100 tip spreader, (5) rail trucks, (6) EKG–8 excavator, (7) EKG–4 (6) M³ excavator.

tons annually and the EKT–12·5 excavators used two to a bench, produce 20 to 25 million tons per annum at Beresovk and Ekibastuz sites. The estimated output in tons-per-man-month from large opencasts using draglines is 1400 to 1600 tons while from continuously operating ERShR bucket wheel excavators it is from 2100 to 2600 tons.

In order to meet the requirements of the chief consumers of opencast coal, the electric power stations, the work routine at most large opencast sites provides for a continuous working year of 360 to 365 days.

Complete mechanization of auxiliary processes is a necessity for efficient working at large opencast sites and in the USSR work is in progress on automatic operational control of equipment and transportation through automatic monitoring.

OPENCAST COAL MINING IN THE USA

Surface coal mining in the USA has a number of problems among which are complying with new legislation on land reclamation after coal extraction, the avoidance of stream pollution and the necessity to reduce costs of production to a minimum, while the overburden to seam thickness ratio is constantly increasing, to retain the hold on a highly competitive energy market. More and better long-range planning and more intensive mechanization is being adopted with a constant upgrading of machine capacity and performance. Teamwork should commence long before the first overburden is stripped and continue throughout coal production and extend into land reclamation.

Before starting or even acquiring a new opencast site two fundamental problems need a solution:

(1) Acquiring and consolidation of sufficient coal reserves to justify the capital expenditure for mining and preparation equipment of sufficient ease of working and quality to render the operation viable.

(2) Gathering reliable information about the coal seam or seams and the mining conditions associated.

The size of the property and the topography influence greatly the size and type of surface operation that can be planned and it would not be wise to choose high-capacity equipment for a mine with limited reserves. Similarly small-capacity equipment may not be economical at a site with large reserves. Information should be accumulated as quickly as possible and getting the information is an engineering and a geological job involving topographic surveying, mapping, and prospecting. If deep mining has been carried out in the area, mine plans may provide a valuable source of information showing how close the old workings came to the outcrop, frequently the dip and thickness of the seam or seams and the geological conditions, particularly the pattern of faulting.

If drilling is resorted to, to ascertain seam and overburden thicknesses, diamond core boring gives valuable information and several coal companies have found it useful to photograph the coal cores to provide a permanent graphic record of the core; this takes less space and the physical characteristics of the coal are often destroyed when the core is split for analysis. Diamond drilling results have improved with the adoption of the wire-line coring technique for coal prospecting in which a retractable inner tube is used. At the end of each run the inner tube assembly is hoisted through the drill-string without the necessity of pulling this out to get at the core barrel and recover the core.

The consolidation of the overburden is a main factor in determining the method to be adopted for its stripping and this can greatly affect the cost of the operation, and is often a main item in determining the economy of the whole process of production. A quick, easy and inexpensive method is the seismic analysis system which is based on the principle that sound shock waves travel at different speeds and along different paths through subsequent materials and this allows a decision to be made whether the overburden can be ripped or whether it will need to be drilled and blasted.

If the site on which the data collected is considered to be viable, the site of the plant must next be considered. Access to rail or water transportation, topography, the size of the building area required, water supply for coal preparation and distance to the coal to be worked are all factors to be considered and a refuse disposal area must also be sited.

Having fixed the position of the preparation plant, the most advantageous point to commence opencast operations must next be fixed, followed by the laying out of a permanent road connecting the opencast site with the preparation plant. This is best carried out by using a map of the area showing the topography and then estimating the cutting and earthworks required on different alternative routes. Good road alignment and good gradients pay dividends in faster transport, lower truck and particularly tyre maintenance costs. A solid well-drained roadbed is a prime essential.

The factors which influence the method of attacking the coal-getting operation are the type, thickness and contour of the overburden, and the thickness and contour of the coal available. If the seam does not outcrop a box-cut will be required to open up the opencast. If the coal does outcrop a considerable distance above drainage and contour mining is proposed, it may be necessary to construct an expensive solid, all-weather road to the coal level. On occasion, it may be more economical to construct a bin, feeder and conveyor to transport coal downhill to the preparation plant and confine dump truck haulage to nearly level roads along the outcrop.

The selection of equipment required is influenced by terrain, coal reserves and type of coal, expected selling price of the coal, type of over-burden, spoil area and tonnage of coal required per shift.

The capacity of the overburden stripping equipment will influence the capacity of the other equipment such as dump trucks and the choice will be influenced by the number of cubic yards of overburden that must be removed to recover one ton of coal, the stripping ratio. Once this economic limit has been established, the stripping unit or units can be selected.

Cross-sections of the proposed stripping area together with major equipment dimensions and ranges are useful in selecting the machine or combination of machines best suited to the conditions. The needs of the specific property in relation to overall efficiency and cost should be the deciding factors rather than in terms of earth moving alone in selecting a machine. First cost is not the only expense involved with an excavator, it must be erected and may require to be taken down and moved to a fresh site later. It is not always possible to amortize the cost of equipment during the life of a site because reserves are limited. A smaller or medium-size machine would be less costly to move to another site.

When the major stripping machine has been chosen, other units, such as drills, face-shovels and dump trucks, should be selected to build up a balanced mechanized production team of machines and any change will generally upset the balance of the cycle of operations unless other changes are also made.

Where items of equipment are working far apart or in different pits in the neighbourhood, radio communication is valuable in improving super-vision and reducing delay from machine breakdown as breakdown crews can be called to a machine immediately trouble is experienced and spare parts from a central stores can be rushed to the job.

In opencast coal mining in the USA rotary dry-type drilling units con-tinue to be most popular for vertical drilling though improved vertical augers are being used. Most heavy-duty drills are crawler mounted for drilling from $5\frac{1}{2}$ in up to 15 in diameter holes, but smaller units are carried on truck frames. Larger holes can be drilled faster than smaller ones and in field studies in identical conditions of overburden a $7\frac{3}{8}$ in diameter hole was drilled at a penetration rate of 70 ft per hour, a $10\frac{5}{8}$ in hole at 90 ft, a $12\frac{1}{4}$ in diameter at 100 ft and a 15 in hole at 110 ft per hour. The life of the bits increased from 6000 ft for the $7\frac{3}{8}$ in holes to 10,000 ft for the 15 in holes.

Large diameter holes permit more effective deck loading, concentrating explosive in the harder layers resulting in better fragmentation and more overburden is broken per hole. Typical drilling results include: in Illinois two-man crews drill 580 ft of $12\frac{1}{4}$ in hole per shift in 75 ft of overburden with 6 ft limestone, 45 to 50 ft of sandstones and shales and 20 to 22 ft of dirt. A 13 in auger drills the upper soft portion; a unit drills 120 ft per hour in overburden containing soft and hard limestone 45 to 55 ft, shale, sandy shales and slag 25 ft to the surface, bit life is 9000 ft; in Illinois increasing the diameter of the holes to $12\frac{1}{4}$ in increased bit life to 10,000 ft; in Ohio changing from 12 in to $13\frac{3}{4}$ in improved the rate of penetration by 20% to 3 ft per minute and bit life was also increased by 20%.

Considerable research has taken place into the use of slurry-type blasting agents which are water resistant and have a high density of about 1·5 with a detonation rate of 17,000 ft per second. Charges of up to 306 lb should be primed with 5 lb of 40, 60 or 75% gelignite.

With references to AN–FO, operators are becoming not only more interested in proper mixing but also in quality, grain size and moisture content and most companies are making their own AN–FO blasting agents.

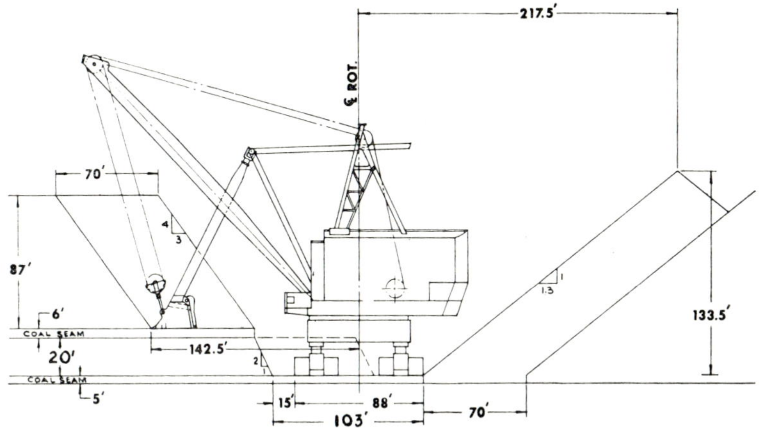

Fig. 62.   Long reach and high discharge are important requirements in two-seam recovery of the Marion type 6360 shovel.

Good records should be kept of all blasts: including depth, spacing, burden and quantity and distribution of explosive in each hole. Type and thickness of overburden, feet of drilling and cubic yards of material broken each day should also be noted. One Alabama site is using explosive to cast 40% of the overburden on to the spoil bank. Three rows of 9 in diameter holes are drilled to provide a 65 ft wide pit. In the first row are 600 lb and in the second and third rows 500 lb of AN–FO primed with 3 lb of gelignite.

In moving the maximum volume of overburden the excavator must not only operate continuously but make each motion perform useful work.

The Marion 6360 shovel (Figs. 1 and 2) at the Captain Mine of the South Western Illinois Coal Corporation with a 180 yd^3 dipper removes overburden from both the Illinois No 6 and No 5 seams (Fig. 62) as it takes a cut along the two mile length of the open pit. As shown the shovel is based on the No 5 seam, and the sequence of operations includes taking a 60 ft wide cut out of the overburden above the No 6 seam, carrying a total width of 90 ft, then removing a 60 ft wide cut of the strata between the seams and carrying a total width of 105 ft, with the monthly duty being the removal of $4\frac{1}{2}$ million yd^3 of overburden, the coal loading rate being 20,000 tons per day or $\frac{1}{2}$ million tons of coal per month.

The method of coal transport is a 240-ton diesel-electric Caterpillar machine consisting of a bottom-dump hopper mounted between two rubber-tyred tractors each powered by a 1000 hp diesel engine which drives a dc generator supplying a pair of traction motors on each of the outer-most axles, thus propelling the truck. The total weight of the empty truck is 190,000 lb and the load is distributed over eight dual-tyred wheels with 36-ply rating nylon tubeless tyres by Firestone. The use of two tractors with dual cabs and controls allows shuttle operation in the narrow pits, and though the machine is 95 ft long it can turn in a circle of 85 ft diameter. Road irregularities are cushioned by eight nitrogen-over-oil suspension cylinders, two per axle, which provide a maximum axle oscillation of 43 in and bogie action between tandem wheels of 17 in. Bottom doors are hydraulically opened to a maximum width of 105 in for rapid discharge. Speed is up to 45 miles per hour on a good road.

Associated equipment includes a Bucyrus–Erie 61R rotary drill, drilling blastholes in the overburden and a 50R for drilling the strata between seams. The 61R drills in two steps, a $13\frac{1}{2}$ in auger to penetrate softer surface strata and a $12\frac{1}{2}$ in diameter rotary rock bit to complete the hole. The 50R drill bores $10\frac{1}{2}$ in holes. Hole spacing is on a 37 ft interval pattern.

Two Marion 181M shovels are used for coal loading, one with a 16 yd^3 dipper and a 45 ft boom, and the other with a 12 yd^3 dipper and a 60 ft boom which is intended to load coal from the No. 5 seam into trucks running on the upper seam.

The specification of the Marion Type 6360 Shovel is given in Table XI.

A Bucyrus–Erie 2550W dragline with a 75 yd^3 bucket (Fig. 63), operating at Ayshire Collieries Corp. in Indiana, has a 275 ft long triangular boom of which the three alloy-steel main-chord members are filled with air under pressure and by reading the pressure gauges the operator can check the safety of these members. Its walking motion combines stepping and sliding movements amounting to $8\frac{1}{2}$ ft. It can dig to a depth of 165 ft and dump material 174 ft from the centre of the machine.

The pit in which the machine is stripping the overburden extends for $\frac{1}{2}$ mile and maintains a pit width of 90 to 100 ft and handles a maximum of 110 ft of overburden in recovering the No. 6 (6 ft) coal. The machine has 74 electric motors. Blastholes are drilled 36 ft apart with a 32 ft burden; three rows are required to give a width of 75 ft for the machine. Six 80-ton Euclid drop-bottom trucks provide transport.

TABLE XI

SPECIFICATION OF TYPE 6360 MARION SHOVEL

Dipper capacity, yd^3	180
Boom length, centres of foot pin and point shaft, ft	215
Dipper-handle length, including dipper, ft	133
Crowd-handle length, effective, ft	102
Stiff-leg length, ft	104
Boom-point sheaves, pitch diameter, in	144
Boom-padlock sheave, pitch diameter, in	120
Hoist cables (four, double-hitch), diameter, in	3·5
Boom-support cables (eight, bridge strand), diameter, in	3·6
Working ranges:	
Boom angle, deg	45
Dumping height, maximum, ft	153
Dumping radius (maximum height), ft	211·5
Dumping height over spoil, ft	133·5
Dumping radius, effective, over spoil, ft	217·8
Dumping reach, effective, over spoil, ft	173·8
Dumping height at maximum radius, ft	113·7
Dumping radius, maximum, ft	219·8
Cutting height, maximum, ft	189·7
Cutting radius at maximum height, ft	218·5
Cutting height at maximum radius, ft	122
Cutting radius, maximum, ft	236·8
Cutting radius at 39-ft elevation, ft	208·5
Radius of clean-up at grade level, ft	156·8
Radius of clean-up at 31-ft elevation, ft	142·5
Length of lower frame, centre to centre of jacks, ft	58
Width of lower frame, centre to centre of jacks, ft	56
Depth of lower frame girders, ft	14
Length over propelling units, ft	103
Width over propelling units, ft	88
Length of each crawler, ft	45
Width of each propelling unit, ft	30
Width of crawler belts, ft	10
Crawler bearing area, total, ft^2	3,140
Crawler bearing pressure, lb per in^2	58·7
Bore of jack cylinders, in	66
Centre journal diameter, in	80
Main swing gear, pitch diameter, ft	55·5
Width of upper frame, ft	70
Depth of upper-frame girders, ft	10
Centre of rotation to boom foot, ft	32
Elevation, ground to boom foot (jack pistons out 45 in), ft	50
Rear-end radius:	
Maximum (61 ft from ground), ft	74
At lower edge of upper frame (39 ft from ground), ft	66·7
Clearance under upper frame (jack pistons out 45in), ft	39
Roller-circle diameter (mean), ft	64·3
Swing rollers (85), mean diameter, in	24

SPECIFICATION—*continued*

Hoist drums, pitch diameter, in 125
Bail pull, maximum, lb 1,700,000
Electrical equipment:

    (Rating, 75°C rise continuous)
Hoist, motors (eight) with blowers, total hp . . . . 8,000 at 230 V
16,000 at 460 V

Swing motors (eight) with blowers, total hp . . . . 5,000 at 230 V
10,000 at 460 V

Crowd motors (four) with blowers, total hp . . . . 2,000 at 230 V
4,000 at 460 V

Propelling motors, total hp 3,200
Steering, total hp 200
AC motor load, total hp 20,000
Total horsepower 33,000

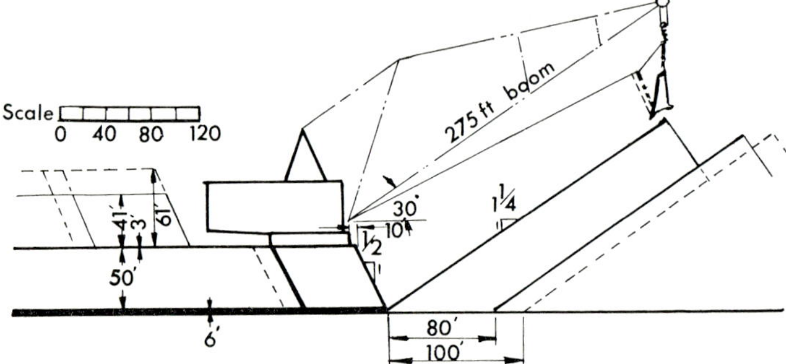

Fig. 63.   A 75-yd Bucyrus–Erie dragline recovers two seams while working to 100–120 ft banks.

Pit width	Depth	Cut	Cut + 25%	Fill
80′	100′	7,280 Ft²	9,100 Ft²	11,200 Ft²
80′	120′	8,880 Ft²	11,100 Ft²	11,200 Ft²
100′	100′	9,100 Ft²	11,375 Ft²	13,500 Ft²
100′	120′	11,100 Ft²	13,875 Ft²	13,500 Ft²

Because of higher speed and greater power, bulldozers are being used to an increasing extent for stripping overburden. The development of the hydraulically operated ripper mounted at the rear of a large bulldozer has increased the range of the scraper for overburden removal. Shales and soft rocks, which in the past could not be loaded by scrapers or rippers, can now be loosened successfully. In one pass an hydraulically operated ripper with three arms, spaced at 53 in can loosen a total width of 53 ft to a depth of 12 in.

The high-speed rubber-tyred tractor-scraper is also finding increasing employment as an auxiliary machine to remove the softer top portion of the overburden. It can also be used to make an opening cut and working bench for a dragline. An advantage of the scraper method of handling spoil is that little extra work is necessary where backfilling and levelling are required. A push-loaded scraper loads faster and carries more because the load is heaped better and packed tighter. It also loads better travelling downhill and the dumping area should be kept in good condition so that loads can be released when travelling at speed.

Box-cutting to open an open pit involves digging down to the coal and then working straight ahead to the boundary of the site. In some instances, where a dragline or a shovel is used to open up a new area along an existing boundary, the spoil from the first cut is placed on the surface next to the boundary line. Overburden removed then advances into the new area and the coal along the boundary line and under the first cut spoil is left *in situ*. Some managements do not consider it economic to rehandle material from the opening box-cut and the material over the coal so that the strip of coal along the boundary is never recovered. However, scrapers or bull-dozers may be used to move the spoil from the opening box-cut.

For handling thick overburden the big dragline is the most popular excavator in the USA but shovels with 75, 115 and 180 yd^3 dippers are also very popular. If only one machine is to be used, the dragline is often chosen, but tandem operation of a dragline and a shovel is often used, the dragline taking the upper section of the overburden and the shovel the lower.

Where multiple-seam mining is required and two or more seams are to be worked, close together, where one alone would not be viable, the choice of mining method and equipment will depend on the dip, the thickness of the coals and the type of rock above and between the seam and the topography.

The wheel excavator has been developed to cut the cost of moving over-burden in 50 to 85 ft highwalls. Three advantages are claimed for the use of the machine: first to handle overburden up to 85 ft thick and place it far enough away to avoid slides; second, to cut the cost per cubic yard below that which can be achieved by conventional machines such as shovels and draglines and third, to leave any overburden not moved by the machine so low in height that the capacity of the shovel or dragline in tandem with it is increased.

A wheel excavator working in 75 ft of overburden in Illinois removes 550,000 yd^3 per month and a 35 yd^3 shovel removes 650,000 yd^3 in making a cut 45 ft wide. The total pit width in the combined operation is 90 ft, the wheel working ahead of the shovel digging 22 ft of surface strata and discharging it 70 ft beyond the shovel spoil.

Another wheel excavator and a 70 yd^3 shovel removes 2·6 million yd^3 of overburden per month in a tandem operation in Illinois. The wheel handles 900,000 yd^3 per month when cutting down 20 to 50 ft of soft material. The tandem team requires a pit 115 ft wide so that the wheel and the shovel can pass each other when the wheel reaches the end of the $1\frac{1}{8}$ mile pit. Normally the wheel leads the shovel by a sufficient distance to allow the drilling and blasting crew to work efficiently. The two machines should advance at the same speed as far as is possible. The wheel maintains a 65 ft bench varying the depth of cut to keep pace with the shovel. The speed of the wheel depends on the depth of cut, and the relative position of the machines can be controlled by varying the depth of cut to be taken by the wheel. The tandem team removes a total of 80 to 110 ft cover and has successfully recovered coal under 130 ft of cover. The average overburden to coal ratio is 16 to 1.

In order to keep the main stripping machine working at full capacity as much as possible the clearing up of the pit should be performed by auxiliary plant such as crawler- or wheel-mounted bulldozers, as well as the motor grader.

The operation of the excavators is the key to their efficient performance. With shovels a full dipper should be the target in each cycle and loading, slewing, dumping and returning for a fresh load must be kept to a minimum.

With draglines the bucket should be loaded as quickly as possible and the bucket teeth must be kept sharp to ensure good digging. Spare sets of teeth should be kept in the stores for frequent changing so that worn teeth can be built up again with hard-surfacing material.

About 13 million tons of coal per annum are obtained by highwall augering and such coal has the advantage of being dry, clean and with a high proportion of large coal but this decreases as the depth of augering increases.

Augers range in diameter from 16 in to 84 in and are capable of a production of $25\frac{1}{2}$ tons per minute and conveyors can be arranged to deliver coal on either side of the auger. A crew of three or four men is usual and is supplemented by trucks and their drivers. Dual and triple-headed augers have enabled thin seams to be won economically since they require no extra men for their operation and recovery is better since the area between the cutter heads is won and the holes are straight.

If augering is to follow and be a part of the stripping operation care should be taken in blasting so that the highwall may be left in the best possible condition since a highwall slide can endanger both the auger team team and their machine. A clean well-drained pit is suitable for augering which should be proceeded with as soon as possible while the highwall is in good condition. The full height of the seam should be taken, the depth of augering depending upon coal thickness and any variation in the dip and strength of the seam. If augering follows stripping a bulldozer and trucks will be the only additional plant required.

In Ohio a compact auger equipped with a 42 in cutting head produces 500 tons per shift while drilling to a depth of 250 ft.

In West Virginia a dual-headed 21 in diameter auger in adverse conditions through subsidence from mining a lower seam produces 150 tons per shift.

In Kentucky a 30 in auger produces up to 460 tons per shift at a two-seam opencast with a 50 ft highwall, penetration averaging 150 ft.

Tractor-mounted rippers for breaking 3 ft and 6 ft layers of shale and then the coal beneath reduce costs and also production of fines in West Kentucky. Scraper-rippers have also been used in similar conditions and two 44 yd units in New Mexico are producing, breaking, loading and hauling 1500 to 2000 tons per day.

In laying out the power supply system for an opencast three factors are paramount. First, the system must be able to supply the equipment without objectionable voltage regulation and not be too large for the load required. Secondly, the supply system must provide adequate protection for personnel and equipment and thirdly, the units in the system must be adaptable to relocation to keep up with changes in the load requirements.

Opencast mining companies can generally get more favourable contract terms if they receive power at higher voltages but this entails the purchase of a transformer to step down the voltage to that required on portable equipment. With single-step transformation primary opencast distribution is usually at 2300 or 4160 volts but with the larger machines now being installed also 6600 and 7200 volts. In two-step transformation the 'super-primary' voltage is 13,000 and there is a trend towards 3-phase rather than single-phase current.

Distribution is often by a 'pole-mounted' high line which is largely standardized with a main line one mile in advance of the opencast and parallel to it and from this pole-line laterals of 1300 to 1500 ft are run to the opencast.

To bring the power factor up to that specified in the power contract, synchronous motors with 0·8 power factor leading are installed on the motor generator sets on large excavators and with this proper correction an overall factor of 0·9 to 0·95 may be achieved which is above the penalty area.

Water should be kept out of the opencast and off the transport roads to keep drainage costs to a minimum. For this purpose streams may be diverted, ditching above the highwall to divert surface run-off from the pit and building flumes to span the pit. Air-cooled diesel engines are more efficient than petrol engines and pumps, diesel driven when electricity is not available, are coming into increasing use.

The opencast coal mining industry suffers from a poor public image and as a result of recent national emphasis on beautifying America, strip mines have received more than their share of criticism. As a result of recent legislation some of the major problems facing the industry today are higher labour costs, supply and power costs, coupled with the market requiring a higher quality product and with general public pressure for stricter legislation requiring more reclamation of disturbed land and closer controls of air and water pollution.

Reclamation in surface mining generally is dealt with in Chapter 14.

OPENCAST MINING OF GERMAN BROWN COAL

In spite of its relatively low calorific value of 1600 to 2900 Kcal per kg, brown coal is the cheapest form of primary energy in the Federal Republic of Germany because of its shallow surface cover enabling it to be worked predominantly by large-scale opencast mining using high-capacity plant to give optimum rate of production. To satisfy the high consumption of coal at the brown coal power plants, output has reached more than 100 million tons per annum and this output requires the removal of more than 267 million yd^3 of overburden which can only be achieved by complete mechanization of both the winning and conveying operations.

In addition to the carboniferous coal seams of the Ruhr deep mining coalfield, Germany has vast Tertiary brown coal deposits, most of them near the surface. The total reserves of German brown coal average some 10,000 million metric tons and of this some 60,000 million tons are within the Federal Republic. About 9000 million tons of this can be won by present opencast mining techniques, 85% of it from the Lower Rhine region in which lie the main deposits, others being near Helmstedt in Lower Saxony, at Regensburg in Bavaria, near the zone boundaries and in Hesse near Frankfurt.

In contrast to deep-mined 'black coal' German brown coal is mainly overlain by loose material such as sand, gravel, clay or loam. Miocene brown coal from the Lower Rhine has a calorific value of 7000 Btu per kg (3182 Btu per lb) with a moisture content of 60 to 62% and ash from 2 to 8%.

Bulges in the underlying rock have caused faults running south-east to north-west and have split seams originally horizontal and from 32·5 ft to 195 ft thick, averaging 130 ft.

The opencast workings are, however, getting deeper and will reach a depth of over 800 ft which, to enable work to proceed, will entail the complete de-watering of the area by deep filter wells 6·5 ft diameter with submersible pumps of 1300 hp with eleven stages raising 3300 gallons per minute against a head of 1000 ft.

Some 550 wells are in operation in the Lower Rhine district to keep the water-table 500 ft below the deepest point of the opencast, the total annual pumping load in the Rhine district being 400,000 million gallons, the water to coal ratio being 15 to 1.

In the districts where opencast mining commenced the overburden to coal ratio is 0·35 to 1 but in new fields it will be $3\frac{1}{2}$ to 1 rising to 6 to 1.

The type of overburden and the relative softness of the coal has enabled continuously operating excavators to be used and these were at first bucket chain dredges mounted on rails or caterpillars with a cutting depth of 130 ft and still in use. They weigh up to 1400 tons with bucket capacities of 3 yd^3 and outputs of up to 70,000 yd^3 per day and consist of a bucket ladder carrying a bucket chain with a bucket every fourth link, which is suspended from a boom by wire ropes controlled by winches.

In the newly opened deposits in less favourable conditions the bucket-wheel excavator without travelling boom has become the dominant type. Transport to the surface is by belt conveyors the buckets doing the digging only. Bucket-wheel excavators are used to win the coal as well as for removing the overburden (Fig. 64a, b), with a maximum cutting height of 163 ft and down to a depth of 82 ft. Conveyor belts up to 86 in in width are used for transport at 780 ft per minute and for transport to power stations rail haulage is used with large wagons, of 128 yd^3 capacity. The overburden consisting of sand, gravel and clay is dumped on outside dumps or close behind the working face or into worked-out pits, overburden spreaders being used with a discharging belt conveyor boom up to 325 ft in length. Transport to the dumping site is by conveyor or rail.

There are three main types of bucket-wheel excavator:

(1) Dredges for high cuts incorporating a crowd action by means of a sliding boom travelling forward and backward.

(2) Dredges for high cuts without crowd action and which have, therefore, to be pushed forward for taking cuts.

(3) Dredges for high and deep cuts without crowd action.

The two last types are more commonly adopted for large machines and consist of a digging unit, an independently movable discharge unit and a telescopic intermediate bridge conveyor connecting the digging and discharging units. The advantages of these machines include reduced weight, simple operation and suitability for automatic control. The bucket wheel carries from six to twelve buckets which empty at the top position and deliver to a separate conveyor, the most modern type is followed by a rotating disc, a set of big rollers or a short belt, all of which throw out the material into a main boom belt conveyor which in turn delivers it to the next flight of conveyor, the tail pulley of which is sited at the slewing centre-line of the upper carriage.

The method of working the opencast depends on whether the different strata comprising the overburden need to be excavated separately. If not, the 'block' method of working (Fig. 64*a*), is often adopted in which the bucket wheel travels ahead and excavates a small face at right angles to the length of the block, the width of the face depending on the length of the bucket-wheel boom which may be 325 ft. For stability the faces ahead of and abreast of the wheel must have a limiting slope and this is attained by travelling into different positions and altering the slewing angle when the wheel changes from one cutting level to another and at all cutting levels the slices taken are deepest directly ahead of the machine and thinnest at the ends of the slices. The angular velocity of the wheel is regulated automatically to the slewing position to ensure constant output. Slewing, direction, reverse and crowd are also automatic where there is no crowd action, movement of the bucket-wheel boom is preceded by movements of the caterpillars on which the machine is mounted. The common working height is 165 ft and the cutting depth 80 ft. The cell-less wheel diameter is 57 ft with ten buckets, each with a capacity of 5 yd^3, and a discharge rate of 38 per minute. The output is 11,600 yd^3 per hour and the overall weight is 7400 tons distributed over three crawler groups each consisting of two caterpillars creating a ground pressure of 20 lb/in^2. The total electrical load is 21,000 hp of which 2350 hp represents the bucket-wheel load.

A tripper unit is fitted to the last movable conveyor carrying the overburden and delivered to a dirt spreader comprising an undercarriage, the slewable middle section carrying the first conveyor flight connected to the tripper crane belt, and finally the superstucture containing the long slewable discharging boom and the counterweight, the whole weighing 2400 tons and mounted on three twin crawlers. The spreader can deal with 11,300 yd^3 per hour.

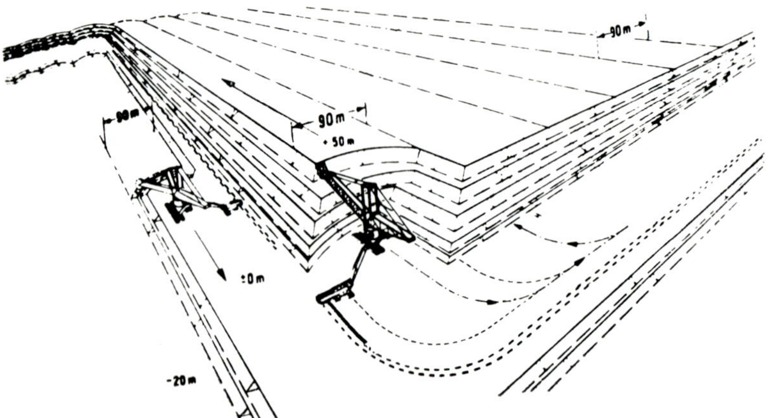

Fig. 64(a). Diagram illustrating the working sections for the block method using a wheel excavator.

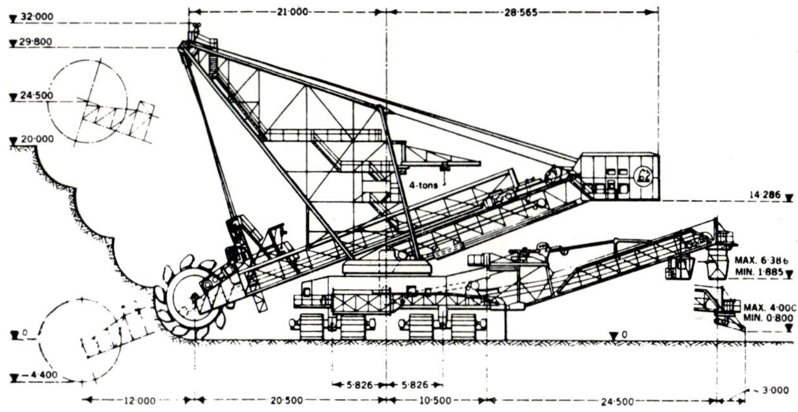

Fig. 64(b).   *Type K800 bucket-wheel excavator manufactured in Czechoslovakia.*

K800 BUCKET-WHEEL EXCAVATOR

Diameter of bucket wheel with buckets	8·5m
Number of buckets	10
Capacity of buckets	800 litres
Theoretical output of excavator (in loosened material)	1440–1920m³/h
Maximum digging force at the bucket tooth	17·5 ton
Maximum reach of bucket wheel centre from slewing superstructure	34·1m
Maximum reach of bucket wheel from slewing superstructure axis ..	20·5m
Maximum horizontal crowd of bucket wheel	12m
Maximum height of bucket wheel axis above ground level at the machine	24·5m
Maximum depth of bucket wheel below ground level at the excavator	4·4m
Slewing of superstructure	± 110°
Slewing of superstructure beyond the limit switch end position ..	± 360°
Loading boom slewing	± 75°
Maximum lift of loading boom ..	6·4m
Minimum curve radius of the excavator when travelling ..	39·7m
Travelling speed of excavator, approx.	6m/min
Maximum operating gradient ..	3%
Width of tracks	2500mm
Weight of excavator	1320 tons
Maximum height of excavator ..	31m
Maximum length of excavator ..	75m
Number of operators	6
Installed output of electric motors ..	1275kW

In the brown coal area of the Lower Rhine district some 130 km² have been taken over for opencasting and 60 km² have been restored and of this area 34 km² have been forested, 16 km² returned to agriculture and 10 km² have found other uses. The landscape undergoes great alteration during opencasting on the 70 km² area in use for mining, involving the removal of agricultural and forestry areas, villages, railways and roads but regulations require the mining companies to restore the landscape as far as possible to its former state.

REFERENCES

'Opencast Coal Mining in Great Britain', W. G. Martin, *Opencast Mining, Quarrying and Alluvial Mining*, Institution of Mining and Metallurgy, 1965, p. 617.

'Planning Large Opencast and Underground Coal Mines for High Productivity', K. K. Kuznetsov, *Mining Magazine*, December 1966, Vol. 115, No. 6, p. 452.

'Lighting of Opencast Coal Mines', W. B. Bell and P. E. Sullivan, *Opencast Mining, Quarrying and Alluvial Mining*, Institution of Mining and Metallurgy, 1965, p. 658.

'Opencast Technique in German Brown Coal Mining', W. Tilmann, *Opencast Mining, Quarrying and Alluvial Mining*, Institution of Mining and Metallurgy, 1965.

'Guides to Efficient Surface Mining', *Coal Age,* July 1967, pp. 234–253.

'Volume Determination in Open-cut Mining', B. Milasovszky, *Mining Magazine,* July 1967, p. 10.

CHAPTER 9

SURFACE MINING OF BAUXITE, CLAYS, CHALK AND PHOSPHATES

BAUXITE

The hydrated oxide of aluminium $Al_2O_22H_2O$, bauxite, supplies almost all the aluminium of commerce. It is like clay in appearance, and varies in colour from dirty white to greyish, and when iron colours it, is yellow, brown and reddish brown and even mottled. Bauxite deposits mostly form on or near the surface of the earth, by decomposition under moist tropical or subtropical conditions, of clays and clayey limestones and igneous rocks with high aluminium silicates content. The aluminium silicates in these rocks decomposed under favourable atmospheric conditions and bacteria played an important part in the concentration of the alumina. The fact that large deposits of bauxite occur in temperate zones in the USA, Provence in France, Hungary and Northern Russia, is, together with other reasons, considered by some geologists as evidence of tropical or sub-tropical climates in earlier geological eras. Horizontal or gently tilting sheets or blankets of bauxite are sometimes found as cappings of plateaux or flat-topped hills which have been levelled by erosion. Most bauxite deposits occur in close association with old land surfaces and the bauxite occurs along unconformities representing long periods during which lateritization took place. Bauxite deposits of comparatively recent age in Guyana, Surinam, Ghana and India, lie on almost level strata which have remained undisturbed for long periods under tropical conditions. Canada, with the Kitimat project in British Columbia, is a very large aluminium-producing country using bauxite mainly from Guyana.

The Demerara Bauxite Co. Ltd, Demba, is a subsidiary of the Aluminium Co. Ltd, of Canada, the principal operating subsidiary of Aluminium Ltd.

Demba mines and processes bauxite and produces alumina in Guyana and exports of metallurgical, abrasive, refractory and chemical grade bauxite and alumina are made to 30 countries.

The opencast mining is centred at Mackenzie on the Demerara River, 65 miles up river from the capital and chief port, Georgetown. The investment of Demba in plant and facilities at Mackenzie totals £25 million, which supports a payroll of 3500 people.

The overburden consists of sediments of a continental deltaic deposit laid down in shallow water, the Berbice formation, consisting of white unconsolidated quartz sands with indurated sand patches, sandy clays, and clays with bands of sandy clays. The formation forms a plateau with an immature drainage system, and a slight run-off since the rainfall percolates through the sand to the underlying clay layers.

Before mining the bauxite, up to 200 ft of overburden must be removed. Trouble arises from the very heavy tropical rainfall of 100 inches per annum, mostly in one short season, which may cause spoil banks to slump and bury drainage ditches and railway tracks when the drainage system is already overtaxed. But the area is practically undeveloped and uninhabited and there are no boundaries limited by agricultural or industrial activities, roads or railways.

At Mackenzie, stripping methods have progressed from hand loading of mule carts in 1917 to monitors, dredge pumps, walking draglines and bucket-wheel excavators. These have already been referred to in Chapter I, and it is intended here to deal mainly with the two last. A combination of hydraulicking by monitors and sand pumping is an attractive method of stripping overburden, where the thickness exceeds that which the draglines can sidecast provided the sand to be hydraulicked is unconsolidated and does not contain clay bands and there is enough area for impounding and settling the sand/water slurry. Under ideal conditions when discharging through a level pipeline not more than 4000 ft long or its equivalent, the combination of hydraulicking and sand pumping can dispose of sand overburden, at a cost, including depreciation, of 10d. to 1s. per bank yd^3. The capital cost is also attractive. To install a 2 million yd^3 capacity overburden disposal system of this type, would cost £300,000 including piping, whereas a bucket-wheel excavator including 6000 ft of conveyor costs £600,000. This plant was chosen however because the overburden was not suitable for hydraulicking, suitable areas were not available and the bauxite to be stripped was 50 ft below drainage level and will, even with the dug method chosen, present a difficult pumping problem.

Railroad mining, and tropical rainfall impose conditions on dragline sidecasting not usually encountered. The dragline bucket must swing over the railway track and not block it with spillage because the shovels load directly from the bauxite face into railway cars close to its base and in addition drainage trenches must be maintained between the toe of the spoil pile and the bottom of the bauxite face. A ledge must also be left between the stripping face and the bauxite face to provide a barrier against any overburden breaking away from the stripping face and falling into the bauxite loading area. Having, in this manner to provide for the ledge, the railroad track and drainage the effective reach of the dragline is seriously reduced. Crushed bauxite becomes stickier each time it is handled, or the solution would probably be dump trucks hauling to a central rail waiting yard which would improve the utilization of the dragline reach. The overburden removed by the dragline is usually clay with an angle of repose of 30° to 35° and the base is saturated and has little resistance to shear or plastic flow and tropical torrents erode the new spoil piles rapidly. Underlying them is a soft kaolinite which also shears readily and is not competent to carry the load of the spoil banks and when the kaolinite clay fails under load, the spoil banks slump. To reduce this trend, a sand base is laid down by the dragline before clay stripping begins.

The draglines operate best from a sand base and where the wheel excavator has cut into the clay horizon the area is backfilled with a foot of sand since operating a dragline on a slippery base is dangerous since the tub tends to spin out of control at the end of a swing, and the machine itself may slide with disastrous results. A disadvantage of dragline sidecasting is that it limits the possible stripped ore reserves, required to provide for a dragline breakdown, to less than the length of the mining face, depending on the dragline's position. The bucket of a dragline is also a rough tool with which to uncover the flat-lying bauxite deposits, so that a bulldozer carries out the stripping clean-up of the top of the bauxite. Four draglines are used, the first, diesel-engine powered with a 9 yd^3 bucket has a 200 ft long boom, the second, also with a 9 yd^3 bucket, has a 220 ft aluminium boom which has given good service though a steel boom has failed. The third dragline has a 235 ft steel boom and a 10 yd^3 bucket, and the fourth is electrically driven, and has a 195 ft steel boom, and a 15 yd^3 bucket. The cost of sidecasting clay by dragline into the mined-out area, has been 9d per bank yd^3 or, including depreciation on a 10-year-life basis, 1s. per bank yd^3.

It was decided that a bucket-wheel excavator was the best machine for stripping, taking into account all the factors concerned, and the choice between the stacker type and the more normal type used in opencast mining in conjunction with a conveyor system, was decided in favour of the latter, since the lowest operating level would be 60 ft above the top of the bauxite, and stacking the material on top of the dragline spoil heaps would have required an extremely long and expensive stacking boom. In addition, the clay under the dragline spoil banks was unstable and liable to slump. The use of the bucket-wheel excavator and conveyor system involves the use of a dragline on the 60 ft, mostly clay, immediately above the bauxite, and the wheel excavator on the overburden above which is mostly fine to coarse grained sands, the dump area for this being beyond the opencast limits. This involves four metre-wide belt conveyor units in series running at 680 ft per minute, the first, face, conveyor is 3000 ft long, the second and the third each 1500 ft long and the fourth, dump, conveyor 1500 ft long, which is swung pivoting about the tail pulley, so that the dump takes the form of a segment of a circle with a mobile double-swing slewing conveyor of 160 ft and a skid-mounted sand-thrower to increase the disposal range of the dump.

The bucket-wheel excavator with seven buckets is capable of digging over a horizontal range of 100 ft to a height of 67 ft. Figure 65 shows the three bench stripping cycle adopted. During the first pass, the wheel excavator cuts an upper bench, 60 to 70 ft in height, in the overburden. On the next cut, a second bench is cut 26 to 36 ft in height below the first bench, the vertical heights depending on the surface contours of the overburden, and so that the floor of the second bench is at the level of the face conveyor. The final 60 ft high third bench is sidecast by the walking dragline.

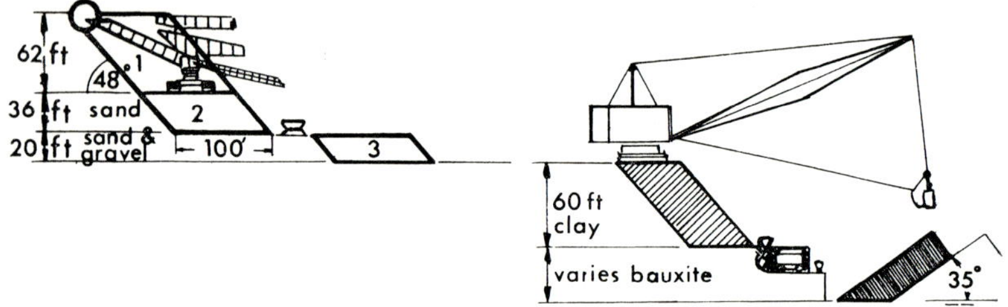

Fig. 65.   Typical stripping section wheel and dragline.

The cost of bucket-wheel stripping and conveyor costs over a two-year period were 9d. per bank yd^3 or 1s. 3d. with depreciation calculated on a 10-year-life basis.

Future stripping plans indicate a trend to increase the number of bucket-wheel excavators in use, and an intention to reduce the thickness of overburden to be sidecast by the draglines to improve opencasting conditions, or to substitute dump trucks for railroad transport from the face, so that the draglines can handle thicker overburden.

CLAYS

A clay is finely divided material with sufficient colloidal matter and clay minerals, with their characteristic affinity for water, to form a plastic mass when they become wet.

Most clays contain a large proportion of hydrated aluminium silicate ($Al_2O_3XSiO_2YH_2O$, where X and Y are both variable), although other clay minerals have a more complex constitution. Clays and clay minerals are the breakdown products of altered igneous minerals, mainly the felspars and ferromagnesian silicates from which they have been derived by chemical weathering. Clay beds therefore occur as water-laid sedimentary deposits.

The amount a clay shrinks when it is dried, depends on the amount of water between the mineral particles or within the molecular structure of the clay minerals, but the amount of contraction is very variable in different clays and more than $\frac{1}{3}$ shrinkage in volume renders a clay unsuitable for brick making or ceramic purposes and preliminary drying and firing tests are necessary to ensure suitability of the clay. To reduce shrinkage pre-burnt material, 'grog', may be mixed with the clay. Ground shales also make very satisfactory bricks if sufficiently weathered to improve plasticity with water.

Bricks are of two types, moulded and wirecut, and there is often a local preference of architects and builders for one or the other, wirecut needing less mortar for bonding. Bricks are dried to drive off excess moisture before firing in a chambered or a tunnel kiln, which may be gas fired. What water is not driven off enters into more permanent chemical combination during firing with the alumina, silica and alkalis to form a silicate glass which binds the unfluxed particles together. Some minor constituents such as soda, lime and even magnesia, are essential for cold-temperature fluxing, and some iron is required to impart the warm red or brown colour.

A carbon-rich clay, 'Knotts', is used by Fletton brickmakers.

Drain pipes are also made of common clays with similar properties to those of a brick-earth.

Refractory clays are also found and are used for furnace and fire-bricks, some being made from bauxite.

China clay

The most important clay, however, with a very important export market is the so-called 'china clay', or kaolin.

In the British Isles, most of this clay occurs in the neighbourhood of St Austell in Cornwall, and some from Dartmoor and Bodmin Moor, but nowhere else outside Cornwall and Devon have viable deposits of china clay been found.

China clay is used in making newsprint and the daily paper contains about one-tenth of its weight of china clay and in periodicals, books, writing paper and wrapping paper it is the chief filler in the wood or other pulp to produce the body of the paper and its smooth surface. The clay is an essential constituent in bodies and glazes of domestic porcelain and other kinds of ceramic ware including porcelain electrical insulators, and it is used in the rubber industry also as a filler.

The paint industry and the cotton and other textile fabric industries, in which it is used as a stiffener, are also important consumers. It is also used as a filler and surface coating agent in linoleum and oilcloth, as a mild abrasive in polishes, cleansing tooth-powders and soaps, as a filler in some wall-plasters and in white Portland and other cements. Specially prepared and of very fine grain-size clay is used for medicinal purposes partly as an absorbent of toxins in the alimentary canal and in disinfectant powders, for cosmetics and in some plastics.

In the 18th century supplies came from China, hence the name, and all world china clay deposits are the product of the decomposition of felspars in granite and this kaolinization is usually effected by downward percolating surface water containing carbon dioxide, organic acids and other substances which leached out the potash in the felspar and left a residue of a hydrated aluminium silicate, $Al_2O_3SiO_22H_2O$, mixed with quartz, mica and other minerals from the granite or by ascending gases and vapours, chiefly carbon dioxide and superheated steam, which emanated from acid igneous magma producing the same effects *in situ* as the surface water and acids by leaching out the potash in the felspar. It is fortunate that the china clay deposits of Cornwall and Devon had this latter origin, since they persist to a depth of over 400 ft below the ground surface, in fact to a depth that has not been bottomed and, at depth, are of the highest quality, whereas those deposits formed by descending waters are in general shallow in depth. The important Czechoslovakian china clays and those of France and China all occur in well-kaolinized granite.

However, all china clay deposits are not *in situ*, some have been transported by water from the kaolinized granite and redeposited some distance away, as those in Georgia and North Carolina in the USA, which are very suitable for paper making but not for high-grade ceramics. Such deposits also occur in Malaya and Thailand. The care taken in preparing British china clay has established its high reputation all over the world.

The most important firm in the china clay industry is English China Clays Ltd, of St Austell, Cornwall, and the workings there extend over 30 to 40 square miles. There are several different operations proceeding simultaneously, the fundamental one being the hydraulicking of china clay from the working face by monitors. The clay then runs in a stream to the bottom of the open pit where the china clay is separated from the sand which is then carried by conveyors to the top of the light grey granite sand pyramid-shaped bank or 'burrow'.

Bucket-wheel excavators are also used.

After passing through various stages of refinement, by pumping the clay arrives at the 'dries' where surplus water is removed by a filter press and the resulting clay is dried in mechanical dryers. The end-product is a very pure china clay for use in the industries already enumerated, some 72% of the total output being exported.

The disposal of waste rock from the open pit bottom and the stripping of overburden from the edge of the pits is carried out by a subsidiary company, the Western Excavating Co. Ltd, which is responsible for all mobile plant in the china clay and building divisions of the group.

A Caterpillar 966A wheeled Traxcavator works in conjunction with a monitor at the bottom of the clay pit; the monitor moves across the clay face and the wheel loader, equipped with a skeleton rock bucket, loads the granite already blasted into dump trucks which haul it out of the pit to a disposal area. Because of the continual flow of water from the monitor, floor conditions are very wet and soft, and it has been found that wheeled tractors stand up to these conditions better than those crawler-mounted because the abrasive action on the track-type loaders meant heavy undercarriage maintenance costs.

The extension and widening of the open pits is a scraping and dozing operation involving excavating the top overburden and moving it to areas where there is no clay underneath. No pit has yet reached the limit of its reserves in depth, even those which have been working for a century. A fleet of wheeled tractor-scrapers handle the bulk of the 4 million yd^3 of material removed annually. These machines include two Caterpillar 657 wheeled tractor-scrapers with a capacity of 32 yd^3 struck and 44 yd^3 heaped and three 631A wheeled tractors. For many years past large burrows of granite sand have been deposited where it was thought there was no clay underneath, but modern methods of prospecting have revealed that some of the burrows cover large deposits of clay and it has, therefore, become necessary to move them. They contain tens of millions of cubic yards of granite sand, and have in some cases to be moved $\frac{3}{4}$ of a mile. A large burrow which rose steeply to a height of 80 ft has been removed recently. The slopes were as steep as 35° in some cases and the material was moved from both the top and the base. Tandem-powered 657 tractor-scrapers loaded from the top with a 631A tractor-scraper and a 619C tractor-loader loading from the base. Because of the nature of the material the scrapers were push loaded by Caterpiller D9G tractors. The machines transported the material 1000 yd over a road with an adverse gradient of 15° and dumped it to form another burrow.

China clay is the best known and most important source of raw material for the ceramics industry. A second material, known as china stone or Cornish stone, has also been quarried and used in ceramics for the 200 years that Cornwall has been so well known to this industry.

China stone is a mixture of equal quantities of quartz and felspar, with up to 15% of other minerals, the felspar acting as a flux, and quartz is also commonly used in the ceramics industry. China stone has the advantages of uniformity and low iron content, and it is because of these that it can be used directly in ceramic bodies.

China stone is quarried from a particular area of St Austell granite and is very uniform with an absence of veins or intrusions of other minerals. The quarried granite is hand sorted into various qualities, depending on hardness, degree of kaolinization or the amount of fluorspar, which is a minor constituent present, and the most expensive grade of stone is called Hard Purple.

Until recently, the fluorspar was regarded as a beneficial minor constituent because of its strong fluxing properties. However, the Clean Air Act demands freedom from fluorine and it was becomingly increasingly desirable to produce a raw material free from this element and advances in mineral dressing technique rendered methods based on hand selection both unsatisfactory and expensive. It was, therefore, decided to install five years ago, a pilot plant to produce a china stone free from fluorine.

After laboratory tests it was decided to adopt a two-stage froth flotation process to remove both mica and the fluorine minerals. This enabled a type of china stone to be introduced into the feed which normally is rejected because it contains brown iron-bearing mica that renders it unsuitable for ceramic use. This type of stone known locally as 'shell stone' is the only material, other than china clay, removed from the quarries.

Since it amounts to about 40% of the granite present, and was previously considered waste material, it is obvious that with the two-stage flotation process in operation, elaborate and expensive selection of the material is eliminated and a pure product can be produced as cheaply as the old china stone. The material so produced has the further advantages of coming from known large reserves of definite composition and containing the same mixture of potash and soda felspars used successfully over a long period. The material has, of course, to be prepared by crushing and grinding for separation by froth flotation.

Following the successful laboratory tests confirmed by the satisfactory performance of the 10 cwt per hour pilot plant, a full-scale plant to treat 5 tons per hour, with provision for ultimately doubling this capacity, was erected, the flowsheet adopted being virtually that of the pilot plant.

A 109 electronic linear programmer and analogue simulator by Measurement Aids, costing some £3500, has been installed by the china clay section, English Clays, Lovering Pochin & Co., the world's largest china clay producer, and part of the English China Clays group, to control its clay blending operations. It will be used to compute optimum blends and to simulate different conditions likely to be encountered. Consideration must be paid to as many as eight types of clay with different chemical and physical properties, and these in the final mixture must be contained within six closely defined limits to ensure the quality of the end products.

Without the services of a computer this requires much time, skill and experience and to obtain a blend which also minimizes costs is well nigh impossible.

Ball clay

The name is derived from the lumps or balls of clay about 8 inches in diameter produced in the beginning of the Devon and Dorset clay mining industry but it now denotes the highly plastic, white burning secondary clays forming an essential constituent in the manufacture of pottery. Possibly the clays have been worked since Roman times at first for local industry, and now for the ceramic industries of the British Isles, Europe and elsewhere.

In 1688 the accession of William and Mary of the House of Orange brought Dutch potters and their technique to England, and by 1720 the skills had become naturalized and Astbury, Twyford and Wedgwood led the expansion of the pottery industry during the end of the 18th century. South Devon ball clay was only used locally until 1693 and even by 1740 only a few hundred tons annually went further afield, but by 1785 over 10,000 tons were sent to Liverpool, London, Bristol and Hull, and by 1900 to Europe and the USA as well.

South Devon produces some two-thirds of the national output of half a million tons of ball clay and about $\frac{1}{4}$ million tons is exported, the remainder is used in the ceramic industry in the Stoke-on-Trent ('Potteries') area, with a small proportion going to individual potters in the UK, and to the foundry, refractory and general filler markets.

The decomposition by metasomatism of the felspathic minerals in granite produces ball clay, the action of steam and carbon dioxide causes the breakdown of felspar into kaolinite, $Al_4Si_4O_{10}(OH_8)$, and this is basically china clay, which, remaining *in situ*, is worked by the hydraulic methods previously described, with ball clay as a secondary derivative of china clay, formed by erosion and subsequent transportation by water to form a sedimentary deposit remote from its original source laid down in the late Eocene and early Oligocene periods, the South Devon deposits occurring in a basin which was originally a valley eroded by fast-flowing streams from the north and north-west. Later it would appear the valley bottom sank below the level of the outlet and a large area of comparatively still water into which streams carrying decomposed felspathic material and small quantities of characteristic secondary minerals discharged, giving ball clay its particular characteristics.

In the centre of the area the lower clay seams reach a depth of over 700 ft and the beds are about 250 ft thick. It is expected that the clay in this area is similar to that at the outcrops and comprises a series of good clay seams with intermediate zones of lignites, sand clays and inferior quality clays (Fig. 66).

For its main use, pottery, the quality of the clay and its chemical and physical characteristics are very important and consistency is essential so that the problem is to produce from a heterogeneous deposit a range of homogeneous products for a large number of different uses. One firm, Watts, Blake & Bearne, control quality by the separation of the clays into four grades, each grade with similar dominant characteristics when used in ceramic compositions.

Ten standard clays and some twelve others, and four standard blends are supplied. To meet final product specifications, deposits must be continually sampled and production work deployed so that the natural clays may be blended to provide the ten standards.

Both underground and surface mining is used to work ball clay. The opencast workings are on the outcrops surrounding the Bovey Tracey

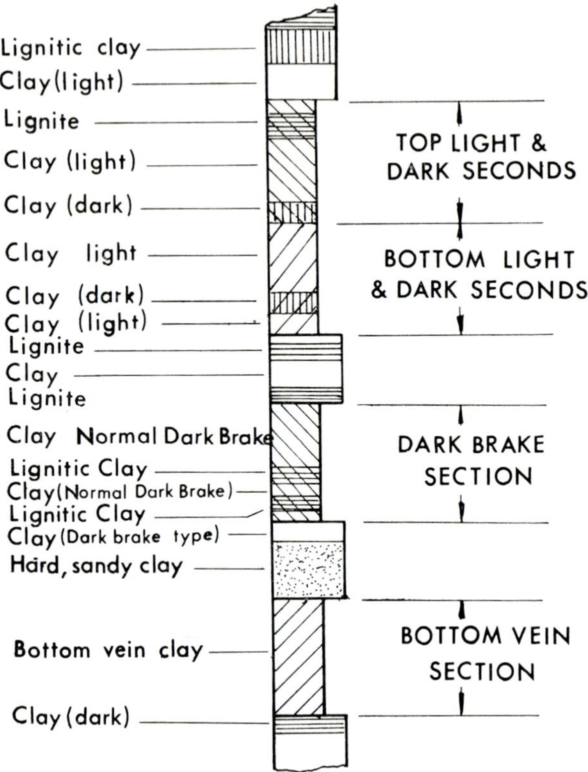

Fig. 66. *A typical section of the seams.*

Basin and both hand mining and mechanical mining are employed. Fully mechanized working gives higher output at higher efficiency, hand working has the advantage that waste material can easily be discarded by the selective extraction of clays, and is, therefore, practised in areas where high-quality clay is associated with a large proportion of lignite waste.

Mechanized extraction by face-shovel working as far as possible selectively, is used in areas where there is much less waste material and differentiation. In both methods, the overburden must be removed using draglines and scrapers, the waste being dumped on waste areas and ultimately used for reclamation work.

In hand working, the face is divided into 10 ft to 20 ft high benches depending on the nature of the seams. The work has to be concentrated, and the labour force is deployed in gangs of three men, each working a section of the quarry similar to the system in operation in the Portland stone quarries.

The benches are divided vertically into sections of about 6 ft by 10 ft convenient for work by one gang. The clay is dug with compressed air powered spades being prised out by the chisel action of the spade and then loaded by hand in a Muir-Hill 10B 3 yd^3 capacity dumper, clay and waste being extracted separately from the top down the bench as the seams are met, the clay being taken by the dumper, driven by one of the gang, to the main covered sheds at the works. The waste lignite bands are dumped on spoil heaps.

Workmen and officials use ladders, for access between benches, which are moved from place to place as required. A temporary road of sleepers is laid on the quarry floor to each working face, and ground conditions are kept as dry as possible by collecting the water made in a sump, in the deepest part of the quarry, and pumping to the surface drainage. Output per manshift is 17 tons by hand working.

There is less lignite in the outcrop areas of the deeper siliceous clay seams and it is only necessary to differentiate the different seams and extract them selectively to provide as consistent a clay as possible. At the White Pit quarry a single bench 10–12 ft high is worked by a Priestman Lion $\frac{3}{4}$ yd^3 dumper. An experienced shovel operator is able to work within the limits of one seam and avoid sending mixed loads to the works storage sheds. In both methods of working regular sampling of the faces is carried out and the results are used in planning the production sequence. A roadway of sleepers is run from the entrance to the face to prevent the crawler tracks of the excavator ploughing the working area. It is planned to experiment with 10 yd^3 dump trucks in the future.

A shredding plant with a mobile shredder is used to give blended clays to the standard products or grades or to other specifications and 77% of the sales are of shredded and blended clays, while a further 15% is dried and pulverized material with the remaining 8% sold as lump clay. Exports total 65% of production.

Reserves of ball clay are estimated to be sufficient for a century allowing for a continual increase in demand and increasing mechanization is inevitable.

Fuller's earth

A further use of a special clay is fuller's earth. In a 112-acre site at Apsley, on the Woburn Abbey Estate, F. W. Berk & Co. Ltd have commissioned a new fuller's earth clay production plant for an output of

1000 to 1500 tons per month. Two grades are produced from clay ground to 98% through 200 mesh Berkbond, containing 5% soda ash for foundry moulding purposes, and Berkbent, with 7 to 8% soda ash added, is used in drilling muds, and as a grouting medium and for specialized civil engineering work.

St Albans Sand & Gravel Co. Ltd, a subsidiary of the Berk Group, is responsible for the extraction of the clay and commenced operations at the end of the property furthest from the plant, the clay being transported $\frac{3}{4}$ of a mile along a road built by the company with a second access road for cars and transport to the plant, the whole of the workings, plant and roads being concealed behind a belt of trees and not visible from the main road running along one side of the area.

Some 75 acres of the 112-acre site are available for fuller's earth extraction, the remainder being left to preserve the tree screen, the processing plant being built on land from which the clay has been extracted, then backfilled and consolidated.

The flat-lying fuller's earth occurs in the Greensand series and is overlain by sand with little topsoil, only 3 in, and the clay thickness is about 2 ft at the outcrop and reaches $14\frac{1}{2}$ ft in places approaching the site of the processing plant. Trial borings over the whole area have proved the deposit and ultimately a maximum thickness of 90 ft of overburden will be encountered in certain sections.

Ahead of overburden stripping trees are removed by the Bedford Estate, and the overburden, depending on its thickness, is removed by a team consisting of a bulldozer, tractor and scraper, and a dragline, the sand being removed in two stages. The upper is removed by a Fowler Track Marshall 70 bulldozer, followed by an International TD 18 tractor and 9 yd^3 scraper combination which remove the sand to a level of 8 ft above the fuller's earth in readiness for the second stripping stage by a Ruston–Bucyrus diesel-engined dragline with a $1\frac{1}{2}$ yd^3 bucket and a 60 ft boom. The dragline stands on the remaining sand layer, carefully removes it, casting aside the 30 to 40 ft wide cut into the worked-out area. The final cleaning of the top of the clay seam is done methodically by hand labour to ensure that minimum contamination is loaded with the fuller's earth, and only a minimum is exposed at any one time.

The clay is loaded by the dragline into AEC 10 yd^3 capacity six-wheeled dump trucks for transport to the processing plant. The trucks are fitted with wooden extension boards, are filled with a 9-ton load in 5 min, and driven by the dragline operator to the plant, the round trip taking 13 min. The bottom 3 to 6 in is left by the dragline and this is carefully cleaned up by hand labour, with the result that little sand gets to the plant. Of the 70 tons of dried clay processed in 24 hours only some 2 tons of waste is returned to the worked-out area. The cut is then backfilled with overburden and reforested with Norwegian spruce.

The clay is discharged at the plant into a feed hopper or into a stock area where 3000 tons may be stored and is reclaimed by a Michigan 55A tractor shovel loading into the feed chute.

CHALK

The Cretaceous rocks of Great Britain are exposed in a broad band from Dorset to Flanborough Head and have a south or a south-east dip with a band running off eastwards along Salisbury Plain to the coast and a narrow strip bordering the Hampshire Basin in the Isles of Purbeck and Wight. The escarpment of the upper dominant portion of the system, the chalk, is to the west overlooking Jurassic or Triassic plains and sinking under Tertiary or more recent deposits eastwards.

The chalk consists mainly of comminuted shell fragments with some Foraminifera and microscopic calcareous algae, laid down in clear water which may indicate arid conditions in the lands bordering the chalk, the depth not being greater than 600 fathoms and probably much less. Sponge spicules and layers of flint are associated with the chalk. The transgression of the chalk sea took place westwards in Upper Cretaceous times when most of the British Isles must have been covered since patches of the Upper Cretaceous occur on the Antrim basaltic plateau and below Tertiary basalts in Mull and Morren.

Chalk has two main uses—for agriculture and for lime and cement manufacture.

In the former field Agricultural Contractors Ltd operate a number of quarries employing a horizontal face working method in the eastern and southern counties and supply chalk for agriculture purposes for all of south-east England to a very approximate line from the Wash to Dorset selling two products—'Carbolim' which is dried by natural processes and contains 6 to 16% moisture depending on the weather conditions prevailing at the time of extraction and delivery, and 'Acolim', which is passed through a rotary drier to reduce the moisture to an average of 0·2%.

In working the chalk the overburden is first stripped to reveal the under-lying chalk, the work being done by an outside contractor. An area of some 5 acres is usually exposed at a time but the total area of some working faces have reached 9 to 10 acres.

Agricultural-type tractor-drawn disc harrows are used for scarifying which produces a fine tilth upper layer of more or less pulverized chalk several inches in depth which is then exposed to the action of sun and wind while it is further comminuted and stirred by further harrowing, the process being continuous throughout the day, weather permitting. For success this method requires essentially that the quarry should be shaped to enable the wind to reach the quarry floor and drive through the tilth so reducing its moisture content, and this becomes increasingly important as the quarry deepens and the floor is progressively lowered. Adequate aeration is a fundamental necessity and when the quarry becomes too deep it has been found possible, when the floor has reached a low level such as 30 ft or more below the surface contours, to open out the site in the direction of the prevailing wind, making a long fairly narrow quarry down which the air can funnel and extend the life of the quarry for several years. Ultimately, the quarry becomes too deep and the air currents do not get down to the floor and, in consequence, the chalk is not properly dried, so that this is the main factor deciding the depth limit to which efficient and economical quarrying may proceed by this method.

At the different quarries various hardnesses of chalk are encountered and this has a major effect on the type of equipment chosen, a Rotovator, for example, is used for relatively hard sections of the floor, mounted behind a tractor and consisting of a revolving shaft carrying special pick tines, power supply being obtained from the power take-off of the tractor. Recently a Conder $\frac{1}{2}$ yd^3 Hydraulic Dipper has been used at another quarry for similar hard chalk. A four-gang disc Model HR 13, employing 36 high carbon heat-treated steel 20 in diameter discs, spaced $6\frac{1}{2}$ in apart giving a harrow width of $8\frac{1}{2}$ ft weighs just over a ton and requires a drawbar pull of 22 hp. For hard chalk discs $3\frac{1}{4}$ in apart are being tried out.

When the chalk is sufficiently dried, it is picked up and transported by M3 roll-over type Land-Levellers, a type of tractor-drawn scraper, which take a cut five feet wide and have a capacity of $\frac{3}{4}$ yd^3. They weigh about $\frac{1}{2}$ ton empty and require a drawbar pull of 27 hp. The fine chalk collected is unloaded on to a grizzly and dumped on to a conveyor belt. A Conder $\frac{3}{4}$ yd^3 capacity scraper Model 20B has also been used successfully at some of the chalk quarries of the group. The Land Utility model of the Fordson Major Tractor is used throughout the quarries. It has a fixed front axle and the engine is a four-cylinder four-stroke developing 30 bhp at 1200 rpm on vaporizing oil, petrol being used for starting only. It has a wheelbase of 6 ft 5 in with an overall length of 11 ft 1 in and weighs 2 tons.

These machines have given excellent service, and five are used with one in reserve so permitting adequate planned maintenance. Six type Vce-tread tyres of six-ply construction are made by Goodyear specially for quarry tractors.

Productivity varies with the weather, and when wet weather is forecast, the scrapers gather the chalk into windrows so that when normal conditions return, a fair proportion of dry material is available. Each quarry has a rain-gauge and readings figure prominently in daily reports. A production of 400 to 500 yd^3 per day can be achieved in good weather from each quarry and peak sales demand occurs in the autumn immediately after harvest. Normal face operations are suspended in winter and the staff of six to eight are employed on major maintenance overhauls and erection of buildings. Each quarryman is responsible for his own set of equipment, tractors, harrows, and scraper, and this has greatly improved the standard of maintenance. One skilled mechanic under the production manager is responsible in the eastern area, but two quarry managers are responsible in the southern area. Labour turnover is small.

The Swanscombe open pit of the Associated Portland Cement Manufacturers in Kent, one of 27 the company owns in Britain, used four Caterpillar DW 21 tractors pulling No. 470 scrapers to remove $2\frac{1}{2}$ million tons of overburden per annum consisting mainly of sand from 2 to 120 ft in depth to enable $1\frac{1}{4}$ million tons of chalk to be loaded by face-shovels into trucks or on to a 2500 yd long conveyor for transportation to the cement works.

The Alpha Cement Works near Rodmell on the South Downs, have used diesel crawler tractors for excavation of chalk. Blasting is said to be unsuccessful generally due to the cushioning effect on the charge. A 95 bhp Fowler Challenger 3 diesel crawler-tractor, equipped with a Bray hydraulic angle-dozer was used on the hardest part of the face, cutting grooves about 20 ft long at right angles to the cliff. The spoil was pushed over the cliff on to the quarry floor 100 ft below, where it was loaded and removed by shovel and dump trucks.

Over 1000 tons per day of middle and bottom chalk were removed from the face by the tractor in an eight-hour working day, excluding time for maintenance and for lunch.

A larger tractor, the Challenger 4 powered by a Meadows diesel engine developing 150 bhp and manufactured by John Fowler & Co. Ltd, has removed from 160 to 200 tons per hour from the face.

Four miles south of the port and town of Larne on the western shore of Larne Lough is Magheramorne quarry of the British Portland Cement Co. Ltd, who built two cement kilns and replaced them with a larger unit in 1934 and added a third kiln in 1954, together with a new crushing and grinding plant to meet a budgeted increase in production to exploit economically the 120 ft thick chalk beds overlain by 320 ft of basalt which had increased rapidly in the direction of face advance.

The main geological features of the area are shown in Fig. 67, one of the main being the basalt cap which was extruded in the Tertiary era under which are the Cretaceous chalk beds and the Greensand. The basalt cap of maximum proved thickness of 320 ft in the lease area, is covered by a shallow soil overburden and there are extensive basalt intrusions into the chalk beds which dip $5°$ to $7°$ to the south-east with a thickness of 120 ft. At the face the intrusions are difficult to negotiate and slow down face advance considerably.

The chalk contains 5 to 7% of flints.

The major quarrying operation is the removal of the 320 ft of basalt overburden, the thin soil cover being bulldozed level to facilitate drilling operations and to provide access roads to the top quarry bench.

Of the five quarry faces worked simultaneously, three are in the basalt, the two lower benches being developed out of a face originally 120 ft high. Annually, some $2\frac{1}{2}$ million tons of waste basalt, for which there is no local market, were removed to uncover 450,000 tons of chalk, about 400,000 tons of basalt going into the worked-out quarry workings and the remainder into the Larne Lough dump.

In the basalt, blasthole drilling is carried out, either with Ruston–Bucyrus 27T well drills with $6\frac{5}{8}$ in forged tempered bits with a penetration

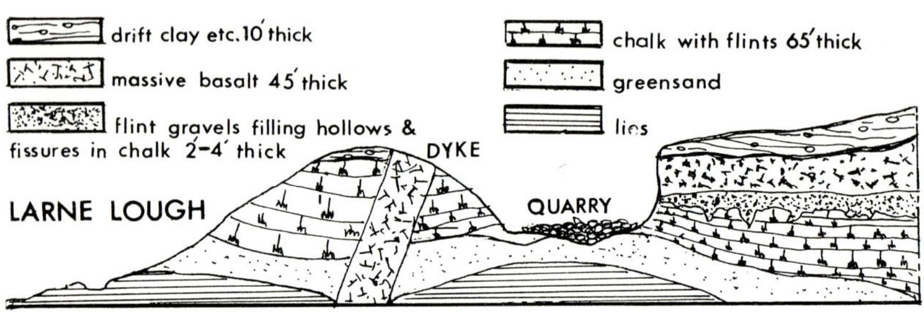

Fig. 67. Geological section of Magheramorne.

rate of about 2 ft per hour, or with electrically powered Ruston–Bucyrus 42T drills, using the same type of bits, $9\frac{1}{2}$ in diameter, which have a penetration rate of $2\frac{1}{2}$ ft per hour. The drills are crawler mounted and are moved under their own power between drilling sites. The holes are flushed with water, and bailed out at frequent intervals, their depth being 66 ft extending 6 ft into the floor of the bench. The $6\frac{5}{8}$ in diameter holes are spaced at 14 ft intervals in one row 14 ft from the quarry face, the $9\frac{1}{2}$ in diameter holes have a spacing of 24 ft and a burden of 24 ft. The blastholes are charged with 100 to 200 lb of blasting gelatine as a base charge and filled up with opencast gelignite to within 20 ft of the collar, Cordtex and millisecond delay detonating relays are used for detonation to reduce ground vibration, this primary blasting gives a stone to explosive ratio of 5 tons per lb of explosive, a maximum of ten holes being fired in a single blast producing some 21,000 tons of basalt. The basalt is brittle and fragmentation is generally good. Secondary blasting when required is by $\frac{7}{8}$ in diameter holes charged with cartridges of blasting gelatine detonated by direct Cordtex line with no delays.

Face-shovels, Ruston–Bucyrus 110B electric with $4\frac{1}{2}$ yd^3 dippers, load into a variety of dump trucks including 18 yd^3 Foden dumpers and 9 yd^3 Euclid, Foden and Aveling–Barford machines, which transport the basalt for tipping into the old quarry or to the Lough basalt tip. A concrete bridge and a road were built to take the largest dumpers: the distance from the face to the tip is about one mile increasing by 250 yd per annum as the tip extends.

The two lowest benches are in the chalk beds, which are hard, fine grained with few fossils but with 5 to 7% of flint inclusions; they have a large number of faults and slip planes which may impede working.

Wagon-mounted drills are used on these chalk benches, compressed air driven Halco–Stenuick machines using $4\frac{1}{2}$ in diameter tungsten carbide insert cross-bits with a penetration rate of 9 to 15 ft per hour, the blast-holes are 63 ft deep extending 3 ft into the bench floor, the holes being air-flushed. Better fragmentation is obtained in the chalk by more closely spaced smaller diameter holes, a single row at 9 ft intervals and 9 ft from the bench face of the $4\frac{1}{2}$ in diameter holes being drilled and charged with opencast gelignite to within 10 ft of the top and fired by Cordtex with millisecond delay detonating relays, the yield of chalk per lb of explosive being 5 tons.

The planned line of advance of the benches is southerly with the benches orientated in an east–west direction.

A Bucyrus–Erie 100B electric face-shovel with a $3\frac{1}{2}$ yd^3 capacity dipper, and a Ruston–Bucyrus 54RB electric face-shovel with a $2\frac{1}{2}$ yd^3 dipper on the lower bench are used to load the chalk into Foden dump trucks of 9 yd^3 capacity and hauled to be dumped directly into the primary cone crusher, the transport distance being comparatively short. Care must be taken that the blastholes in the chalk do not penetrate too deeply into the greensand below which is soft, or bad quarry floor conditions result in bad weather.

The primary cone crusher is a 30 in McCully set to crush down to 6 in and is driven by a 250 hp GEC induction motor which has a maximum capacity of 250 tons per hour.

At the bottom of the old quarry workings a drainage pond has been made into which the run-off seepage water collects and is pumped by two 8 in centrifugal pumps mounted on a floating pontoon to the main works with a feeder tapped off for flushing the drill rigs.

The annual production of cement is 300,000 to 350,000 tons, 75% of which is used in Northern Ireland and distributed by bulk transport in 8-ton capacity tankers, 20% is bagged and 5% shipped to the United Kingdom.

PHOSPHATES

Of all the chemicals necessary for plant growth, and no animal or plant can exist without phosphorus, compounds containing available phosphorus are liable to be deficient.

Phosphorus is quite common in nature but is frequently in a form inaccessible to plants. Animals eat plants and so obtain phosphorus to build bone, but before calcium phosphate in bones can be taken up by plants the bone must be ground to a fine powder as bone meal and even then the plant roots extract it slowly and with difficulty. The chemist, however, has converted insoluble natural phosphate into a soluble superphosphate. A ton of wheat takes 47 lb of nitrogen, 18 lb of phosphoric acid and 12 lb of potash from the soil and unless replaced its fertility is reduced.

Natural phosphates occur as rock phosphates such as phosphorite, phosphatic limestones, guano and bone beds and particularly the mineral apatite, $3Ca_3P_2O_5CaF_2$, which occasionally contains chlorine in place of fluorine (chlorapatite). Apatite is a constituent mineral of almost every kind of igneous rock but only in pegmatites associated with deep-seated alkaline intrusions is it sufficiently concentrated for quarrying profitably as in Ontario and Quebec and at Kola in the USSR.

Phosphate rock occurs also in beds of marine origin interstratified with limestone, marls, sandstones or shales. Most accessible to European users are the extensive beds of fossil phosphates from the base of Eocene in North Africa, particularly Morocco. These deposits probably derived their material from decaying organic matter accumulating on the sea floor where the organic phosphate reacted with calcium carbonate or lime to form concretions of nodules of calcium phosphate rocks, the important constituent being collophane or collophanite $Ca_3(PO_4)_2H_2O$.

In 1964, some 20% of the world's output of rock phosphate amounting to some 10 million metric tons was produced by Morocco where production of phosphates began soon after World War I and exports now go to Europe, Asia, America and Africa, and amount to about 40% of the import requirements of these continents. Increasing interest and use of fertilizers and phosphorus and phosphorous derivatives has led, since then, to further expansion of the Moroccan industry. The mining and marketing of phosphate in Morocco is controlled by the Office Chérifien des Phosphates, which is a state-owned company with monopoly powers but managed as a limited liability company with all stock in government hands. It constitutes the largest firm in the country employing 15,000 people.

The deposits are of sedimentary origin with continuous beds separated by bands of waste material. The reserves of high grade are very extensive and practically inexhaustible since a number of beds extend over large areas of Morocco and there are six main regions on a line running northeast to south-west across the country.

The centre of production at Khouribga, some 90 miles from the port of Casablanca and 2600 ft above sea level, is the source of over 70% of the exported ore. The beds are of Eocene origin and are covered by 100 ft of overburden decreasing in thickness towards the east. Four beds are separated by limestone and marls and the top bed is too thin to be viable and operations are limited to the second bed (called No. 1) which has a content of 75% tricalcium phosphate $Ca_3(PO_4)_2$ and a thickness of 3 to 16 ft averaging 6 ft, the phosphate rock being a dark friable material forming a dark, rather compact sand when broken with a density of 1·7 and a moisture content of 12 to 15%.

Recently a rich deposit of phosphate rock averaging when dry 80 to 82% $Ca_3(PO_4)_2$ has been worked at Sidi Daoui 18 miles east of Khouribga and is one of the richer sources of P_2O_5 in the world having upgraded by the removal of calcareous material through weathering and surface water infiltration.

At Youssoufia, 100 miles to the west of Khouribga, is the second main producer where three main beds are recognized interbedded with limestone of a very friable nature although only the top bed is being worked 5 to 8 ft in thickness and 70 to 72% of $Ca_3(PO_4)_2$ when dry, run-of-mine rock contains only 12% of moisture.

The phosphate won in Morocco is worked by both opencast and underground mining, some 70% by the latter and 30% by opencast.

The deposits are only viable for opencast working where the overburden is less than 49 ft thick and the hard limestone needs drilling and blasting before stripping. A Joy rotary drill, track mounted, drills 7 in diameter holes on an $11\frac{1}{2}$ ft grid pattern at drilling speeds of an average of 45 ft per hour. AN–FO blasting agent is used initiated by detonating fuse and about 12 tons of explosive are used for each primary blast.

After blasting the overburden is removed by Bucyrus–Erie type 9W walking draglines with 10 yd^3 capacity buckets and 163 ft booms working on the surface of the overburden and swinging the broken waste across the 130 ft wide open-cut. The worked-out area into which the overburden is dumped is not levelled to restore the land to a flat surface, but is left in the ridge and furrow condition of unreclaimed opencast workings.

The removal of the overburden uncovers the friable phosphate bed which is excavated by draglines working on the phosphate top surface. The excavators, which work in pairs, are usually Bucyrus–Erie 110B crawler-mounted draglines with 5 yd^3 buckets and $97\frac{1}{2}$ ft booms which discharge into a mobile screening plant to remove some of the waste product. The phosphate is then transported in 15- and 30-ton capacity Euclid trucks and dumped into storage bins for transport by rail or conveyor to the central treatment plants, production from each open pit usually averaging 2500 to 3000 tons per day.

Where the overburden is only 3 to 7 ft thick and does not require blasting, Euclid TS24 wheeled motorized scrapers with tandem engines of 336 and 227 hp, are used to uncover the phosphate bed and may be assisted by a bulldozer when necessary.

REFERENCES

Minerals in Industry, W. R. Jones, Pelican Books, 1943.
'Stripping Operations of Demerara Bauxite Co. Ltd, British Guiana', W. Forbes, *Opencast Mining, Quarrying, and Alluvial Mining*, Institution of Mining and Metallurgy, 1965, p. 693.
'Production of Defluorinated Stone', W. Windle, *Mining and Minerals Engineering*, May 1965, p. 338.
'Phosphate Mining with Wheeled Tractor Scrapers', *Mining and Minerals Engineering*, May 1967, p. 194.
'Phosphate Production in Morocco', *Mining and Minerals Engineering*, January 1967, p. 21.
'Ball Clay Production in South Devon', *Quarry Managers' Journal*, August 1954, p. 301.
'Clay Handling at Marston Valley', *Mine and Quarry Engineering*, April 1962, p. 176.
'Goonvean China Clay Pit', *Mine and Quarry Engineering*, November 1958, p. 476.
'Fuller's Earth at Woburn', *Mining and Minerals Engineering*, December 1964, p. 142.
'Quarrying Chalk', *Mine and Quarry Engineering*, September 1953, p. 321.
'Excavating Chalk by Tractor', *Mine and Quarry Engineering*, April 1956, p. 153.
'Quarrying at Magheramore', *Mine and Quarry Engineering*, April 1963, p. 142.

CHAPTER 10

SURFACE MINING OF GOLD, PLATINUM, URANIUM AND GEMSTONES

Gold, platinum and diamonds are resistant to chemical weathering and are often recovered from detrital deposits.

GOLD

Surface mining in the manner of placers and dredger washings supply half of the gold won outside South Africa and Canada and one theory is that the so-called 'banket' reefs of South Africa originated as placer concentrations in Pre-Cambrian Bench deposits and are therefore 'fossil' placers, subsequently covered by later deposits and folded into an immense synclinal fold. It is agreed that the 'banket' conglomerate and other associated rocks were so transported by water but another theory is that the gold was deposited later, redistributed by wandering hot solutions infiltrating the conglomerate which deposited also the pyrite generally associated with the gold as well as minerals such as pyrrhotite and cobaltite.

Gold placers may also accumulate through wave and current action producing beach or marine placers like those of Nome in Alaska where gold is recovered from the present beach and from raised beaches now 47 and 70 feet above sea level.

Gold dredging in Rhodesia

Gold dredging is being carried out on the River Sanyatti in Rhodesia some 235 miles from Salisbury, close to the Kariba Dam, where gold-bearing sand and gravel is being won at a rate of 300 yd^3 per hour in free-running sand, 210 yd^3 per hour in free-getting gravel up to a stone size of a 7 in diameter sphere and 62 yd^3 per hour in gravel contained in crevices and between large rocks and boulders down to depths of 30 ft.

The dredge used (Fig. 68) is a 10 in Hydrojet mounted on a multiple pontoon hull made by Hymatic Dredges Ltd, owned jointly by Simons–Lobnitz Ltd, of Clydeside and Hays Wharf Ltd, of London. The jet pump principle is used for dredging free-getting material for depths down to 200 ft. Dredging depth required decides the number of pontoons required to form a rigid watertight platform, the pontoons are of all-welded mild steel and are securely coupled and braced together. The engine room and operating control panel, mounted forward to provide a full view of the dredging operations, are accommodated in a deckhouse.

The ladder, gantry, fairleads and bollards are all included in the hull arrangement, the final depending on the operating conditions and specific user requirements. The control panel contains controls for the sideline and ladder winches, engine speed indication and gauges.

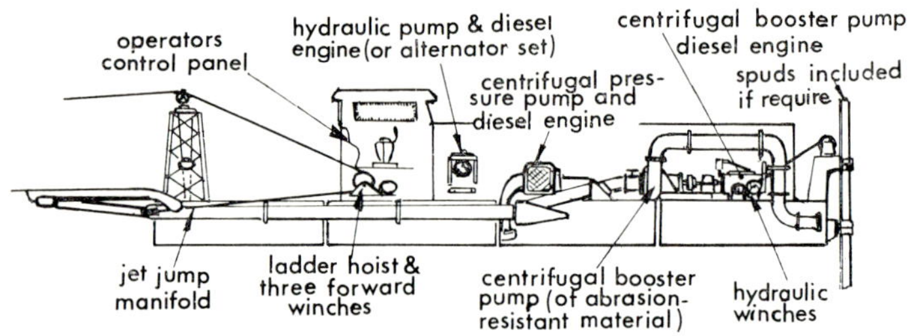

Fig. 68.   *General arrangement of the Hymatic Hydrojet dredge.*

When tougher materials are encountered a cutter is fitted and cutter controls are also mounted on the panel. The dredging ladder is carried in a well between the forward pontoons and consists of a pressure pipeline, jet pump manifold, suction pipe, replaceable wearpiece, discharge line, a fully braced and rigid steel frame, special flexible rubber hoses and pivot arms. The size and power of the pressure pump and power unit depend on conditions, a centrifugal pump supplies clean water at high pressure to the jet pump which is capable of lifting a 40% solids mixture in free-running sand. Where required, a Simons–Lobnitz cast steel dredge pump capable of discharging material to a distance of about a mile may also be included, the power unit must be of sufficient capacity to maintain continuous running under arduous tropical conditions. The power supply to the dredge can be diesel-electric, or electric power from a shore station, the various units being driven by individual electric motors or a diesel hydraulic drive may be used.

Open pit gold mining in Australia

Mount Morgan Ltd's mine was discovered in the 1880s and has produced some 45 million tons of material by underground and open pit methods, the original company amassing in profits and liquidation capital distribution a total of £10½ million. The present owners took over in 1929 and have since worked the deposit by open pit mining. Ore reserves

exceed 10 million tons assaying gold 2·7 dwt per ton and copper 1·02%
and assays of gold 3·13 dwt have been obtained.

Mining of the main ore-body has proceeded to 850 ft and selected gold-
bearing overburden containing 90% silica and one to two dwt gold per ton
is used as a flux, the total weight of material mined amounts to $2\frac{1}{2}$ million
tons per annum of which just under a million tons is of milling ore.

Bench height varies from 76 to 125 ft and width 70 to 90 ft. Above the
first five benches material is hauled up an inclined roadway to ground
level, below this level ore is transported by diesel-engined trucks to underground crushers and hoisted to the mill through the main inclined shaft at
45° to the surface.

Loading is by three diesel-engined shovels with a $1\frac{1}{2}$ yd^3 dipper,
three $2\frac{1}{2}$ yd^3 dippers and one 5 yd^3 dipper electrically driven together
with bulldozers and front-end loaders for cleaning up and auxiliary
duties.

Diamond T 20-ton capacity diesel-engined six-wheel trucks are used to
transport ore and waste.

For drilling in ore only diamond drills are used, churn drills and wagon
drills are used in waste only with hammer drills for secondary blasting to
reduce to a size suitable for the three 42 in by 30 in Hadfield crushers,
which are located underground and at the mill.

Placer goldmining in Alaska

The Yukon Consolidated Gold Corporation of Canada operates five
dredges and a bulldozer-sluiceplate operation in the Klondike River
region of the Yukon Territory under sub-arctic conditions, operations
being restricted to the period between April and November.

The gold occurs partly in gravels and partly in a deeply weathered bedrock. The gravels are covered by a deposit of Aeolian origin locally known
as 'muck'. All the superficial deposits are frozen perenially and must be
thawed before removal by a combination of solar thawing and hydraulicking. The gravel and the bedrock are thawed by the circulation introduced
into the ground under pressure.

Three distinct gold-bearing gravels are worked in the Klondike district.
The Hill or White Channel are now sited 200 to 300 ft above the present
stream level and are up to 200 ft thick with gold values mainly near
bedrock.

The Bench gravels are composed of quartz pebbles and boulders with
fragments of the underlying schist, sand and silt. Some have proved to be
profitable when worked by bulldozers pushing the gravels into sluice
boxes for gold recovery. A $2\frac{1}{2}$ yd^3 dredge works these gravels.

In the beds of the present streams are the Creek gravels similar in composition to the foregoing Bench gravels and all but one of the dredges are
concentrated in these gravels.

The Dominion Government grants mining claims and leases under the provisions of the Yukon Placer Mining Act and full size creek claims amount to 500 ft along the valley by 1000 ft on each side of the base line or from the centre of the valley if there is no surveyed base line. Full size hill and bench claims are 500 ft along the valley by 1000 ft at right angles to it, all claims being on a ten dollar annual renewal fee together with assessed representation work of 200 dollars per annum.

The recent development of the placer industry is due to the development of the successful method for stripping the frozen muck which overlies the gold-bearing gravels and the use of water under pressure instead of steam for thawing the gravels and, also, the increased price of gold. A four-year programme of project boring was embarked on under expert supervision to furnish reliable figures estimating yardage and values of each dredging area, and equipment was ordered on these estimates.

The gravels are overlain by a deposit of barren frozen 'muck' from 10 to 65 ft deep, difficult and expensive to thaw and dredge. Hydraulic stripping removes it more cheaply than by thawing and dredging. On the creeks dredging entails cleaning the surface of brush, trees, old buildings and machinery and then removing as much 'muck' as the available runoff grade will permit; this work was formerly done by hand but now by bulldozers drawn by tractors. Water supply for stripping is obtained from ditches or by pumping from local streams, a high efficiency 10 in centrifugal pump at 3000 US gallons per minute against a 150 ft head has become standard for this purpose and each pump is driven by a 150 hp 2300 V synchronous motor, the unit being correct to supply a $3\frac{1}{2}$ in nozzle monitor, the casings being strong enough for two pumps to work in series and so generate double pressure and parallel operation of pumps is also possible.

Hydraulicking is used for stripping, using water under pressure through a series of pipes leading to a series of No 2 Giants so placed that all surrounding 'muck' can be reached with a minimum transporting of Giants. A large area must be treated at one time to give the sun's rays time to thaw the surface. Water duties at Dawson vary from 8 yd^3 to 15 yd^3 per Miner's Inch Day (MID), a flow of $1\frac{1}{2}$ ft^3 of water per minute for 24 hours. Giants should be placed for 'muck' stripping a radius of operation for a distance in feet equal to $1\frac{1}{2}$ times the water pressure in ib/in^2.

Thawing with water at ambient temperature follows completion of the 'muck' stripping and is accomplished by injecting water into the ground through pipes driven into the ground by hand to bedrock as the water thaws the ground ahead of them. The water flows through a 12 in gate valve into a flanged pipe ranging in size from 11 in to 8 in with 6 in outlets with 6 in valves distributing the water through header pipes to the ground pipes.

Dredging of the ground follows complete thawing and the dredging season is short, from May to November.

Areas which are too small or too shallow or have a very uneven bedrock are mined by what is known as the 'bulldozing' method. The ground is first cleaned and then 'muck' stripped off by hydraulicking and the gravel is generally shallow enough to thaw naturally after the 'muck' overburden has been removed so no thawing by water is required.

When the gravel is ready for gold recovery a sluice box with angle-iron riffles is built with a wide dump box at its head and into this the bulldozer pushes the gravel. The gravel is well washed before reaching the riffles. When the gradient is insufficient to allow the tailings to run away it is necessary to have a bulldozer stack them as often as required.

Each operation originally had its own camp but it has been found more economical to maintain fewer camps and transport the men to and from the job.

Electric power lines of 33,000 V deliver power to the dredges, to Bear Creek, the main operating headquarters, and to the city of Dawson, secondary lines as 2300 V. A hydro-electric plant for electric power supply is situated on the Klondike River which has three 5000 hp Francis type IP Morris turbine units supplied with water from the north fork of the Klondike River with a head of 220 ft and a 16-mile canal of 10,000 Miner's Inch capacity was constructed to supplement the supply.

PLATINUM

This metal was first discovered in 1735 by the Spaniards in Columbia where it occurred with gold in river gravels but was afterwards developed in 1778 until in 1883 rich platinum-bearing alluvial deposits were discovered in the Urals which became the main producer of 93% of the world supply, Columbia providing the remaining 7%. In 1923 International Nickel Co. of Sudbury, Ontario, produced it as a by-product from nickel ores and its share of the market increased rapidly.

Platinum occurs in intimate association with the platinum group of metals which includes also palladium, osmium, iridium, ruthenium and rhodium and it is also found alloyed in addition with gold and iron. Platinum is used in jewellery, particularly as a setting for diamonds, for dental, electrical, chemical purposes as crucibles, foil and wire, and as a catalyst in the manufacture of sulphuric acid, in the synthesis of ammonia from nitrogen and hydrogen and in the hydrogenation of many organic compounds.

It has been found economic to mine it in peridotites in the Transvaal.

The price of platinum in the free world was raised at the beginning of February 1967 from £35 15s to £39 per oz; this was not unexpected since the price had not been increased for two years by Rustenberg in South Africa and International Nickel of Canada, the two main producers, the other being the USSR. The free market depends on the USSR for the bulk of its supplies although Russia's potentiality and production capability are unknown and when she withdrew from the market in 1966 it sent the free market price rocketing to £57 to £59 per oz, equivalent to a premium of £15 to £18 per oz. The non-Communist world has been able to get most of its supplies at the official market price but Rustenberg and International Nickel have been hard pressed to equate supply with demand, and Rustenberg with its very large platinum reserves is the sole Free World producer capable of stepping up production specifically to meet increases in demand since supplies from Inco are bound up with nickel production and sales. The modest increase in official platinum quotations implies no departure from the policy of price stability followed by both western major producers. In Rustenberg's view, however, the shortage of platinum is likely to persist and in addition the higher price may reflect the special risks attached to Rustenberg's costly expansion programme, having regard to the uncertainties surrounding the USSR production potential and sales policy.

The primary deposits of Sudbury and the Urals and the Bushveld of South Africa are related to igneous rocks of the noritic type. Some 90% of the Urals production is dredged from alluvial deposits where the metal has been concentrated in stream gravels derived from weathering of the olivine-rich basic igneous rocks, dunites. The platinum deposits of Choco in Columbia, of Ethiopia, Alaska and Sierra Leone are also of placer type originating from basic igneous rocks.

Although placer mining is carried out throughout the countries of South America, the greatest output is from Columbia, gold being the principal product with platinum in appreciable quantities, Choco being one of the three most important areas. In this area a hand-operated easily transportable Ward drill is used for prospecting particularly in coarse gravel. Bucyrus and Keystone track-mounted power churn drills are used where ground conditions and objectives permit. Formidable natural difficulties must be faced, the lack of roads necessitating buying special types of aircraft, including tri-motored Worthrop aircraft capable of a payload of six tons. This adds considerably to the cost of all operations. In addition, tropical rainfall causes great and rapid fluctuations of water in rivers.

In Columbia over the years the major problems have been legal in nature involving title ownership and dredging rights in navigable rivers, even where titles conferred before the passing of a new law were upheld in the Courts. Inspectors are appointed by the Ministry of Mines to supervise the dredging, recovery, yardage measurements and reports. The only dispute of consequence involved the inclusion of transient river sand in yardage calculations for determining royalties.

There are two governing factors in the evaluation of drill-hole samples— the platinum or gold content recovered and the volume of the material from which it was derived. The platinum or gold can be weighed and is, of course, definite, but the determination of the true volume of the rest of the material involves many variables. There is a theoretical volume of core rise for each foot of drive of the drill casing which is a function of the outside diameter. Experience has proved that even the application of correction factors to adjoining areas can be dangerous. In Choco, where positive corrections up to 42% were allowed based on dredging experience, these same factors applied to deposits 10 to 15 miles away did not stand and the resulting recoveries were very disappointing.

However, negative corrections should always be applied.

The cost of equipping a property in South America can be very high in an isolated location where it is necessary to construct an air strip and fly everything in. It costs twice as much for a new dredge to dismantle, transport to the port, ship it, transport it to an airfield and to fly it to site, as the cost of a second-hand dredge where it had been working previously. The power costs for dredging in the Choco, where some diesel power is used in conjunction with hydro-electric, range from $\frac{1}{2}$ to one cent per yd^3, power costing 0·62 cent for electric and 2·6 cents for diesel power per kW hour at Choco. The dredge at Choco had jigs installed for platinum recovery which improved recovery so jigs are to be installed on further dredges at Choco.

The platinum produced in Columbia is sold in the USA to refiners through brokers.

URANIUM MINERALS

Radium has proved of inestimable use to science, particularly medicine, and in helping to unfold new conceptions concerning the nature of matter culminating in the making available of nuclear energy and the atomic bomb. It is extracted from uranium ores, but even after the uranium minerals have been extracted from the large quantities of the other minerals with which they are associated, only about one part radium is present in three million parts uranium so that for every gramme of radium produced, nearly three tons of uranium are obtained. Pitchblende, after its velvet-black or pitch-like lustre, is one form of uranite and the main ore of uranium and is extremely complex—a uranate of uranyl, lead, often thorium or zirconium and the metals lanthanum and yttrium and containing the gases nitrogen, helium and argon up to 2·6% and other elements may also be present. Pitchblende occurs in pegmatites, granites and in certain metalliferous veins containing the minerals of tin, copper, lead and silver, where it was deposited by mineralizing solutions. Uranium is also a by-product of gold mining of the South African Rand and one of the most important radium-uranium deposits in the world is that of Chinkolobwe in the Congo which for many years gave Belgium a natural monopoly until the discovery of the now famous deposits at the Great Bear Lake in the North-Western Territories of Canada within the Arctic Circle which are in shear zones traversing metamorphosed lavas and sediments near their contact with intrusive granite.

From the prospector's point of view, uranium is probably the most important of all the metals in the second post-war years. Government encouragement in Australia has been given to prospectors, a uranium-buying pool was established and the Atomic Energy Commission was instituted.

The Mary Kathleen Project opencast uranium mine in north-west Queensland was developed after the ore-body was discovered in 1954 to supply 4500 short tons of uranium oxide to the United Kingdom Atomic Energy Authority which was completed in 1963 and during the contract some 5000 short tons per day were obtained from eight benches within a restricted surface area. During the contract 7·8 million short tons of material were mined of which 2·9 million tons were ore. Selective mining was used because of the very irregular ore occurrence. Sampling of all blastholes was carried out by the use of a geiger probe and each face was mapped geologically. On these results firing instructions were issued to the mine foreman, the objective being to separate the ore from the waste.

The mine and its township were placed on a care and maintenance basis in 1964.

Opencast uranium mining in Sweden

At Ranstad, Sweden has erected a plant for the production of 140 metric tons of uranium oxide, U_3O_8, per annum, the raw material being alum shale which is part of the Cambrian-Silurian stratified rock series of the Billigen mountain which consists mainly of sandstone, alum shale, limestone and claystone. The uranium is confined to the alum shale $11\frac{1}{2}$ ft thick and contains approximately 300 grammes of uranium per metric ton. The mine will ultimately have a capacity of 800,000 metric tons and for the first 15 years opencast mining will be used and then deep mining. The bench height in the uranium shale is $11\frac{1}{2}$ ft, while the overburden height averages 40 ft, giving an overburden ratio of $3\frac{1}{2}$ to 1. The alum shale bed is virtually horizontal, extends over a wide area, the uranium reserves reaching one million metric tons.

The decision to adopt opencast mining at first was made after diamond drilling and seismic prospecting and a trial tunnelling programme had revealed that opencast was more economic for at least the first 15 years.

Since the uranium shale with shallow overburden has a restricted extension, it was decided, however, to site the primary crusher underground at a level suitable for future underground mining.

The main factor influencing the selection of the best mining area to commence operations was the quantity and distribution of the overburden. The diamond drilling and seismic geophysical investigation results were used to produce maps showing the depth of earth, rock and earth-plus-rock overburden within the whole of the potential opencast mining area. These were used to construct a 'cost map' to show the relative stripping costs in the area and at Ranstad the rock overburden has to be drilled and blasted before it can be removed, making the costs per rock removal three or four times that for earth removal so that the total stripping cost depends on the ratio of rock-to-earth overburden depth. When preparing the 'cost map' of a total stripping undertaking costs for each 300 ft by 300 ft unit area were first made and then transferred to the 'cost map'. The selected opencast mining area was 435 acres, sufficient for a production of 800,000 tons per 15 years. The top 10 to 50 ft of overburden, averaging 24 ft, was composed in descending order of peat soil, gravel and moraine sod.

The next 0 to 26 ft averaging 10 ft was limestone, these overlying 8 ft of waste shale above $11\frac{1}{2}$ ft of uranium shale. The stripping ratio thus averages $3\frac{1}{2}$ to 1.

The mining method uses one walking dragline for all stripping, and winning the uranium shale by benching at, normally, only one working place using a face-shovel. It is the method used in a number of British opencast mines including ironstone open pits in Northamptonshire and Lincolnshire.

The pit is opened up by a box-cut alongside the north-west border of the area, that is close to the plant and the mill and all future workings will be to the rise since the dip is towards the north-west and this facilitates drainage of the site which is important as the area is marshy, and a permanent pumping station has been installed near the tunnel entrance to the site. In addition, the cost of haulage roads are minimal but the quantities of overburden to be handled are greater than would have been the case if starting along the outcrop, south-east side of the site. As the mining advances, the central haulage road on the bottom level will gradually be extended through the site. Figure 69 shows the mining method adopted. Stripping of the overburden is proceeding on one side of the central road at the same time as the benching of uranium shale is proceeding on the other side. When the fronts reach the limits of the area the machines return to the central road where a new cut is started. If the dragline reaches the road having finished a cut before the benching on the other side is finished the dragline must wait because haulage on the bottom would be stopped. If, however, the benching advances faster than the stripping, the benching may continue on the same side of the road as the stripping without any difficulty. All stripping is done by the dragline without rehandling and the uranium shale is hauled by diesel direct from the shovel to the underground primary crusher.

The walking dragline used for stripping overburden is electrically driven and has a jib of 195 ft, a jib angle of 40° and medium-type buckets of

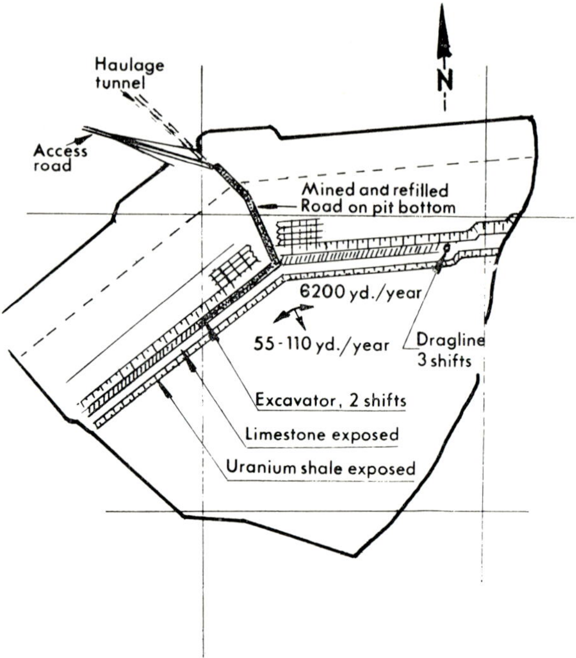

Fig. 69.   Plan showing mining method (Olsson).

14 and 16 yd^3 capacities, the dragline weighing 660 tons. On full production it works three shifts per day on five days per week. It removes from 2 to $2\frac{1}{2}$ million tons of overburden per annum and has a dumping radius of 165 ft which allows a cut width of 65 ft minimum, necessary for truck hauling on the bottom. As shown in Fig. 70, the overburden must be drilled and blasted before removal by the dragline and the holes are drilled vertically or slightly inclined and the blasted rock will be directed towards the waste side. The dragline moves on the pile of blasted rock after it has been levelled by a bulldozer and if necessary covered with peat soil and earth without rock pebbles; the length of the jib required is reduced when the rock pile is used as the base for the dragline. This method is used at several British open pits except that the blasting at

British open pits is not heavy and the surface is almost in the same condition as before blasting. The rock overburden at Ranstad is much harder than the limestone in most British open pits and it is necessary to use fairly heavy charges for fragmentation for easy loading and also some 15% of the rock is thrown directly on to the waste heap without dragline handling.

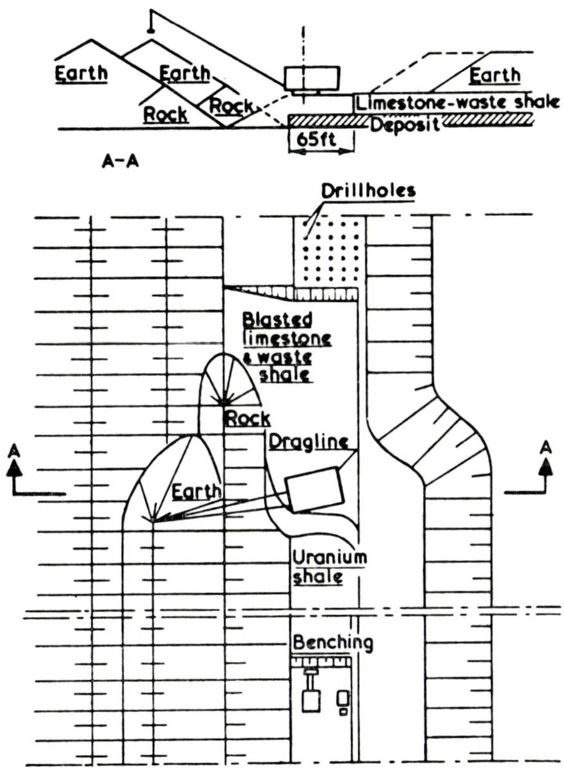

Fig. 70.   Plan detail of stripping method (Olsson).

Figure 70 also shows the distribution of rock and earth in the waste heap; the rock is dumped in advance of the earth, the rock at the bottom and the earth at the top which has the advantages that the rock allows drainage through it and the earth is comparatively easy to restore into forest or farming land.

The bucket factor of the dragline is 0·93, the operating efficiency 0·73 and its capacity some 490 yd^3 on rock and 790 yd^3 in favourable conditions with an overall average of 450 yd^3 bank measure per hour. There are two crews of two men trained for dragline operation and they are also in charge of the drilling and blasting, changing over jobs each week.

The rock overburden is blasted with $3\frac{1}{8}$ in diameter holes spaced 10 ft by 10 ft and from 9 to 38 ft deep, averaging 20 ft, drilled by two Atlas–Copco 101 rotary drilling machines, electro-hydraulically operated on 500 V ac power taking about 100 kW and rotating at a maximum of 250 rpm and a thrust of 4 tons, the feed length being 20 ft, penetration rate being $3\frac{1}{2}$ to

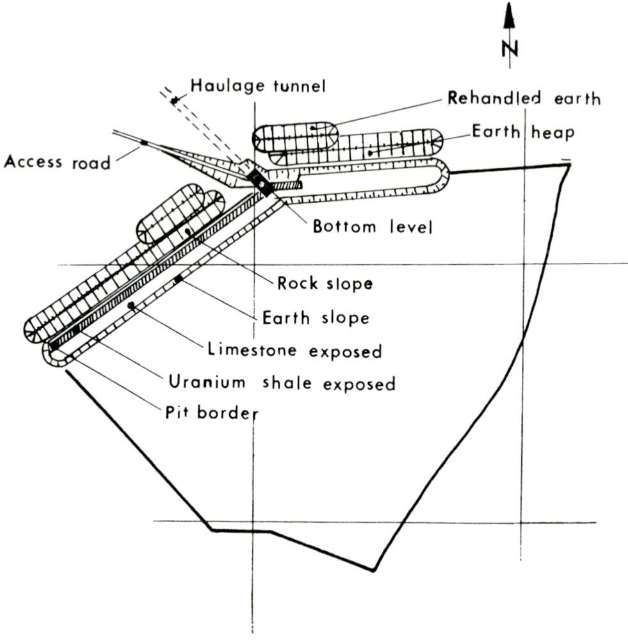

Fig. 71. *Plan drawing of box-cut excavation during the development period (Olsson).*

5 ft per minute. They are crawler mounted. The holes are charged with AN–FO prepared on the site, with a slurry type explosive, Reolite, at the bottom of the blastholes.

The consumption of explosive during the blasting of the box-cut (Figs. 71 and 72), which was done without mucking, was higher than for blasting during production. The width of the exposed uranium shale in the box-cut is 82 ft. In addition to this, a constant width of 65 ft of limestone surface must be exposed on the advance side and from 40 to 80 ft on the waste side depending on the total overburden depth, giving a total exposed width of limestone of 187 to 227 ft. The overburden from the initial gullet is heaped outside the mining area so that the uranium shale along the border limit is buried deeply.

The benching of the uranium shale involves drilling, blasting and loading by a 5 yd^3 face-shovel into rear dump trucks of about 20 tons capacity. The hydraulic-electro rotary drills by Atlas–Copco used for overburden drilling will also be used for drilling the shale.

The transport distance from the shale face to the primary crusher varies between 1500 and 3000 yd. The gradient of about 1 in 250 is towards the crusher and the first part of the journey to the central road is on a temporary road on the mine bottom built from plant waste and levelled by bulldozer to last a maximum of two months. The central road, including the tunnel, will have a long life and is of first class quality with an asphalt

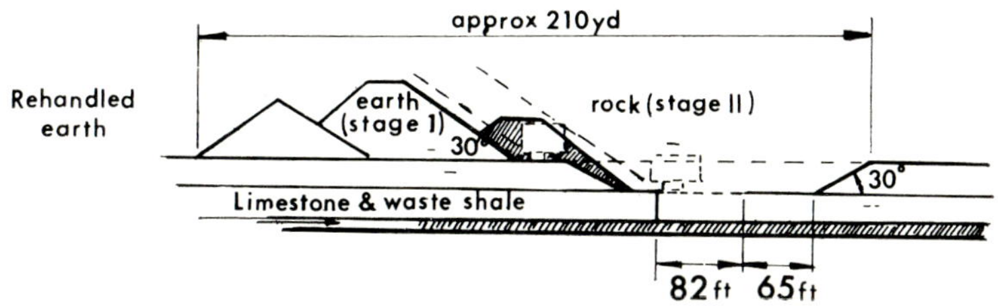

Fig. 72.   Section of box-cut excavation (Olsson).

surface constructed for heavy traffic at high speed. Two tunnels are used separately for loaded and empty trucks, are 17 ft 4 in wide and 18 ft high to take 40-ton capacity trucks, if required, and are lighted by sodium discharge lamps. A permanent pumping station is established at the tunnel entrance where the lowest point in the uranium shale is located.

A flexible electrical power supply has been provided so as not to interfere with mining operations, three portable transformer stations are used to supply the machines through trailing cables, the overhead lines feeding these stations can be disconnected when required.

The dragline and shovel work on 3000 V and other machinery on 500 V, the incoming voltage to the transformers being 10 kV.

GEMSTONES

The three cardinal virtues of the gemstones are beauty, durability and rarity and the greatest of these is rarity. Durability is linked with hardness and chemical inertness and all true gemstones are stable and do not deteriorate with age except the pearl which is not mineral but organic. Pearls steadily deteriorate with time and go dead and may even disintegrate. All gemstones are also hard. If the diamond is rated at 10 in hardness, rock crystal is 7, ordinary window glass is 6 and most semi-precious stones are 7 or 8 and the precious stones $7\frac{1}{2}$ to 8.

Beauty is enhanced by cutting and faceting and such cutting is a very skilled technique although it only dates back some three centuries.

The diamond is the top gem and in ancient times and the Middle Ages, India was the only source of diamonds which were obtained from placer deposits in recent or fossil river gravels. They were found in Brazil in 1732 as alluvial products and quickly flooded the market, were government controlled and became less important.

In South Africa the first diamonds were picked up by Boer children in 1867 and for some time were not even recognized as diamonds, and also from river gravels. Later, more stones were found on De Beers' farm on a dry plateau where the absence of water hindered development. The friable yellow soil was then found to contain diamonds in quantity and this in turn gave place to an unweathered compact rock known, from its colour, as blue ground, which also contained diamonds in abundance and is the source of the present supply which is mined, with some recovered by alluvial mining. The blue ground, or kimberlite, is a magnesia-rich ultra-basic volcanic rock with olivine, pyroxene, garnet and other ferromagnesian minerals and formed the filling of an ancient volcanic neck. The neck was followed down to 600 to 1300 ft by open pit mining and also by underground mining to 3000 to 4000 ft. Recovery is by passing crushed blue ground over greased plates to which the diamonds stick.

A faintly bluish tinge (blue white) is the most prized colour in diamonds with quite colourless stones next but the slightest yellow tinge is a serious defect, although cloudy and black varieties known as bort have a great industrial demand particularly for diamond drilling among a host of other uses.

A significant proportion of the world supply of diamonds is won by alluvial mining from recent and ancient gravels and raised benches in Africa and Russia.

The ruby and the sapphire are important gems with a composition of aluminium oxide, Al_2O_3, a hardness index of 8 and specific gravity of 4.

Rubies are recovered by placer mining and the most famous ruby mines are at Mogok in Upper Burma, where the gems occur in metamorphosed crystalline limestone and in alluvial deposits in the same distrcit.

Sapphires are found in Siam, Ceylon and Kashmir, where they occur in pockets of china clay and are of large size.

Another important gem is the emerald which is a silicate of beryllium and aluminium, $3BeO$, Al_2, O_3 $6SiO$ with a hardness of $7\frac{1}{2}$ and a specific gravity of 2·7. Emerald with its green colour is rarely unflawed but when it is, and is of good colour, it ranks with the most precious of gems. In micaceous schists the emerald mines of Upper Egypt were worked under the Pharoahs and ultimately rediscovered, but the main supply now comes from Columbia, Equador and Peru.

Jade is not a gem but highly prized in its carved form, the work of

Chinese craftsmen throughout the ages. True jade, nephrite, is a silicate of magnesium and calcium of the amphibole group. Jadeite, very similar in appearance but generally more granular, is a silicate of sodium and aluminium related to the pyroxenes. Both are tough, have a hardness of $6\frac{1}{2}$ and take a high polish and both are used for Chinese carvings.

Open pit diamond mining in Tanzania

Dr John Thorburn Williamson, a graduate of McGill University, Montreal, left Canada in 1933 at the age of 26, to take an appointment as geologist with the Loangwa Concessions (Northern Rhodesia) Ltd, and, in 1936, with Tanganyika Diamond and Gold Development Co., which was formed by South African interests to acquire Mabuki diamond mine and other interests including the Kisumbi Mine gravels. In 1938 he obtained an option over these two properties and reworked the pan and jig tailings to finance further prospecting and in 1940 he discovered diamonds at Mwadui. He delimited the kimberlite pipe and the diamond-bearing area around it as swiftly and secretly as possible and after pegging out his prospecting claim of $4\frac{3}{4}$ sq miles had accurately delimited the kimberlite pipe but also the diamond-rich areas some distance from it on to which diamond-bearing ground had been eroded.

He incorporated a private company in 1942, Williamson Diamonds Ltd, with himself as sole governing director and general manager. He died in January 1958 and the company and the mine are owned and operated on an equal basis by the Tanzania Government and De Beers Consolidated Mines Ltd, with the Anglo–American Corporation of South Africa as consulting engineers.

The mine is in the Chiefdom of Mwadui of the Lake Province on Lake Victoria, 4000 ft above sea level. Bulk supplies reach the mine mainly by rail from Dar-es-Salaam.

Mwadui lies within an area of 10 sq miles fenced off for security reasons and unauthorized entry is punishable by a £1000 fine although employees pass freely in and out on passes in accordance with the security regulations. The company must provide housing accommodation because of the isolation for 2000 employees, which, with families, means a population of 4000 of which some 600 are European, 150 Asian and 3250 African.

Illicit diamond dealing is always a problem requiring very adequate security organization and this is in the hands of the Diamond Protection Division of the government police stationed at the mine, both European and Asian officers. Since mining is entirely mechanized, diamonds are not visible until the treatment plant is reached since initially the distribution of the gems is one in 16 million by weight. The plant areas are therefore double-fenced with trained dogs patrolling the space between to warn of anyone trying to enter the area other than by the guarded gate. In the

final section of the processing plant where secondary and tertiary concentration is carried out mechanically and the final concentration is hand-
sorted, only four Europeans are employed.

In the kimberlite magma it is generally assumed that diamond crystals
were formed where it was rich in carbon and under high temperature and
pressure. The kimberlite was probably brought to the surface by eruption
in Cretaceous times along the kimberlite pipes together with volcanic ash
and rock debris. The pipe at Mwadui, surrounded by a crater, occurs in a
strongly fractured zone and in a granite country rock and is the largest yet
discovered, being 5000 ft by 3500 ft, covering an area of 250 acres. Relatively little erosion has occurred, probably less than 100 ft and backfilling
of the crater by shales and kimberlite has taken place perhaps at different
times.

The quartz gravels and calcareous and siliceous minerals overlying the
pipe and country rock were first sampled by 50 ft long sample trenches at
200 ft intervals on a grid down to the granite and at each foot horizon
samples were washed in diamond pans and evaluated. Sampling was by
5 ft^2 shafts at 500 ft intervals on a grid and at smaller intervals where
more information was required. Around the periphery of the pipe nine
exploratory shafts have been sunk into the blue ground to a depth of 200 ft.

Mining operations commenced in 1940 when equipment was difficult to
procure so the stripping of the overburden and the excavation of the
diamond-bearing ground was done manually by pick and shovel, African
labourers carrying the diamondiferous gravels in flat tin pans on their
heads to the diamond pans where hand screening at $1\frac{1}{2}$ in was carried out,
the diamond pans being sited near the diggings so that the tailings could
be run into the mined out areas. Later when more capital was available, face
excavators of $\frac{3}{8}$ and $\frac{5}{8}$ yd^3 capacity and some flat-body diesel trucks were
used. The shovels dug the quartz gravels overlying the country rock,
blasting not being required, and the gravel was transported to the central
pan plant and later 8-ton capacity diesel dump trucks were obtained and
the black-cotton overburden was stripped with a Carryall scraper drawn
by a D8 crawler tractor.

The capacity of the pan plants was increased to 700 tons per day in
1949 and motorized scrapers bought to speed stripping of the overburden
and to build haulage roads and to establish gravel stockpiles for treatment
during the rainy season when the workings become flooded and transport
on the black-cotton soil with tyred machines becomes impossible. The $\frac{3}{8}$ yd^3
shovels were replaced by $1\frac{1}{2}$ yd^3 shovels now used to load the shallow
quartz gravels overlying the country rock, the output being added to the
ground from other sources. The black-cotton topsoil which carries high
values in places is now treated in rotary scrubbers.

The yellow ground deposits are won by benches 24 ft high using $2\frac{1}{2}$ yd^3
Ransomes and Rapier electric shovels and prior blasting even though the

deposits are weathered. Ruston–Bucyrus 2TRT churn drills mounted on trucks and driven by diesel engines are used to drill $6\frac{5}{8}$ in diameter at a penetration rate of 10 ft per hour. Holes are overdrilled by 5% in order to ensure a level floor, the spacing and burden of the holes depending on the depth of the holes. Millisecond delays are used to ensure maximum fragmentation and to reduce overbreak and for blasting to a free face holes are drilled and detonated to the pattern shown in Fig. 73a, which

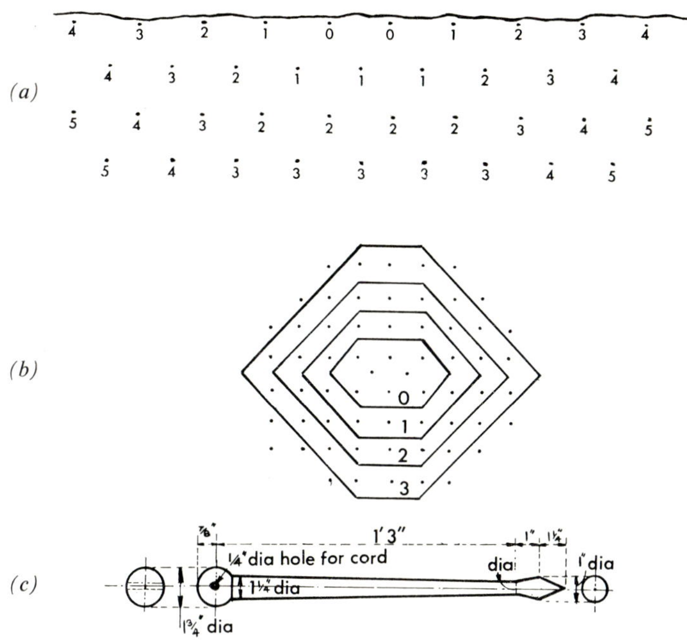

Fig. 73. *(a) Drilling pattern used for face blasting showing sequence of firing; (b) box-cut blasting pattern used to open up a new bench in a working floor; (c) hardwood lowering rod used in charging the blastholes.*

indicates the sequence of firing and the layout also gives a minimum forward throw. If the number of rows exceeds three the loading ratio must be increased to overcome the resistance on the back rows. The best ratio has been found to be 5 tons of rock per lb of explosive, preferably Dynagel but this is not often available so 60% ammon gelignite in 6 in by 9 in cartridges which weigh $12\frac{1}{2}$ lb each are used. Primers are lowered to the bottom of the hole by two lengths of Cordtex threaded through them, long enough for a detonator to be attached at surface, a loading rod (Fig. 73c) is then used to lower the rest of the charge so that it is in contact with the Cordtex. A steel gauge $6\frac{1}{2}$ in diameter is lowered down the hole immediately before charging to ensure that the hole has not swollen or collapsed in the interval between drilling and charging.

The box-cut is used for opening up a bench in the floor of a working as shown in Fig. 73*b*, where the figures show the sequence firing and is not dependent on a free face as the blasting action is vertically upwards and only slightly outwards and sideways. The $6\frac{5}{8}$ in diameter holes should be so spaced that the conical shape of the ground blasted by each hole overlaps that of adjacent holes and holes should be overdrilled by some 75% of this spacing, since little breaking takes place at the apex of the cone. The central holes with 0 delays must be charged with a continuous explosive column whereas the succeeding holes have the explosive in decked charges to ensure good fragmentation.

The weathered kimberlite and silcrete is loaded by a Ransomes and Rapier W.150 walking dragline with a 6 yd^3 bucket. This overlies a layer of shale over the pipe and varies in depth and because of the uneven floor is better worked by a dragline which can operate from level ground above the face and the cut can vary from a few to 60 ft and eliminates truck transport on the uneven floor. It has a capacity of 230 tons per hour.

The depth of blastholes is determined by sampling and blasting is kept well ahead of the dragline and the holes are not overdrilled because the shale under the silcrete is soft and the 6-ton dragline bucket is able to remove it without over-blasting.

The dragline loads into a mobile hopper mounted on 90 lb rails, the bottom being fitted with a Shure feeder which discharges on to a 42 in boom conveyor which in turn delivers to a 25B Telsmith gyratory crusher mounted on its own bogies on the same railroad. This primary crusher reduces to 6 in and delivers to a 30 in belt conveyor running parallel and near the rail track which is in sections on flat steel pads to follow the dragline and mobile hopper and crusher unit when they move into a new cut.

The gravels loaded by the small face-shovels into 15-ton capacity Euclid diesel rear-dump trucks are discharged to a dumping point feeding a 10,00 ft long belt conveyor which delivers the ore from all the separate workings to a 1800-ton capacity storage bin at the treatment plant. The dumping point consists of a 20-ton capacity concrete hopper with a 36 in steel apron feeder belt with a variable speed delivering to a vibrating grizzly with 6 in bar spacing, oversize being crushed to -6 in by a 13B Telsmith gyratory crusher which joins the undersize on the trunk belt.

The $2\frac{1}{2}$ yd^3 electric face-shovels deliver the yellow ground ore to Euclid trucks which dump into a steel hopper mounted over a 40 ft long by 36 in wide steel apron belt feeding a vibrating grizzly with 6 in bar spacing, the oversize again being crushed down to -6 in in a 60 in by 48 in Traylor jaw crusher delivering with the undersize to the trunk conveyor which is 24 in wide running at 500 ft per min and in four sections. The Euclid trucks have an average go and return round trip of 4000 ft.

During the rainy season a battery of pumps mounted on rafts is used to de-water the workings. The storms are short and heavy at several days' interval and the pumps are connected by armoured hoses to the water mains on the benches above and the rafts rise and fall with the water in the workings and continue to pump until the sumps are dry. The method has the advantage that the suctions are short so that priming difficulties are few.

Diamond-bearing gravels are also found off the S.W. African coast along a thousand mile area near the Orange River. The diamonds are in a mixed alluvial deposit of sand, gravel and boulders in water at depths of 40 to 100 ft. The Collins group have a concession of 384 miles extending from low-water mark to 3 miles out to sea. A large number of other companies have also taken up concessions and there has also been some Russian interest in the area. The South African government propose to extend the 3 miles territorial limit to 12 miles.

The Collins organization began prospecting the area in 1961 using teams of coastal surveyors equipped with Tellurometer and Hydrodist electronic distance and position locating equipment, progressed with the marine prospecting units up the coast plotting to an accuracy of one fathom which, together with aerial photographic surveying, produced accurate mining maps of the ocean bottom and the exact areas of diamond-bearing gravels trapped in old marine terraces and submerged river beds. The first sea-going unit was launched in 1962 and operations were so successful that further dredges were planned. The gems are recovered from the sea floor by airlift pumps with water jets to loosen the host material. The dredged material is forced up large-diameter rubber pipes, oversize is screened off and diamonds recovered by heavy medium separation followed by screen washing and conditioning with caustic soda and fish oil. Diamonds are separated on greased belts.

REFERENCES

'Gold Dredging in Rhodesia', *Mining and Minerals Engineering*, October 1964, p. 72.

'Gold and Platinum Dredging in Columbia and Bolivia, South America', P. H. O'Neill, *Opencast Mining, Quarrying and Alluvial Mining*, Institution of Mining and Metallurgy, 1965, p. 156.

'Operations of the Yukon Consolidated Gold Corporation, Canada', W. H. S. McFarland, *Opencast Mining, Quarrying and Alluvial Mining*, Institution of Mining and Metallurgy, 1965, p. 180.

'The Mary Kathleen Uranium Project', A. Nelson, *Mine and Quarry Engineering*, February 1960, pp. 46–54, March 1960, pp. 94–101.

'Selective Mining Practice at Mary Kathleen Uranium Mine, Queensland', D. L. Munro, *Opencast Mining, Quarrying and Alluvial Mining*, Institution of Mining and Metallurgy 1965, p. 681.

'Ranstad—A New Swedish Opencast Mine', G. Olsson, *Opencast Mining, Quarrying and Alluvial Mining*, Institution of Mining and Metallurgy 1965, p. 711.
'The Williamson Diamond Mine', G. J. du Toit, *Mine and Quarry Engineering*, March 1959, pp. 98–103, April 1959, pp. 146–153, May 1959, pp. 194–200.

CHAPTER 11

SAND AND GRAVEL

This branch of the surface mining industry has expanded by leaps and bounds in the past decade through the boom in building and road making and particularly the increased construction in concrete. The output increased from 80,546,799 tons in 1961 to 100,994,589 tons in 1964.

METHODS OF WORKING SAND AND GRAVEL

The methods of working sand and gravel depend particularly on the position of the water-table either naturally or after it has been lowered by pumping. The main methods adopted are:

(1) Wet pits using draglines.
(2) Wet pits using pumps.
(3) Dry pits using draglines or shovels.
(4) Hand methods.
(5) Offshore and river dredging.

In 1949 the Hoveringham Gravels Ltd group was inaugurated with the purchase of a gravel quarry at Hoveringham in the Trent Valley which had already been operating for some ten years, and since that date seven further gravel pits have been acquired by the group extending from Nottingham in the south-west to six miles down river from Newark in the north-east ranging in output from 40 to 200 tons per hour.

The overall cost of prospecting in the Trent Valley for sand gravel averaged £6 per acre.

The deposits worked belong to the last phase of two periods of glaciation, the Pennine drift and the Eastern Glaciation with the alteration of the Trent Valley system caused by the melt water from the snow and ice in interglacial and postglacial periods. The deposits worked by Hoveringham are Flood Plain Terrace gravels in the newer drift of Pleistocene age while old Pleistocene gravels are worked by other firms up river from Nottingham and a Beeston Terrace deposit also exists. The overburden thickness averages 6 ft over gravel beds from 9 to 30 ft thick. The gravel size varies from $\frac{1}{4}$ to 3 in with some larger sizes but not exceeding 5%, the proportion of sand $-\frac{3}{16}$ in to gravel being 1 to 1. Some 80% has been eroded from the Bunter Pebble Beds and are mainly of quartzite, the pebbles being well rounded and of high crushing strength and are generally less than $\frac{3}{4}$ in in size. Hard well-rounded pebbles of vein quartz $\frac{1}{2}$ in to $\frac{3}{16}$ in are also present and some of flint of angular form. Erratics include sub-angular pebbles of Carboniferous and Liassic limestones, chert and flint, Millstone grit, Carboniferous and Keuper sandstones, granite and basic igenous rocks which may be well rounded to angular showing conchoidal fractures with hardness and texture also varying widely. Such gravels make excellent concrete with a high degree of workability with the lowest possible water-cement ratio.

D7E Caterpillar tractors and 9 to 11 yd^3 scrapers of 125 to 160 hp are normally used to remove the topsoil which is then stored in an area convenient for ultimate resoiling during reclamation. When it lies above water-level the overburden can also be removed by tractor and scraper down to a limit of 6 ft deep, deeper than this excavators should be used. Overburden removed by excavator is spoiled into a worked-out area well clear of the line of the new face.

Wet pits using draglines

The raw gravel in the Trent Valley is below the water-level and extraction is by Ruston–Bucyrus 38RB dragline although at the different sites bucket sizes and machines vary from $\frac{3}{4}$ yd^3 to $2\frac{1}{4}$ yd^3 depending on the method of dealing with the overburden, the depth of the gravel bed and the rate of output required. One site is worked by a gravel pump but the dragline is generally preferred since the gravel pump, though efficient in ideal conditions, generally incurs a high maintenance cost in replacing wearing parts and the distance a pump can deliver often entails two pumps in series or, as an alternative, conveyors to transport the gravel to the processing plant. If the material is free running, 300 yds can rather exceptionally be achieved, but delivery ranges are often much shorter and blockages are often a fairly common occurrence resulting in delay and loss of production.

When extraction is by dragline, the handling of the wet material presents problems. Buckets may be drilled to allow some drainage, but sand and small size stone may be washed out below water level and lost and a more usual method is to dump the material on one side and let it drain and then reload it sufficiently dry to allow it to travel on feed conveyors.

A danger in wet working is that good quality material may be left behind since it cannot be seen, or that clay is picked up with the gravel and faults are also difficult to work round when they cannot be seen.

Access from the wet face to the process plant must be kept moderately dry for transport.

When the total depth exceeds 10 ft it is common practice to pump the face dry, but in the Trent Valley depths may exceed 20 ft and the problem is greater but improvements in de-watering pumps allows the face to be sealed off in areas of reasonable size. Drainage of gravel is no longer required and the clay bottom can be seen and avoided and access roads remain dry.

For transport from face to processing plant, several alternative methods are used. Diesel locomotives hauling five or six 2 yd^3 tubs allow drainage of gravel en route to the processing plant, and wet weather does not interfere with production. Labour to maintain track and equipment is, however, high as is the initial capital cost for locomotives and tubs and it is inflexible and being less used in the Trent Valley.

Where large areas of water are available, 200-ton capacity barges and tugs may be used to advantage to transport the ballast from face to plant, but double handling in loading and unloading is required increasing the labour cost.

At the larger operations where mobile plant is moved from site to site, the dump truck such as the Aveling–Barford S.L. 13½-ton capacity dump truck, has the advantage of flexibility and can be used for removing overburden and feeding the hoppers of the processing plant. Bad ground conditions can prove adverse and seriously affect efficiency, but a well-made hard-core main road suiting the type of truck used to the conditions, should overcome these adverse conditions. Where distances are considerable and ground conditions bad, belt conveyors prove relatively efficient and easily coupled lightweight sections reduce the time and labour required for repositioning.

Wet pits using pumps

Gravel excavating by pump is used at the Shepperton pit of W. J. Lavender Ltd, which started in 1937 when 150 acres were purchased of which 68 acres have been worked, so that at 1000 yd^3 per day production considerable reserves remain.

Gwynnes Pumps Ltd designed and installed the 200-ton per hour plant which excavates the gravel by pontoon-mounted pumps, delivering to barges, supplying a shore installation. The material is pumped from the barges to the treatment classification and storage plant, occupying an area of 14·6 acres.

The stripping of the topsoil from 18 in to as much as 12 ft in places, is the first operation in working the deposits, and is carried out with a Ruston No. 4 dragline with ½ yd^3 bucket and a 30 ft boom, and powered by a Ruston–Hornsby 2VQBN diesel engine of 38·5 hp, running at 1000 rpm.

The dragline weighs 16 tons and is carried on crawlers 12 ft long and 24 in wide.

The spoil is loaded by the dragline into bottom-emptying hopper barges of 65 yd^3 capacity 53 ft long, 16 ft wide and 6 ft 6 in deep with a 6 ft draught. These barges are self-propelled by hydraulic jets which operate by means of a diesel-driven pump ejected forcibly through swivel pipes at the stern of the barge. When filled with overburden the barges are moved out from the banks and their load dumped over a worked-out section of the gravel pit.

The gravel bed varies in thickness from 25 to 35 ft and is won by means of two pontoon-mounted 8 in Gwynne Invincible gravel pumps which are completely lined with renewable manganese steel components and have manganese steel impellers to resist abrasion. The pumps are each capable of lifting 125 tons of sand and gravel per hour and are claimed to be able to deal with spherical stones up to $7\frac{1}{2}$ in diameter. The pumps are belt driven from an 80 hp Davey, Paxman and Co. Ltd, of Colchester, diesel engine mounted on each pontoon. An Ace friction hoist driven by a Lister 6 hp diesel engine is used to raise or lower the suction pipe. It is more economical and convenient to use diesel power on the pontoons than to carry long trailing cables on buoys out from the banks.

The barges into which the sand and gravel is loaded for transport to the processing plant on shore are similar in size and construction to those used for spoil removal but they are not self-propelled nor fitted with bottom-opening doors as the ballast is removed from them by pumping (Fig. 74). Each pontoon is served by three barges, one being filled, one in transit and the third being emptied at the shore processing plant. They are towed to and from the pontoons by an ex-Admiralty tug driven by a Ford V8 engine or a Bantam pusher tug of 30 bhp of length 23 ft, beam 8 ft 6 in and depth 4 ft with a 3 ft 4 in draught. The tug is of all steel welded construction and weighs $5\frac{1}{4}$ tons. The keel plate is $\frac{1}{4}$ in thick steel plate and the side plating, casing and forward bulwark are of $\frac{3}{16}$ in thick steel. The 30 bhp Lister three cylinder 31PMGR diesel engine running at 1200 rpm is fitted with a 2 to 1 reverse and reduction gear and electric starting.

These boats have been specially designed for work at gravel pits with special attention to parts subject to abrasive wear from fine sand in suspension. They are recommended to be used for pushing rather than towing by making the barge and the tug a single rigid unit which saves labour on the barge, saves power since the tug is not pulling the barge through the propeller's wash and saves wear and tear on barges. The tug brings the loaded barge in and it is moored to the wharf, which is constructed on steel piles and carries the pumps for unloading the barges which are hand warped along the wharf as one end of the barge is emptied. After completing the unloading, the barge is filled with water to lower it in the water and allow it to be positioned under the pontoon pump delivery pipe.

For unloading a similar 8 in Gwynne pump is used, belt driven from a Fuller 100 hp electric motor through a Fluidrive coupling to provide infinitely variable speed control and start-up against the load. It is a Vulcan Sinclair fluid coupling of the scoop type. The input shaft drives an impeller which acts as a centrifugal pump transmitting the power to a runner which acts as an oil turbine driving the output shaft on which it is mounted, the power being transmitted by a vortex ring of circulating oil.

Fig. 74. A barge moored alongside a pontoon being filled by a gravel pump. Excess water weirs over the side of the barge.

A hand screw control operating a scoop controls the amount of oil in the rotary reservoir and hence in the working circuit. By means of the coupling the pump may be operated between 590 and 720 rpm, the electric motor running at 740 rpm.

As the sand and gravel dries out in the barge, water is added to it by an 8 in Gwynne KI type horizontal split casing double entry pump driven by a 35 hp motor so that the water acts as a transporting agent; it requires 75 to 80% water with 25 to 20% sand and gravel. The barges are emptied in about 20 minutes.

A $2\frac{1}{2}$ in Gwynne pump driven by a 5 hp motor and capable of 150 gal per min is used for priming and to supply clean water for flushing to keep the pumps free from sand and grit.

Dry pits using draglines or shovels

In dry pits, the dragline is often preferred for gravel removal and as the machine stands on the deposit and it may be free flowing, the toe slope should be 35°. However, if the sand and gravel is hard packed, the slope may be increased to 45°. This angle of repose will affect the length of the boom of the dragline since if this is too short it will be impossible to reach the bottom of the deposit. In other respects the operation of the machine follows standard practice.

Where the excavator can stand on the floor of the deposit a face-shovel may be used to load the material into dump trucks running on the floor of the pit. The working face may be practically vertical and the shovel may work the deposit up to a thickness of 60 ft. In certain rather exceptional conditions blasting may be required to break up the deposit and bring it down to floor level for loading.

The Cleish Sand and Gravel Co. Ltd have been producing sand and gravel since 1959 from a valley deposit some two miles east of the Crook of Devon in Kinross-shire, Scotland, which was originally worked as a wet pit but is now worked dry since the water-table was lowered by pumping. The deposit is 40% sand and 60% gravel and substantial reserves exist beneath the present working. A dragline with a $1\frac{1}{4}$ yd^3 capacity bucket and a 55 ft long boom is used to remove the overburden, approximately 9 in deep, over a considerable area of the property. Also used intermittently is a Smith 21 machine with a $\frac{3}{4}$ yd^3 bucket and a 40 ft boom.

The water-table was lowered by means of a 1200 gallons per min Harland pump which exposed a dry face from 15 ft to 28 ft in height.

The water rises 3 ft overnight if the pumping is stopped and in very wet weather as much as 6 ft.

The sand and gravel is loaded into a portable hopper through a grizzly bar screen which separates an oversize of $+8$ in material. The hopper delivers to a 24 in wide Meco conveyor driven by a 10 hp Brook motor which delivers in turn to a second main rising conveyor 100 yd in length, also driven by a 10 hp Brook motor, which feeds the treatment plant through a Niagara screen 10 ft by 4 ft, the oversize $+1\frac{3}{8}$ in passing to a 36 in wide picking belt for the removal of the felspar. The remaining aggregate on the belt passes on to a bar screen to by-pass the $-1\frac{1}{2}$ in from the subsequent jaw crusher. Oversize from this screen passes to a Kue–Ken 24 in by 10 jaw crusher to give a $-1\frac{1}{2}$ in product, which passes to a Huwood 24 in wide conveyor and also receives the $-1\frac{7}{16}$ in undersize from the bar screen as well as the $+\frac{3}{16}$ in material from the bottom deck of the Niagara screen. The third product from the primary screen is the $-\frac{3}{16}$ in sand fraction and this is passed to a No 1 static cone in the sand section of the plant. Two jigs are used to remove any deleterious material from the sand and the gravel.

Haulage from the open pits is by contract, by an outside firm and the sand and gravel is sold by weight over an Avery weighbridge of 30 tons capacity. The plant has a capacity of 80 tons per hour of gravel and 40 tons of sand, the principal products sold being coarse and fine sand, $1\frac{1}{4}$ in, $\frac{3}{4}$ in and $\frac{1}{2}$ in gravel, and special sizes are supplied to particular requirements.

Hand methods

Except for a limited local market the small sand and gravel pit worked by manual labour and loaded into small tubs and skips of 1 to $1\frac{1}{2}$ yd^3 capacity running on 2 ft gauge rail track of 30 to 35 lb per yd flat bottom rail in 18 to 22 ft lengths spiked to 4 in by 4 in by 2 ft 6 in long wooden sleepers at 3 to 6 ft centres, is less used. Single line working with loops or 'passbyes' for full and empty skips to pass each other may be adopted and the track is temporary and is moved and added to to follow the working face and connect it with the processing or lorry loading hoppers if the ballast is sold as won.

Offshore and river dredging

Now mainly used for gravel working from rivers is a standard crane or a dragline fitted with a grab, the machine being mounted on a pontoon, the sand and gravel being dumped into barges moored alongside or into the pontoon itself for transport to the stacking area.

Bucket-ladder dredges (Fig. 75), for digging underwater deposits which are not free running and need to be dug by the buckets are also mounted on pontoons. In many cases the pontoons are moved by mooring lines to the shore which enable the pontoons to be moved along the cut. The machines may be powered by steam, diesel, petroleum or electricity, but the latter involves waterproof trailing cables from pontoon to shore with earth leakage protection. The ordinary navigation channel dredger used at river mouths and estuaries to maintain navigation channels which tend to silt up are similar but larger examples of the bucket-ladder dredge, the vessels having crew accommodation, the sand and gravel reclaimed being loaded into barges, and in the case of the Thames estuary is sold direct without processing as 'Thames Ballast'.

The multi-bucket excavator can be used in both wet and dry gravel pits. Suspended on one side of the machine there is a long boom or jib fitted with endless chains carrying a line of digging buckets which travel outwards from the excavator along the top of the jib and return on the underside, digging the deposit from the working face and discharging near the foot or hinge-pin of the jib into a second series of buckets which convey the gravel to a loading point which may include a complete screening plant with hoppers for different sizes discharging to trucks or wagons. A further jib and a series of wire hawsers enable the angle of the digging jib to be varied for varying depths.

As in the United Kingdom, the demand in West Germany for sand and gravel for constructional concrete has increased by over 400%. Previously bucket dredges, floating as well as mounted on rails, or trucks or tower drag scrapers consisting of a main tower, a mobile tail anchor, and a digging bucket had proved adequate for requirements, the maximum depth of working under water being 20 to 45 ft. Draglines and face-shovels were also used for dry working. The increased demand for sand and gravel and the scarcity and high cost of land for their production was met by dredging, a fact made possible because the Rhine Valley between the Vosges and the

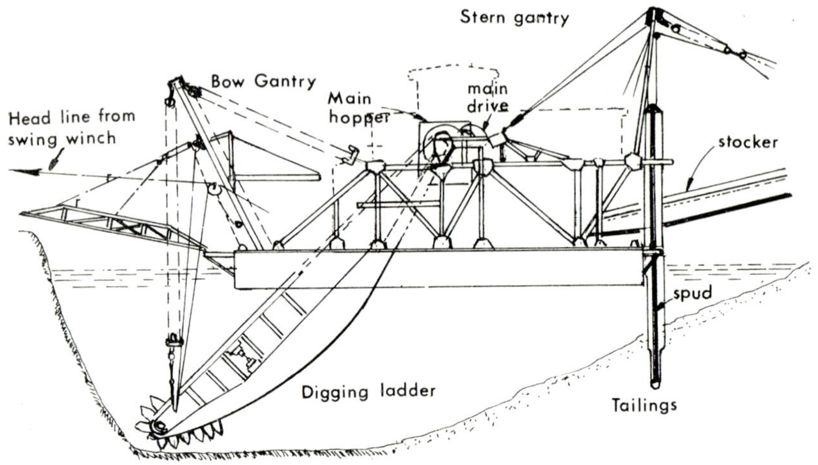

Fig. 75.   Details of a California-type bucket-ladder dredge.

Black Forest is a rift valley which has dropped to depths of 7000 to 9200 ft below the surrounding country by step-faulting and during this tectonic movement the mid-Oligocene sea invaded the valley and partly filled it with marine sediment. The deposits to be dredged are those of the Holocene and gravels of the Würm Glaciation, while gravels of the Riss Glaciation overlie Pliocene gravels and sands on both sides of the area are being worked. Molasse strata of Oligocene period are to be found below the alluvials and gravel beds at depths of 260 ft.

As the rift valley was filled from the south towards the north, very coarse material, up to 12 in diameter is to be found near Basle in the south becoming gradually finer towards the north until very fine sand below $\frac{1}{8}$ in diameter predominates between Mannheim and Mainz in the north.

All these circumstances drove owners of gravel pits to seek to exploit methods of working deposits to be found at greater depths and the necessary co-operation and technical knowledge was provided by two manufacturers of cranes and material-handling equipment and culminated in 1955 in the production of the floating deep grab dredge. Now more than 50% of the Rhine Valley sand and gravel production is extracted by such equipment and these dredges have also been sold to other countries including the United Kingdom.

The main sections of the modern floating grab dredge include the pontoon substructure, the grab, the supporting structure for the winch unit and the grab winch unit with gear and hoist motors. The size of the pontoons depends on the capacity of the grab and the load factor. The pontoon may have a well through which the grab may be lowered or raised, or they may be arranged in the form of a rectangle for this purpose and two parallel pontoons joined by the superstructure may be used for particular purposes.

The design of the grab is orthodox but is specially adapted with interlocking teeth for dredging sand and gravel.

Grabs have increased from 1·3 to 7·8 yd^3 in approximately $\frac{1}{2}$ yd^3 rises, the largest now being equivalent to a hoisting capacity of 20 tons.

At first the supporting superstructure was a luffing jib which was tilted forward to dump into a bin or hopper but as grab capacity and weight increased the luffing jib was replaced by an overhead gantry system on which the winch unit travels and in turn this has been superseded by a portal frame carried by two parallel pontoons at a distance apart allowing the passage of barges to be loaded by the grab, whose winch unit is located on the portal. In another model a slewing crane is placed on the pontoon to increase the area of operation of the grab under water like the portal.

A fundamental part of the design in its relation to capacity is the special winch unit system with its reduction gears and motors. Greater dredging depths to 135 ft necessitated higher hoisting speeds to reduce the unproductive portion of the dredging cycle and hoisting speeds of 250 ft per minute have been adopted and the problem of the washing out of fine particles is not serious since a closed bucket with a good standard of maintenance probably loses less material than a drag-scraper for example. Even higher hoisting speeds are stated to be possible. The hoisting system has two identical independent winches each with a grooved drum, a totally enclosed helical gear in roller bearings running in an oil bath, coupling, brake and slip-ring motor, each winch catering for half the hoisting capacity, each being separately removable for repair.

The grab is lowered or raised if both motors are turning in the same direction at the same speed; if the closing motor and its winch are working alone, or with a speed differential with reference to the other motor, the grab is opening or closing. These functions need accurate selection by the electrical control gear which also controls the closing of the grab under water, limit switches establish the depth of operation and also the upper limit of grab travel. All operations are automated to maximize productive capacity, relieve operator fatigue and guarantee depth of operation in accordance with legal requirement or geological conditions. Complete automation is forecast in the near future.

Dealing with capacities of 350 yd^3 per hour from floating grab dredges led to problems of transport. The first solution was by the use of floating conveyor systems connecting to fixed shore conveyors and, later, bottom-dumping barges were also used.

Sand and gravel dumped by the grab have a high water content and the pontoon is generally at a lower elevation than the main shore conveyor leading to processing plant to which the sand and gravel must be delivered. A de-watering screen is, therefore, provided between the hopper into which the grab dumps and the floating conveyor loading point. This reduces the moisture content, allows the full capacity of the conveyor to be utilized up to slopes of 15°, and facilitates the manual picking off of clay, wood and other garbage.

Where bottom-dumping barges are used, the sand and gravel is dumped near the shore and recovered by a slewing crane. Alternatively, a suction pump or a second dredge with a limited working depth may be used for recovery. A further development has been that of equipping the floating dredge with a complete washing, screening and crushing plant. The disadvantage is that stocking is not possible and the single grades of sand or gravel must be disposed of as produced or dumped back into the water.

Floating grab dredges are anchored and moored in the same way as bucket dredges, by cables on winches connected to the four corners of the dredging unit, the winches being electrically or manually operated, in the former case controlled by the operator of the grab winch unit on the portal or pontoon. The cables are fixed to ground stakes on shore or to anchors in the water.

The greatest wear occurs to ropes but wear also occurs to the teeth and bearings of the grab and the bearings of the winches. Essential spares are, therefore, one grab, one winch drum, one hoist motor and electrical contacts and other parts.

Capital costs of pontoon dredges including assembly are £12,200 for a 2 yd^3 grab, £15,200 for a 3·3 yd^3 grab, £25,900 for a 5·2 yd^3 grab, and £38,000 for a 7·8 yd^3 grab. For the largest grab cost per yd^3 varies from 4·6d. at 33 ft depth to 9½d. per yd^3 at 132 ft depth.

Ridinger floating grabbing plant

The plant is electrically driven, power being supplied from a Mercedes-Benz six-cylinder turbo-blown diesel engine driving a 235 kVA alternator giving 3-phase 400 V supply at 50 cycles with a transformer providing a 110 V supply for control circuits and for lighting.

The plant is mounted on five pontoons rigidly and accurately bolted together, each pontoon divided into watertight compartments with sealed inspection covers, the overall dimensions of the pontoon assembly being 61 ft 4 in (18·7 m) by 30 ft 10 in (9·4 m) with draught at 1 ft 4 in (40 cm). The diesel-electric power plant is carried on the rear pontoon which also carries an 800 gallon diesel-oil storage tank.

As access to road tankers is restricted, a 3000 gallon storage tank was installed at the end of a concrete access road some 700 yd from the limit of grab operation and from this fuel is fed by a $2\frac{1}{2}$ in diameter pipeline and a pump, the line having flexible hose connexions at locations where a change of angle can occur as the floating plant and shore conveyors are movable.

The grab of 3·4 yd^3 (2·6 m^3) capacity is of the four-rope type. The grab winches and motors are carried on a platform which moves with the jib to ensure constant rope tension. Luffing of the jib is by an hydraulic ram and is confined to the fore and aft centre line of the plant. The grab delivers to a 7·8 yd^3 (6 m^3) surge bin with a grid to separate oversize material and rubbish picked up by the grab, which is discharged through a motor-operated sliding gate, controlled from the operator's elevated cabin with a wide view of the equipment and the controls grouped within easy reach and in telephonic communication with the processing plant on land. From the surge bin, the undersize material is delivered to a double-decked de-watering screen, the upper deck fitted with hook-sided meshes, the lower comprised of 15 wedge-wire panels with 1 mm wire spacing, the moisture content being reduced to 25%. Three floating conveyors each 20 in wide and 65 ft long take the discharge from the screen, each being separately powered from the grabbing plant pontoon and being sequence operated with an emergency switch at the transfer point. They are connected together by slewing rings and hinges which allows vertical and horizontal relative movement, the latter to some 270°. The third conveyor delivers to a land-based conveyor system feeding the processing plant.

Special safety measures are adopted, including handrails and toe boards, particularly over outer hoops round access ladders, walkways along both sides of the floating conveyors and sliding handrails at transfer points, emergency stop buttons and a foot-operated emergency cut-out in the control cabin together with fire extinguishers and roped life belts. Checks for any leak of water into the pontoons are made by lifting the inspection covers and access to the floating conveyor walkways is from a 120 ft aluminium boat with a 10 ft by 4 ft drop ramp for loading, which is used to transport heavy repair equipment. Mooring ropes from the winches on the pontoon, consisting of $\frac{1}{2}$ or $\frac{5}{8}$ in old excavator ropes, run to mooring points.

The plant was manufactured by Ridinger in Germany to work a deposit which had already been worked by face-shovel to about one foot above the winter water level and the floating grabbing plant was installed to excavate the remaining deposit consisting of zones 3 and 4 sand with a comparatively small proportion of −3 in gravel down a varying depth to the underlying sandstone and clay.

Grabbing at a depth of 30 ft the hourly output is 180 tons in a steady flow, without surges which could inconvenience the processing plant on

Fig. 76. *The hold of the 1200-ton Pen Avon dredger being filled with gravel from the Solent. The dredger cost £250,000 and is equipped with a 155-ft pipe for dredging the sea bottom and is lowered by remote control.*

land and in shallow depths a pause of 10 seconds or so above the water level for the grab to drain enables the water content to be below 25% to reduce the load on the de-watering screen and to prevent spillage from the conveyors.

Sand and gravel is also won in the United Kingdom by sea-going dredges generally provided with centrifugal pumps and suction pipes capable of loading cargoes of 500 tons in two hours which is off-loaded at the wharf by grabs. In the English Channel the gravel is rounded and angular and consists entirely of flint and the grain size of the material on the sea bottom is said to be related to the current in the water above it. Sand and gravel is won all round the coasts of Britain, for example about 1,000,000 tons annually from the Bristol Channel, 1,750,000 tons off the Isle of Wight and 1,250,000 tons from the southern North Sea. Dredging companies in the Thames Estuary are prepared to dredge in 120 ft of water and make a round trip of up to 150 miles, a modern suction dredge costs some £300,000 and carries 2000–3000 tons of material. The tonnage of gravel won from off-shore deposits around Britain amounts to about 10 million tons per annum with a cash value ex-wharf of some £10 million.

Figure 76 shows the hold of a 1200-ton dredger, the Pen Avon, being filled with gravel and water during operations in the Solent. The vessel cost £250,000 and is equipped with a 155 ft pipe which is lowered on to the sea-bed by remote control.

ARCHAEOLOGICAL EVIDENCE OF GRAVEL BEDS

British archaeologists are introducing systemized planning to keep ahead of mechanization and save sites throughout Britain from destruction due to mass building programmes which need in increasing quantities sand and gravel generally laid down by ancient rivers which were often also the favourite haunts of early man. The gravel of former lake shores and river banks also offered one of the best media for preserving evidence of man's occupation and this is the evidence now threatened with destruction by mechanized diggers such as draglines and shovels.

The organization and planning of digs, as well as the correlation of potential sites with commercial quarrying plans, are being handled by the Council for British Archaeology. Many sites are first detected by air reconnaissance of crop marks and the Council recently completed a pilot scheme in the Welland Valley area between Bourne and Peterborough. As a result valuable data on life over 5000 years from Neolithic to Saxon times was obtained including a mile long avenue of sacred Neolithic sites and wood and bone implements.

A small 'family cemetery' of the early Bronze Age contained the skeletons of two adults and three children with a finely decorated beaker, a flint knife and bronze ear-rings.

Excavations are now planned in the Avon–Severn Valley, in the Nene Valley in the East Midlands, in the Upper Thames Valley and the Trent watershed area.

REFERENCES

'Gravel Operations in the Trent Valley, Great Britain', W. Whiteside, T. D. Paterson and I. H. Lean, *Opencast Mining, Quarrying and Alluvial Mining*, Institution of Mining and Metallurgy, 1965, p. 57.

'Elements of Gravel Pit Design', D. A. Webb, *Quarry Managers' Journal*.

'Gravel Excavating by Pump', *Mine and Quarry Engineering*, December, 1953, p. 448.

'Use of Modern Floating Grab Dredges for Sand and Gravel Extraction in the Upper Rhine Valley', F. Konz, *Opencast Mining, Quarrying and Alluvial Mining*, Institution of Mining and Metallurgy, 1965, p. 20.

'Underwater Winning of Sand and Gravel', D. Lester, *Quarry Managers' Journal*, November, 1966, p. 477.

'Sand and Gravel Dredger', A. A. Raymond, *Mining and Minerals Engineering*, March, 1967, p. 91.

'Aggregates in Concrete', D. R. Sharp and B. W. Shucklock, *Quarry Managers' Journal*, May 1966, p. 181. *See also* Appendix III.

Saga Pit and Quarry Textbook. Published for Sand and Gravel Association of Great Britain by Macdonald and Co. (Publishers) Ltd, London, 1967.

CHAPTER 12

ALLUVIAL MINING

MARINE DEPOSITS

Examples of alluvial mining methods have already been cited for the surface mining of minerals such as placer gold, diamonds and sand and gravel. It has also been advanced that the discovery of new mineral reserves is not keeping pace with the increased demand for minerals necessitated by an increasing world population with a rising standard of living. It should be borne in mind, however, that four-fifths of the earth's surface is under water and that virtually all our present minerals are won from the remaining one-fifth, that is 'underground'. It is a safe inference, therefore, that much mineral wealth exists beneath the surface of the sea, both in the sea water itself and in deposits that may or may not outcrop on the sea-bed.

Manganese and phosphate nodular deposits have, perhaps, attracted the most interest to date since large tonnages are reported on the floors of the Pacific, Atlantic and Indian oceans. Although the presence of these deposits has been known for nearly a century, their wide distribution was not, perhaps, fully appreciated until an International Geophysical Year expedition began a study of the floor of the south-east Pacific Ocean. At various points in the South Pacific between Tuamota Archipelago and the coast of South America deep-sea camera observations disclosed an extensive area in which the sea floor is covered with a loose layer of nodules having rich concentrations of manganese, iron, aluminium, magnesium, copper, cobalt and nickel, the percentages being respectively 20, 15, 5·5, 2·3 and 0·5 each for the last five.

The deposits range from 120 to 15,000 tons per square mile and extend over millions of square miles of the Pacific. They are also believed to exist over large areas of the Atlantic.

To work these deposits the first technical difficulty would be to bring the material from the average of 12,000 ft depth though photographs have been obtained of it at 18,000 ft in the Atlantic east of Bermuda. Drag-bucket dredging might be feasible but slow and costly, and hydraulic dredging delivering 5000 tons of nodules would be more economical with a capital cost of £20 million although other authorities would increase this to £25 million for mining and processing.

Because of the intimate association of the various elements in the nodules a further problem would be their separation and it is considered that flotation as a means of benefication would not be successful. Preliminary hydrometallurgical tests indicate that chemical methods could recover the manganese.

The phosphorus nodules, which contain as much as 30% tricalcium phosphate, are formed in a similar manner but at lesser depths. These have been located in the Pacific, Atlantic and Indian Oceans at depths from 400 to 8000 ft.

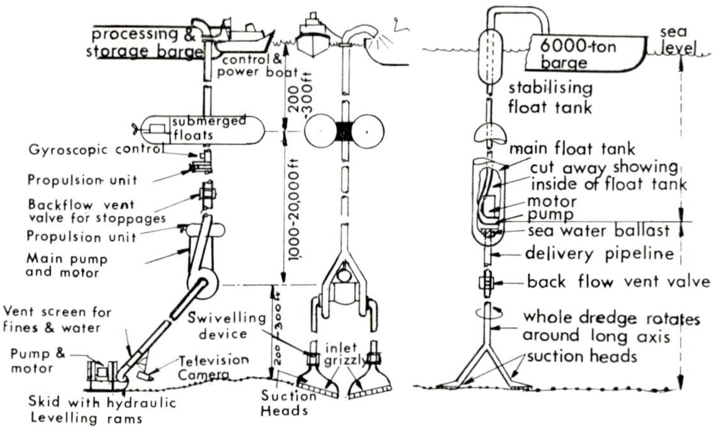

Fig. 77.   Two proposed layouts for dredging deep-sea manganese modules (Mero).

Recently, however, material dredged from comparatively shallow depths has aroused interest in producing and processing these manganese nodules. The concretions occur in oceanic sediments as microscopic grains of iron oxides and manganese dioxide as spherical nodules 1·6 in diameter or as a surface coating of rocks and stones. Enrichments of up to 40% manganese, 1% cobalt, 1·4% nickel and 1·8% copper have been known but not in the same nodule. Two processes have been designed to win these deposits: the first is deep-sea drag-dredging and the second hydraulic dredging, two systems of which are shown in Fig. 77. These methods are considered capable of producing at some £1 2s. 6d. per ton.

The ownership of such minerals raises an interesting question of law. The present law is vague outside the 3 or 12 mile limits or the territorial water limits with reference to ocean beds and what lies on or under them.

The rule of first come first served might apply, but this is probably politically naïve.

The establishment of an international ocean law is really required, probably *via* The Hague, to determine national boundaries in respect of the ocean floor and to avoid legal complications which may now be foreseen before political complications are engendered.

The rules that apply to the outer continental shelf of the United States, that is beyond the states' jurisdiction, with offshore petroleum and sulphur in mind, are contained in Public Law 212 Outer Continental Shelf Lands Act passed in 1953. If exploration is successful, the Bureau of Land Management puts the area up for bid. Bids are in terms of cash, bonuses, lease and for royalty, and the leases are for specific minerals only.

According to the Geneva Convention of 1958, international law gives a country sovereignty over its continental shelf in terms of benthic sea life and minerals, the shelf being defined as limited to the sea bed and subsoil of the submarine areas adjacent to the coast but outside the area of the territorial sea to a depth of 200 metres and similar areas adjacent to the coasts of islands.

On the continental shelves, economic minerals similar to those occurring on adjacent lands also occur at depths down to 100 fathoms where the shelf terminates in a steeply dipping continental slope leading to the deep sea. It is entirely fortuitous that the shelf is at present submerged and the location of the edge of the sea is only temporary, speaking in a geological sense.

Echo-sounding, using the gas-exploder or the sparker and a low frequency of 150 to 300 cycles per second is used for initial bottom surveys and records the sub-bottom down to depths of 1500 ft. Distinction between sediment and bedrock and even sediment type can be determined from the continuous record of the instrument. The area to be worked is gridded with traverse lines at regular intervals to give a systematic areal coverage.

Physical sampling follows and this may take the form of dredging, drop coring or drilling. Divers have been used to collect sea-bed samples as have submarines to a depth, it is reported, of 15,000 ft. The drill used is a modified version of the churn drill and the Bank a manually operated drill has also been used. These drills are used in depths down to 50 ft and great care is required to minimize sampling errors, a grade factor being used where the same type of drill is used in the same type of ground. Recently air lifts, suction pumps or venturis for sludge removal in conjunction with large-diameter casing up to 24 in and a rotary drill stem have been used. If the water is shallow, the vessel is positioned by four or six anchors or stakes driven into the bottom while in deeper water dynamic positioning is used in which a series of outboard engines is mounted on the corners of the vessel and position is maintained by automatic centring in a circle of sonar reflectors positioned around the drilling target either at the bottom or suspended on taut wire buoys. In the Mohole deep-water drilling prospect the ship was maintained within 100 ft of the vertical projection of the drill-hole collar over 2 miles below.

Accurate positioning of the dredge both for sampling or for mining is necessary and by triangulation from the shore, this can be correct to ± 6 ft. Radio contact may also be maintained by shore tracking stations. Out of sight of land, electronic position locators such as the Loran, Shoran and the Electronic Position Indicator systems are used, the latter accurate to ± 200 ft in 275 miles. Sun and star sights give accuracies of $\pm\frac{1}{2}$ mile.

The Decca Navigator Co. has produced a position-fixing system known as the Hi-Fix which is particularly suitable for the Dinosaur automatic grab dredge as the aerial can be fixed immediately over the grab. The path of the grab is plotted with spot soundings by a stylus on a chart. An accuracy of ± 4 ft is claimed at a distance of 100 miles. The Hi-Fix co-ordinates have been linked *via* a small computer to a pair of Voith–Schneider propellers by which the dredge can be held exactly in position or moved in any direction irrespective of wind or currents with an accuracy of ± 2 ft at a range of 30 to 40 miles thus dispensing entirely with moorings or anchors of any sort. A similar system has been used successfully to maintain off-shore drilling oil platforms in position.

Systems of dredging used for mining alluvial deposits include bucket-ladder dredges for depths to 150 ft, draglines for depths to 50 ft, grab bucket or clamshell dredges for depths to 200 ft, a hydraulic dredge with cutter head for depths to 120 ft, a suction or hopper dredge for depths to 150 ft, hydraulic deep sea dredges, proposed for depths to 15,000 ft, and airlifts for depths to 100 ft.

The first, the bucket ladder dredge, consists of an endless chain of steel buckets which dig continuously into the deposit below the level of the floating dredge. Bucket capacities of such mining dredges vary from 2 to 20 ft^3, resulting in monthly capacities of 60,000 to 500,000 yd^3.

Electric or diesel-electric power or steam may be used and Ward–Leonard controls, eddy current couplings and silicon rectifiers are used where electric power is available.

Grabs have outputs of 800 to 1000 tons per hour. A modern dredge of this type is the Dinosaur (Fig. 78), designed for the mining of alluvial tin in south-east Asia. The dredge has twin 6 yd^3 buckets giving a monthly output of 308,000 yd^3 in 135 ft of water at a power cost of 1·4d. per yd^3. The crane structure is mounted on bogies running on a circular track for accurate pattern dredging and after each operation the crane is rotated through 12°, thus positioning the grab automatically for the next dip.

The concentration plant is carried on the dredge and is specifically designed to give a high recovery of fine tin. The power cost of the recovery plant is $3\frac{1}{2}$d. per yd^3.

Hydraulic dredges are used both for the removal of overburden and for the mining of unconsolidated sand and gravel deposits. Cutter heads of the rotating hollow-bit type sited directly in front of the suction pipe are employed to break up the ground and direct the flow of solids to the pipe, the pipe and the rotary shaft being attached to a rigid ladder which is raised and lowered in front of the dredger. A rotary suction pipe with the cutter head attached to it is also used.

The dredger is usually anchored at the stem by hawsers or by spuds driven into the sea bottom and the dredger is swung in an arc with centre

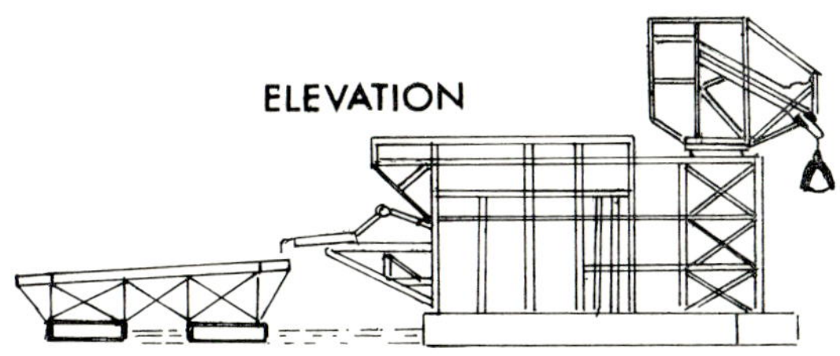

Fig. 78.   *The Dinosaur grab bucket dredge.*

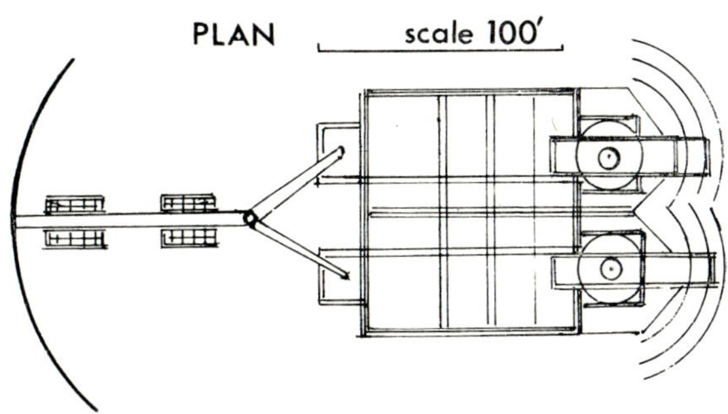

at the point of anchorage, to give a maximum sweep to the cutting head. Water jets are used to agitate the solids when a cutter is not used and jet venturis are used in conjunction with skin divers in California and Colorado.

Shell sand 10 miles offshore in Faxa Bay in Iceland is used for the manufacture of Portland cement. A hydraulic suction dredge without a cutter head with a 24 in diameter suction pipe 160 ft long is used with the maximum limit of dredging at 140 ft with hydraulic controls to maintain the suction on the bottom with a 6 ft rise or fall of the dredger. The hydraulic pump deals with 8000 tons of water per hour carrying 3 to 5% of solids. The dredger is a sea-going vessel of 1100 tons gross weight and carries 1000 yd^3 of material which it transports and pumps, using the dredge pump, to a stockpile $\frac{1}{2}$ mile from the quay.

ALLUVIAL TIN

Some 70% of tin in the form of cassiterite is won by alluvial mining.

Cassiterite (tin oxide SnO_2) occurs sparsely in quartz veins traversing the margins of a special type of granite which also carries tourmaline and sometimes topaz. It is mined together with wolfram within the granite and immediately outside and still further out it is found with arsenical pyrite and copper pyrite in the tourmaline-bearing 'peach'.

However, most cassiterite is recovered as grains, often very small, and pellets from placer workings in weathered soils or river gravels where, as a resistant heavy mineral, it has been concentrated from the wash of veins and lodes in the country rock of the neighbouring hills. The main producers of alluvial cassiterite are Malaya, Burma, Thailand and Indonesia which together account for over 60% of world production of tin.

Malaya is by far the most important producer of tin and within the Malayan peninsula the Kinta Valley is the most productive area.

The high rainfall, 120 in per annum, and the topography of this region provide the supply of surface water required for hydraulic mining. The Kinta Valley runs north and south, is about 25 miles long and 14 miles wide at the southern end. Both eluvial and alluvial deposits are found, the latter predominating. The deposits lie on eroded limestone with ridges which act as ripples to concentrate the cassiterite and other heavy minerals brought down by rivers and streams.

The deposit varies considerably in places but is usually a relatively soft, fine sandy clay with varying amounts of gravel and pebbles ranging from white to brown in colour where iron is present. The cassiterite is dispersed in the clay and is from $+\frac{1}{4}$ in to -300 mesh BSS with the majority in the size range 60 to 150 mesh. The viable portions of the deposit in which the cassiterite is dispersed occur erratically and the economic cut-off or 'break-even' grade depends on the prevailing metal price, the depth of the deposit, the nature of the ground and the valuable minerals present. This grade varies, but has been shallow sandy ground with $\frac{1}{4}$ lb of cassiterite per yd^3, while 0·4 lb per yd^3 may be regarded as the average limit for profitability and if restoration is mandatory this may rise to 0·6 lb per yd^3 as the marginal limit.

Pipelines for water supply at pressure to hydro-electric power stations and particularly for hydraulicking are 18 in to 45 in in diameter though supplies fluctuate through the smallness of catchment and storage areas.

Pressure is usually about 150 lb/in^2, distribution pipelines are 8 to 36 in diameter, while an 8 in diameter pipeline generally serves each hydraulic monitor with a $1\frac{1}{2}$ to $2\frac{1}{2}$ in nozzle, depending on the type of ground being worked.

When sufficient natural head is unavailable engine- or motor-driven pumps may be used to impart hydraulic head to pond or stream water with the pump installed as near the workings as possible to reduce pipe friction, the pressure being from 75 to 85 lb/in^2. Locally made 10 in by 12 in pumps driven by 200 hp engines or motors are required, each to service three 2 in diameter monitors at 80 lb/in^2.

The advantages which accrue from hydraulic gravel-pump surface mining include the sufficiency of land area for tailings disposal; the same equipment can be used to mine to varying depths such as deep pockets; complete extraction of values can be realized as the bedrock can usually be cleaned up; selective mining is possible for balancing outputs and leaving uneconomic ground *in situ*; working faces are continuously visible and capital cost of equipment required is less than for most alternative methods of mining. Paddocks are required to segregate tailings away from mining faces. The slope of pit sides is generally about 45° for safety. Monitors are advanced with the faces and ditchlines are constructed in the floor of the paddock about 4 ft wide at a gradient of 1 in 40 to conduct at high speed to prevent redeposition of the slurry produced by mining, to the pump sump otherwise a build up of material would occur in the paddock. The gravel pumps used are of the vertical type (Fig. 79) in which the motor frame of the prime mover and the pump casings are rigidly joined by a fabricated cage and the rotor and pump impeller are direct coupled, the unit so formed being suspended during operation with its axis of rotation in the vertical plane. The prime movers are usually electric motors, the pumps single stage centrifugal with some diesel-engined units. They have the advantages of no power loss through being direct driven; requiring no foundations, being suspended from a tripod, and when suspended by a chain block can be raised or lowered as required to follow variations of water level or flooding; are mobile and flexible and can be easily and quickly installed and they operate on very low suction head. The size of pumps used are 7 in or 8 in diameter inlet driven by 100 and 150 hp motors which usually supply three 2 in monitors when 8 in and two monitors by each 7 in pump for heads of 60 to 80 ft.

Each paddock would carry a 30 in water supply line which serves eight with 2 in nozzles and four with $1\frac{3}{4}$ in nozzles requiring four 8 in pumps to elevate the slurry carrying 8 to 10% solids to the treatment plant.

The disposal of tailings from the concentration plant is severely restricted by mining laws because of the sluggish flow of most of the rivers so that large areas of land, extensive dams and systems of ditchlines are required to conform and produce a virtually clear effluent.

Tailings are being used increasingly to reclaim the mined land, particularly adjacent to towns where the leases often stipulate that the land must be left filled, levelled and drained on the completion of mining operations. Since this must be carried out at the lowest possible cost the pulp density of slurry pumped to the area being filled is increased to 15% with this end in view.

Average tin-bearing ground in Malaya yields 0·3 to 0·6 lb per yd^3 or about 0·01% to 0·03%.

In palong mining (Fig. 80) the ground is again reduced to a free-running state as a slurry by agitation from high-pressure water jets ejected from monitor nozzles supplied by a centrifugal pump. The slurry bearing the tin ore flows by gravity from the working face along natural channels in

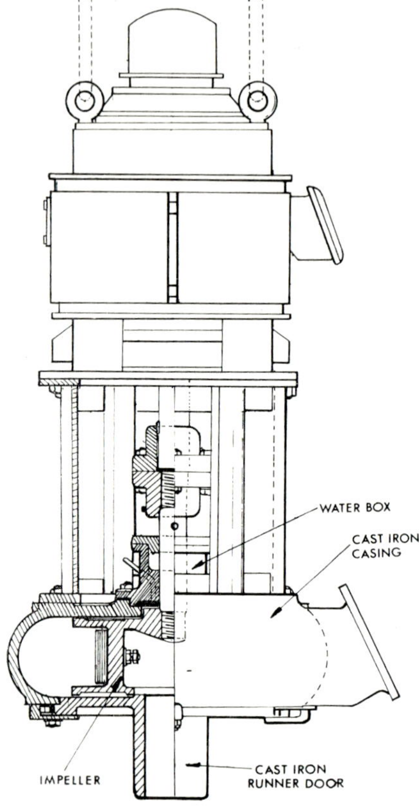

Fig. 79.
Eight-inch vertical pump (Tyrrell).

the bottom of the mine to the lowest point—the pump. The slurry is then elevated by a gravel pump to the top of the palong which is an inclined flume usually erected on trestles but sometimes on suitable sloping ground. A few mines use jigs instead of palongs but despite its crude appearance, the palong has the advantage of acting as a conveyor for the tailings in addition to its primary purpose as a separator.

The gradient and flow of the palong are so arranged that the heavy tin concentrates are trapped by a series of lateral baffles of varying depths.

The water, containing light clay and sand in suspension, passes over the baffles and is discharged from the lower end of the palong to the tailings area. When sufficient low-grade concentrates have been trapped, the delivery from the gravel pump is stopped and the concentrates are washed by a stream of clean water and the concentrates are then removed to a separate plant for final and efficient cleaning. It is usual for the palong to be built with two flumes so that a continuous sequence of gravel pumping and washing can be maintained.

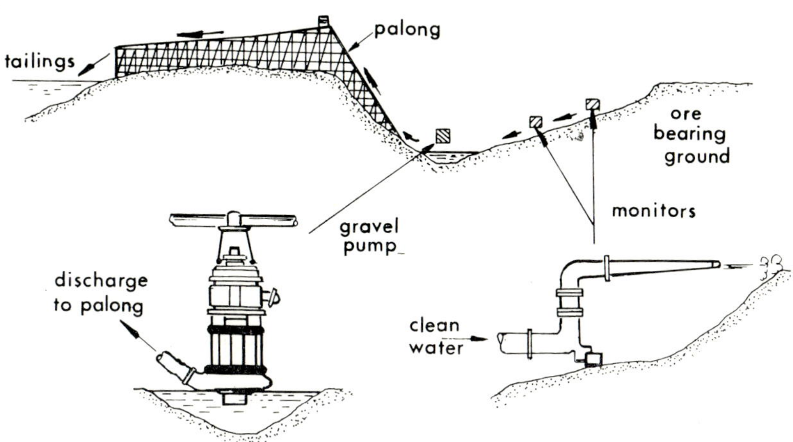

Fig. 80. *Gravel-pump tin-mining method indicating the three main operations—monitoring, gravel pumping and separation (Tyrrell).*

OFFSHORE SULPHUR MINING

Sulphur is one of the most used elements in modern industry and few manufacturing processes do not use it in some form or other and this follows from the fact that sulphuric acid is the most important single commodity in the chemical industry. The acid is obtained from three sources—from pyrite, FeS_2, from sulphur recovered from smelter and other industrial gases and from native or natural sulphur. In Europe, most of the sulphuric acid is made from pyrite but 60% of USA production is derived from native sulphur.

Native sulphur is formed at and near the craters of volcanoes especially where hydrogen sulphide and sulphur dioxide are being emitted since when these combine sulphur and water are deposited.

The principal sulphur deposits of the world are in the USA round the Gulf of Mexico and in Sicily where they occur in sedimentary rocks in close association with limestone, gypsum and anhydrite and gaseous and solid hydrocarbons such as bitumen. Sulphur is the product of local reaction between escaping oil or bitumen and the gypsum or anhydrite deposits. The most productive deposits are those in the cap rocks overlying great plug-like intrusions of rock salt, known as salt domes, which are common in Texas and Louisiana.

The Sicilian sulphur deposits cover several hundred square miles. They are interstratified with limestone, gypsum, bituminous rocks, clay and sandstones of the Tertiary age and average about 25% sulphur.

Above a single salt dome in Texas is as much as 45 million tons of sulphur and in these salt domes the sulphur in these deposits is mined by the so-called French or Frasch process involving the introduction into the sulphur beds of superheated water at a temperature of 175°C, which is higher than the melting point of sulphur, 115°C, the molten sulphur being conducted into bins where it solidifies. The superheated water is under pressure and compressed air at 400 lb/in^2 is forced down an inner pipe to push the sulphur to the surface.

In water 50 ft deep is located the Grand Isle sulphur mine seven miles off the coast of Grand Isle, Jefferson Parish, Louisiana, in the Gulf of Mexico, where in 1956 Freeport Sulphur Co. acquired the sulphur rights under the area. Construction of an offshore sulphur mine (Fig. 81) was commenced in 1958 and production of sulphur began in 1960, the estimated expenditure on the project being £10 million.

The sulphur deposit in over 200 ft of limestone, was discovered in 1949 when the Humble Oil and Refining Co. were drilling at depths between 1813 and 2073 ft. The deposit is the third largest sulphur reserve in the USA covering several hundred acres with its north-east, east and south boundaries not yet defined, ten prospect holes encountering sulphur in eight holes.

Sulphur deposits on the Gulf Coast occur in a cap-rock mantle of calcite, anhydrite and sometimes gypsum covering the apex of an intrusive mass of salt. The sulphur-bearing limestone has a thickness of 220 to 425 ft at a depth of 1800 to 2500 ft within the present boundaries of the site and the sulphur content varies from 15% to 30% and is not particularly high but the thickness of the ore-body is much greater than that in other Gulf Coast deposits. The sulphur occurs in typical blue-grey to black moderately hard fine-grained fractured limestone, varying from dense to cavernous, porous and highly permeable which has been recemented by white secondary calcite and sulphur. Traces of gypsum and lenses of marl and shale are also met in the gangue.

The sulphur occurs in crystalline masses, in vugs, in disconnected veinlets and is disseminated throughout the rock. The sulphur is discoloured since traces of oil also occur. Cavities up to 18 ft in vertical extent were discovered and were generally confined to the sulphur horizon. The cap rock, which is barren of sulphur, varies in thickness up to 250 ft and averages 100 ft. It is permeable, porous to moderately dense medium grey limestone with white secondary calcite cementing small cracks and fissures.

Rock salt occurs below the anhydrite to an unknown depth and the anhydrite, which underlies the sulphur horizon and varies from 29 to 190 ft in thickness, has the typical saccharoidal texture found in Gulf Coast salt domes, dark grey in colour and dense, except for occasional fractures which sometimes contain oil and sulphur.

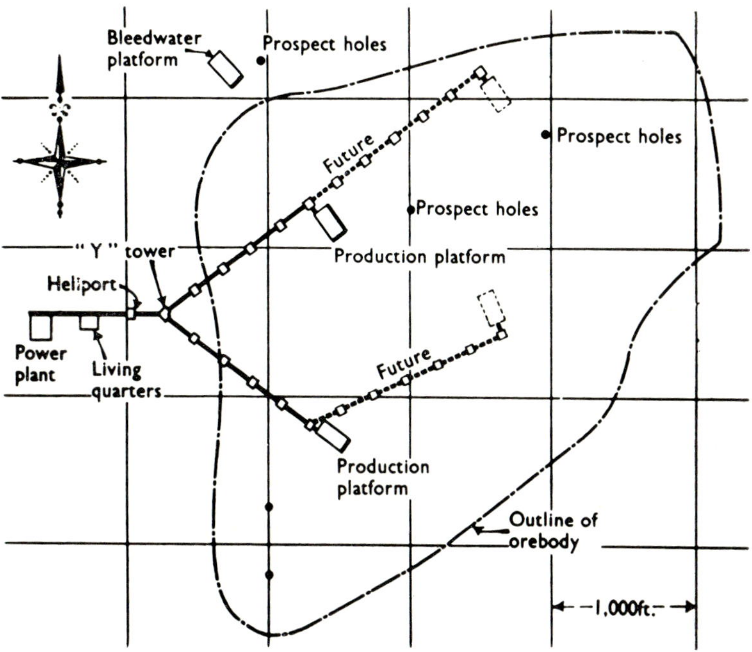

Fig. 81. *Layout of Grand Isle sulphur mine facilities in relation to the deposit.*

In the Frasch sulphur extraction process, the wells which are drilled into the sulphur horizon are equipped with concentric pipes (Fig. 82) and water under pressure heated to some 320°F is injected, the hot water percolating through the formations and gives up heat gradually to them. When the enclosing rock and the sulphur reaches a temperature exceeding 240°F the sulphur melts, separates from the rock and flows downwards by gravity to the bottom of the well where the sulphur separates from the water, its specific gravity is double that of water. The accumulation of melted sulphur rises in one of the concentric pipes and is brought to the surface by means of an air lift. An impervious formation covers the saturated limestone cap rock, thus creating the equivalent of a pressure vessel and to avoid a build-up of pressure a volume of water equal to the volume injected must be withdrawn. The hot injection water is lighter than the formation water and collects along the higher contours of the dome. Waste or bleed water is removed through wells which are drilled along the deeper flanks of the dome where the heavier cold water lies.

The Grand Isle venture is unique because the mine location is in 50 ft depth of water and extremes of weather in the Gulf imposed problems of design and foundation construction. Formerly, the Frasch process had been

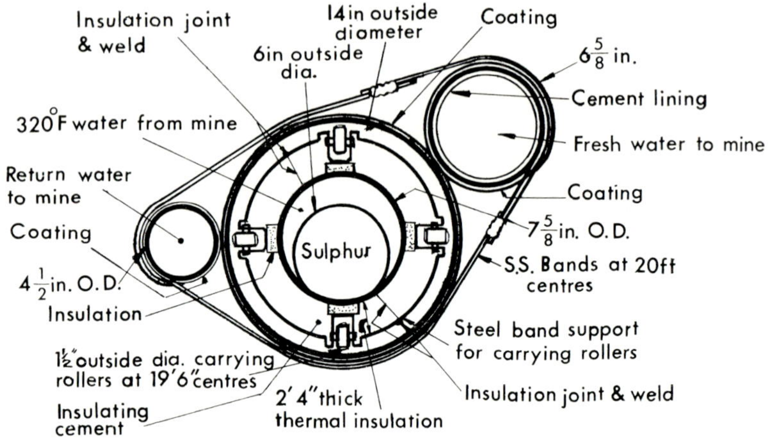

Fig. 82.   *Arrangement of the pipelines for sulphur extraction.*

dependent on fresh water in large quantities but since the nearest unlimited source of fresh water was the Mississippi River, 25 miles away, means had to be found to reduce these requirements.

A further problem was the necessity for a safe all-weather system of getting the sulphur in-shore for storage.

Pile-supported platforms are used to carry the major units of the installation some 60 ft above water level and are connected by 200 ft long bridges between units (*see* Fig. 81) the larger platforms support the steam boiler plant, warehouse and shop, bleeding wells and disposal system, office and living quarters and two drilling and production stations. The bridge spans of 200 ft are supported on small platforms or towers 50 ft by 50 ft on four piles and carry pipelines and roadways connecting the larger platforms; they are 15 ft wide by 15 ft deep, the pipelines are carried on the bottom chord of the bridges and the 12 ft wide roadway on the top is designed for a 7-ton moving load enabling rubber-tyred vehicles to be used for personnel and materials handling.

When the sulphur is removed at most Frasch sulphur mines subsidence occurs through weakening of the rock since the overlying sediment offers little resistance. The size and shape of the subsidence basin depends upon the volume and rate of removal of the sulphur, the character and strength of the overburden and the depth, thickness and extent of the sulphur deposits and the time which has elapsed since sulphur mining commenced.

Because of the subsidence the production platforms are not permanent and they will be salvaged and re-erected at their original elevations when some 10 to 15 ft of subsidence has occurred and provision has been made in the tower legs to enable the bridges to be jacked up and levelled when necessary. The power plant and the living quarters are near the edge of the anticipated subsidence basin where subsidence should be less.

Steaming of a number of wells simultaneously is required in the Frasch process and it is not possible to predict the length of time a well will operate satisfactorily and how much it will produce so that further wells must be in course of drilling and equipping in advance of their actual need to ensure that the capacity of the heating plant is utilized continuously. The wells will be drilled directionally from a multi-well platform so that the capacity of the heating plant is utilized continuously, so that the well-heads are concentrated in a small area, and the well bottoms are widely disseminated over a considerable area. Twelve wells can be operated simultaneously from each platform. The true vertical depth of the wells is about 2200 ft but a horizontal deviation of as much as 1250 ft is used and expected to be economical, the maximum angle of drift being 35° from the vertical and the maximum build-up of drift angle is 5° per 100 ft.

Each platform has openings to drill from 36 surface locations on 11 ft centres but only 12 will be in use at any one time.

When a well has exhausted its reserves of sulphur, the pipe will be salvaged and the bottom portion of the well plugged and abandoned, but the conductor pipe and surface casing will be used again and another hole will be side-tracked in a different direction. The drilling of 108 wells will be carried out from each platform since each of the 36 surface locations will be used three times, the bottom hole spacing being 175 ft. Since the area drained by any one well is limited it is anticipated that four platforms will ultimately be required to give complete coverage of the dome and each platform will be rebuilt and relocated one or more times during the life of the project. About three-quarters of the top deck, measuring 224 ft by 116 ft, is occupied by drilling equipment and auxiliaries, the remainder comprises crew changing-room, tool house, pipe threader and pipe storage racks and the centre bay of the lower deck contains equipment for operating the producing sulphur well-heads and piping. The pumps, shale shaker de-sander and mud pits for the drilling mud plant are arranged along one side of the platform and the relay or well control station is on the lower deck beneath the pipe reworking shop.

The drilling equipment is of the light-duty oilfield type, a conventional 120 ft high by 24 ft base derrick being mounted on a double substructure, one sub-base being movable longitudinally along the platform while another can be moved at right angles to the length of the platform so that the derrick and drilling machinery can be easily and quickly moved to any location. Permanently mounted on the derrick floor is a 300 hp model F–30–D Ideco double-drain draw-works capable of drilling to 5000 ft. The pipe racks are located on either side of the derrick to avoid turning the derrick round when drilling near the end of the platform. Under the derrick floor is the 16 in blow-out preventer. The bottom deck is 60 ft above sea level, the top deck 75 ft, the rotary table 94 ft and the crown block of the derrick 223 ft. Drilling mud returns through troughs permanently installed in the main derrick substructure and beneath the platform floor. Two mud pumps are provided, a 300 hp 8 in by 14 in, and a 150 hp 6 in by 12 in standby unit. De-sanding equipment is provided and the mud storage capacity totals 750 gallons in two tanks. Centrifugal pumps are provided for mixing, transfer, supercharging and de-sanding.

REFERENCES

'Mining Offshore Alluvials', M. J. Cruikshank, *Opencast Mining, Quarrying and Alluvial Mining,* Institution of Mining and Metallurgy, 1965, p. 125.
'Valuation of Alluvial Tin Deposits in Malaya', J. K. Broadhurst and D. J. Batzer, *Opencast Mining, Quarrying and Alluvial Mining*, Institution of Mining and Metallurgy, 1965, p. 97.
'Gravel-pump Mining in Malaya', P. J. Tyrrell, *Mine and Quarry Engineering*, August 1953, p. 267.
'An Offshore Sulphur Mine', *Mine and Quarry Engineering*, April, 1960, p. 53.

CHAPTER 13

POWER SUPPLY IN THE SURFACE MINING INDUSTRIES

Surface mining machinery has increased in size and power to enable deeper deposits to be exploited economically at high rates of production. Most of these machines are electrically operated and this requires carefully planned electrical distribution systems and supporting services such as an increasing degree of automation to reduce the expensive requirement of manpower, planned maintenance of equipment to reduce the chance of breakdown of machines which with larger machines and an increasing degree of mechanization and concentration can have a catastrophic effect on output and communication.

DISTRIBUTION SYSTEMS

As in deep mining, the electrical distribution system at surface mines must be designed to advance progressively as the high wall in the working face advances and the conflicting requirements of reliability, transportability, safety, capacity and flexibility must be balanced realistically. At the maximum end of power requirements, are individual machine loads of up to 20,000 kVA and distribution voltage of 35 kV, so that large surface mines may have electrical loads equivalent to those of towns of medium size.

The electricity supply boards, both in Great Britain and other countries, are anxious to supply the loads of surface mining industries and tariffs are usually of the maximum-demand type, the normal for heavy industry users such as deep mines. However in view of the fact that operations may be adversely affected by inclement weather, the maximum tariff is not necessarily a fair one as very high maximum-demand charges can be incurred during a very severe winter when output would be low. Special ceiling prices or alternative tariffs for surface mining consumers are common so that ceiling costs for power are known in even the worst conditions and can be budgeted for accordingly.

A separate high-voltage distribution line feeds most quarries and surface mines since these are generally located far from distribution mains. The selection of the primary distribution voltage is important and the economics of each case require consideration of such factors as voltage regulation, problems of insulation and necessary flexibility. Primary distribution voltages of 33 kV are common and 60 kV has been adopted for large loads. To quarries with medium or smaller loads supplies may take the form of a duplicate 11 kV supply and less frequently a 33 kV supply has been provided.

The power supply to a large surface mine must be secure and certain for the following reasons:

(1) Loss of production may be very expensive because of the magnitude of the project and the high capital value of the installed equipment which must have a high load factor to be viable.
(2) Operations may be continuous round the clock so that lost output cannot be recovered.
(3) The very security of the project may depend upon vital services such as power when say continuous pumping is necessary.

To secure this continuity of supply as economically as possible it is usual to provide:

(1) Duplicate feeders to the project preferably by separate routes.
(2) In the project a ring-main system with sectionalizing ring-main circuit-breakers where applicable.
(3) For vital services local standby supplies.

If the supply is by overhead lines, hazards such as lightning and hurricanes may need consideration, the stringing of shielding overhead earth wires, surge diverters and rod gaps and the routing of the feeders to avoid exposing, should be carried out in conjunction with the power company or authority.

If the project is remote or if the capacity of the local utility station is comparatively small, it may be necessary to set up a power station on the project. The type of prime mover then depends upon the type of fuel available locally, the supply of cooling water and site conditions. Where the source of the power has to be moved frequently the prime mover is then generally a diesel engine using distillate-type fuel; it may also be used for driving stand-by power supplies for emergency use.

In the United Kingdom electricity in quarries and other surface mines is controlled by the Quarries (Electricity) Order 1956 and an important regulation is 7 (2) which specifies that metallic coverings of cables shall have a conductivity throughout of not less than half that of the largest conductor enclosed and must be suitably specified when being ordered.

When agreement is reached on terms between the quarry operator and the supply authority, the latter's estates department commence negotiating way-leaves and statutory consents that will be required for the line that is to supply the power. As this may take a considerable period, such projects must be planned well in advance to ensure that power is available when required.

The determination of the location of the main incoming substation must take into account the centre of gravity of the surface mining load and its future movement and extension, although it is often possible to locate the main incoming substation permanently throughout the life of the project. If it has to be moved, careful planning is required to prevent disruption of power supply. The shorter the distance of the approach of the overhead supply lines to the main intake substation the lower will be the cost but the substation location must be suitable to present and future requirements of the project.

The distribution from the main to the distribution substations is dependent upon the magnitude of the load, the method of working adopted, and the degree of security required. Where it is required to transmit large loads over long distances it is generally more economic to use overhead lines than insulated cables but they are more vulnerable to lightning and shielding by overhead earth wires may be required where the isoceraunic level is high. Good voltage regulation is of great importance since low voltage can result in overheating of electrical equipment and reduction of load becomes necessary. Some form of compensation is requisite, usually the provision of on-load tap changers in the supply transformers.

The maintenance of the power factor of the installation as near unity as possible reduces considerably power cost and where large motor-generator sets are used on excavators, synchronous motors are usually employed in the Ward–Leonard control system with automatic control of the excitation to provide a constant power factor of unity or, often, over-compensation is employed to improve voltage regulation. Conveyors, however, are generally driven by induction motors with a good starting torque but having a power factor less than unity. However, power-factor correcting capacitors may be connected across their terminals for compensation.

Alternatively, the load as a whole may be compensated by the installation of an automatically controlled high-voltage capacitor bank at the supply point.

It may be necessary to install several distribution substations and if practicable one or more ring-main systems may be employed to advantage, giving greater security of supply, increased flexibility, particularly where periodic advances of substations is necessary, and better voltage regulation. The ring-main may be sectionalized by the use of ring-main circuit-breakers equipped with directional-type protective devices, so that a faulty section may be isolated without interfering with the supply to the remainder of the system (Fig. 83). Where parallel advance working is adopted the security of supply is of great importance and a system of 'leap-frogging' is often adopted; Fig. 83 shows such a system in which a 33 kV overhead line forms a ring-main around the mine. One distribution substation is located at each end of the mine and ring-main circuit-breakers

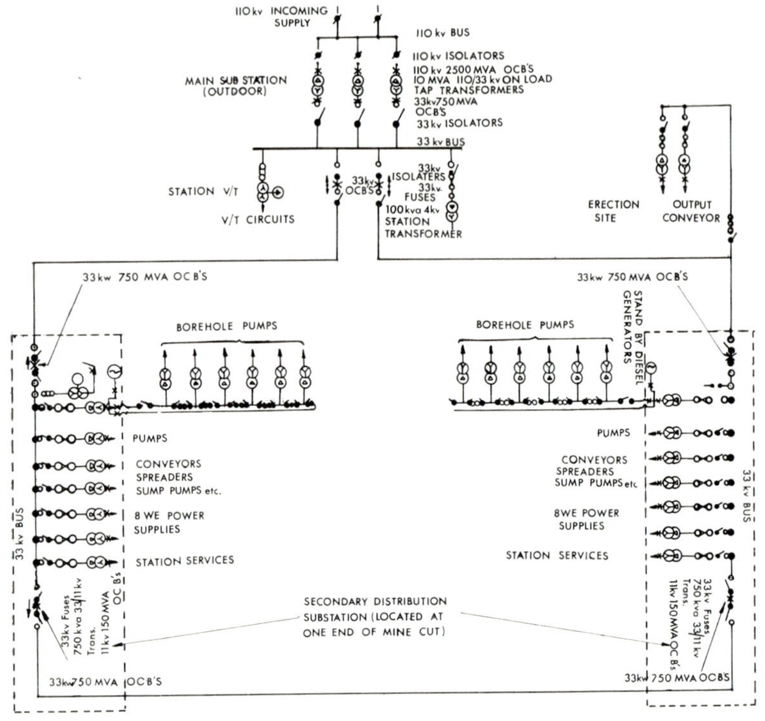

Fig. 83.   *Typical schematic electrical distribution for adopting the 'leap-frogging' method.*

of the outdoor type are located at both ends of the substation and are connected to a 33 kV overhead busbar. Feeder transformers of 1750 kVA are supplied from the overhead busbar through ganged, air-break, disconnect switches and 33 kV liquid fuses to eliminate the high cost of circuit breakers.

In some cases, the substations are linked in the form of a ring which is normally kept open about midway round, each half of the ring being fed from each of the two main outgoing feeder switches in the main substation. For overhaul or maintenance or in the event of a fault the ring can be closed at one point and opened at another thus enabling switchgear and cables to be taken out of service without interruption of supply. High-voltage cables are normally laid 2 ft 6 in down at least and low-voltage cables 1 ft 6 in to 2 ft down. High-voltage cables are laid directly in trenches or ducts.

The field substations should all be similar in design to make plant installation easy and reduce construction times to a minimum. The switchgear for such a project often 11 V is usually of the outdoor type requiring protection in accordance with the legal requirements of the Mines and Quarries Act 1954 and its regulations and in addition reasonable protection from flying stones.

Many factors will influence the choice of the high-voltage switchgear, but in Great Britain it is recommended that it should be of standard outdoor type to the Electricity Board's normal specification. The switchgear should be readily assembled and stand on a concrete plinth, the protection incorporated in the high-voltage circuit-breakers is usually of the high rupturing capacity (HRC) fused type which, with the failure of one fuse, will cause the whole switch to trip. These circuit-breakers are usually connected to the transformers, while the incoming and outgoing sections controlling the ring-main are normally of the solid-link type which do not incorporate any protection.

The high-voltage link switches normally incorporate built-in earthing devices making the earthing of cables easy for repair work to be carried out. The feeder switches on the main panel on the incoming side incorporate overload and earth-leakage protection in accordance with legal requirements.

Cables for surface mining have to encounter adverse conditions and over a considerable period mass impregnated non-draining cables have been adopted. Oil-filled paper insulated cables are an alternative for level runs but are not suitable where vertical runs occur.

Recently the development of polyvinylchloride (PVC) insulated cables has simplified cable design and a PVC insulated, PVC sheathed low-conductivity armoured and PVC-served overall cable has been found very suitable for surface mining work.

Copper-cored cables are used for electrical installation but with the dramatic rise in the price of copper, aluminium is now being used to an increasing extent. However it is not without its problems, for rapid oxidation makes jointing difficult and if stranded aluminium is used sweated joints must also be used to ensure that all the strands are incorporated. However, with solid-sectored cables crimped connexions, which are cheap and quickly completed, can be made and the cables are much lighter than copper-cored cables. For the smaller sized cables to 0.0145 in^2 copper cables are still cheaper, but solid-sectored aluminium-cored cables are available from 0.0225 in^2.

An important safety factor in the pit distribution installation is the method of earthing the star-point of the transformers. The practice followed by electrical supply authorities is often followed and in many countries there are statutory requirements to fulfil. The contact resistance between a machine and the earth is important if an earth fault occurs and where a machine is travelling on a hard dry rock surface, dangerous shocks can occur in these circumstances to persons coming in contact with the machine.

Figure 84 shows the power supply system at an opencast coal site in the USA.

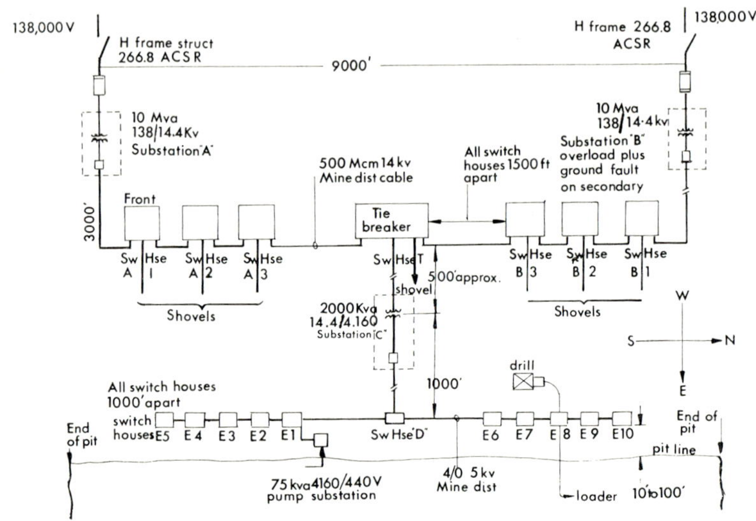

Fig. 84. The power supply system at an opencast coal site in the USA. Power system for mine employing 180-yd shovel features parallel substations which step voltage down to 14,400. Secondary substation supplies 4160-V power for drill and loading shovel.

AUTOMATION

Automation means a number of different things in different circumstances and automation control circuits tend to be very complex. The initial control systems pioneered in the gas, oil and petrochemical industries were simple closed-loop systems used for regulating individual temperatures, flows, pressures and *p*H values among others. These control loops, however, contained the essential requirements of automatic control measurement comparison of the 'measured value' with the 'desired value' and the regulation of the process by the out-of-balance signal to reduce the difference between the two to zero.

In order to improve upon this static mode of control it was realized that it was necessary to know much more about how process parameters affected each other not only in magnitude but as a function of time, that is in order to optimize process performance the dynamic relationship between the operating variables had to be established.

The high speed and versatility of data-logging equipment enabled process measuring points to be scanned at rates of several points per second; the value of each point being printed out or stored on a punched card or tape. As the reliability of the electronic computer improved, particularly as a result of changing from thermionic valves to transistor circuitry, its potentiality as a master controller was realized. If the operation of the process could be simulated mathematically by expressing the relationship between the operating variables and process performance in the form of mathematical equations, then for any measured set of operating variables the computer could solve the equations and thus predict the plant's performance.

Because of the high speed at which these mathematical operations are performed, the computer generates a whole series of values for the variables and solves the equation in each case and so discovers the set which gives the optimum performance and since it next automatically adjusts the set points of the conventional process controllers acts as a super-operator.

Saving of manpower is one of the key advantages of automation and the most common and effective way of doing this is by the centralization of measurement displays and controls enabling one man to control many sections of the plant from a central control room. The use of industrial television has widened considerably the field of application.

An example of automatic control is that of the Waterswallows granite quarry of Hughes Brothers Ltd, which comprises two plants, both of which are controlled from one control room by one man.

The main stone supply for each plant is drawn from seven 80-ton storage bunkers containing 2 in, 1 in, $\frac{3}{4}$ in, $\frac{1}{2}$ in, $\frac{3}{8}$ in, and $\frac{1}{4}$ in screenings, the bunkers each having two outlets, one for each plant. The first plant produces coated granite only, from aggregates derived from the main source.

The second plant processes mixed chippings for surface dressing, the granite being supplemented by three grades of limestone drawn by feeder conveyors from three pairs of storage bunkers.

All the aggregates are weighed cumulatively in a weigh hopper mounted on load cells and conveyed to the batch hopper. The aggregates are brought to the mixing temperature by a heater before they are charged into the mixer. The additives, binder and flux are proportioned by flowmeters, fillers are pre-weighed and added to the ingredients in the mixer.

The first plant offers, in addition, the selection of five fillers comprising four types of bitumen binder and two types of flux. The granite aggregates discharge from their bunkers directly to the weigh hopper. When the correct weight has been reached the weigh hopper discharges into the conveyor serving the mixing line and the material is fed by the batch hopper to the heater. Heated aggregate is discharged into the mixer by a swinging chute. Four of the fillers are weighed in a common hopper to which they are discharged from their separate bunkers. They are taken by conveyor to the mixing plant and fed through a chute to the mixer. The fifth filler is fed by a screw feeder from its own bunker to a weigh hopper over the mixer. The four binders and two fluxes are pumped through the flowmeters to the mixer, no flux being sprayed on top of the binder. No provision is made for storing the mixed product, which discharges directly into the delivery vehicle. The hopper and mixer doors and heater swing chutes are operated pneumatically, all storage hoppers are fitted with high and low bin level indicators and photocells are used to monitor the burner flame and supervise the positioning of vehicles under the mixer chute.

Production is in two-ton batches and up to ten can be processed in sequence under automatic control.

The weight indicators, the heater timers and the filler and binder delay timers are housed together with the various switches and potentiometers for selecting and pre-setting the weights of aggregates and fillers on the control panel.

The mimic diagram on the panel shows by signal lights the selection of ingredients made for a specification and the presence of correct weights of materials in the weigh hoppers. The opening of the bunker chutes and the doors of the weigh hopper, batch hopper and mixer are indicated also on the mimic diagram.

When a wagon is in the correct loading position, as indicated to the operator by the lighting of a lamp on the mimic diagram, he depresses the auto-start button and the selected aggregates start feeding in sequence to the weigh hopper. On reaching the pre-set batch weight, the doors of the weigh hopper open and the material passes on to the grading conveyor and on to the batch hopper, heater and mixer. Meanwhile, the first filler is being weighed and fed to the mixer chute, the second filler is weighed and discharged to the mixer and binder filler and flux are metered and sprayed into the mixer and, when the mixer time has elapsed, the product is discharged into the waiting vehicle. As soon as the aggregate weigh hopper is empty, the next batch is initiated and the cycle is repeated automatically until the pre-set number of batches have been delivered.

The second plant differs from the first plant in that it provides for three additional limestone aggregates and only one type of filler. Aggregates are weighed cumulatively in one weigh hopper into which the limestone is fed under time control after the granite has been discharged from the storage bunkers.

Isolated from the rest of the plant and overlooking the mixing building, the control consoles for both plants are housed side by side. A clear view of vehicle loading and other operations is given by large sloping windows and one man operates both control systems and can keep both mixing plants in continuous operation.

PLANNED MAINTENANCE

The benefits to be derived from planned or preventive maintenance are as follows:

(1) The elimination of avoidable breakdowns of plant and machinery with a consequent reduction in loss of output, cost of spares, materials and labour.

(2) Improvement in the safe working of equipment since the breakdown may cause damage and injury to personnel.

(3) Improvement of supervision by day-to-day control of the work of mechanical and electrical engineering staff.

(4) Provides a basis for determining the staff required for the proper functioning of the planned maintenance, extension, development and withdrawal sections of the work of the engineering department.

(5) Aids in deciding the correct level of essential stores and equipment to be carried at the surface mine and also at the machinery and equipment pool operated by headquarters of the operating company so that excessive capital is not sterilized in the unnecessary duplication of stores or machines.

(6) To improve former practice when maintenance often consisted largely of repairing breakdowns generally in conditions of great urgency and stress in the shadow of heavy production loss. In such circumstances, proper spares were often not available and a patched up job had to be done to allow production to be resumed at the earliest possible moment. In these circumstances, fitters and electricians did an excellent job of work often in very trying circumstances but excessive labour was often required, interruption of other work occurred and output was lost. The basic idea of planned maintenance is to detect wear and incipient breakdown of plant and machinery before it occurs by systematic regular inspection, and the withdrawal of machines and equipment, before the time elapses in which records indicate wear or breakdown is liable to occur, for thorough overhaul and reconditioning in adequately equipped workshops to the same limits of accuracy as adopted by the manufacturers originally.

Large quarrying groups and opencast operators are now paying increasing attention to the organization of their plant maintenance programmes but visits to service plants of manufacturers of surface mining equipment are still common. Cracked castings and shafts that have worn right through bearing housings are commonplace. It is surprising that such abuse of machines is tolerated when the results from planned maintenance reported by reputable enterprising firms are studied. In one operation conducted by the largest firm in the United Kingdom not only did planned maintenance virtually eliminate plant breakdown, but it allowed the number of tradesmen in the maintenance team on this plant to be reduced from 23 to 10.

An incentive scheme for allowed times for each job, was undoubtedly an important factor in the scheme for it led to the remaining ten earning an average weekly bonus of £2 10s 0d each person.

It was found that far more breakdowns were attributable to neglect of relatively simple precautions, such as lubrication and nut tightening, than is generally realized, particularly in those operations where the management have not bothered to scrutinize their plant maintenance programme critically. The use of work study personnel to organize the programme has often proved rewarding and the first requirement is an accurate and up-to-date compilation of the location, and details of all plants and equipment in and about the surface mine, since it is the basis on which planned maintenance schemes are founded. On the inventory will be details of the regular systematic withdrawal for thorough overhaul, the cost of this and of any other repairs carried out as a result of inspections.

For quick identification all plant, both mechanical and electrical, should bear a letter and a number to a code which indicates the class of plant to which it belongs and care should be taken that movements of plant are noted and the inventory kept up to date. It is then decided which equipment is to be maintained and the amount and type of inspection required to keep it in full running order.

A list of inspections is then drawn up which gives clear but concise instructions of the examinations to be carried out and such lists are generally designed to cover the daily, weekly, monthly, quarterly, half-yearly and yearly examinations respectively for each type of equipment in use. Those at longer intervals are internal and progressively more comprehensive, while the daily and weekly examinations are external and semi-external in character. Since lubrication is of such prime importance in plant maintenance, lubrication check lists may also be issued.

The length of time to be allowed for the inspection of each item of plant for the different periodic inspections should then be set and also to ensure that the work load is spread fairly over the inspection team.

The scheme for planned maintenance operations, therefore, depends on the following factors:

(1) The period required for each individual examination and whether external or internal.
(2) The frequency of inspection of each item.

(3) The number of men required for each inspection.

(4) The period during the 24 hours when the item is available for inspection. This may only be in a holiday period.

(5) The location of the equipment on the project and the requisite travelling time for a tour of inspection, to be taken into account.

(6) The number of shifts per week allotted to planned maintenance. Schedules are generally calculated on the basis of a five-day week which reduces overtime charges and also liberates staff for special duties which can only be carried out at weekends. Generally each maintenance man is given a work load equivalent to 75% of his effective shift time. Work sheets for daily and weekly examinations are prepared and inspections occurring at a frequency of more than one week will be regulated through a filing system.

A good record system is essential for success. A plant inspection record card and a maintenance record card together with a description of the item, its code number, the date of its initial installation and the day of the week on which weekly examinations are carried out, is provided for each item of plant.

COMMUNICATIONS

Efficient communication is a requisite particularly in the case of a large surface mine with mobile equipment operating over a wide area which needs control and supervision.

For this duty VHF radio-telephone equipment is suitable. A central transmission-receiver must, however, be located in such a position that radio transmission and reception is good throughout the life of the project. The first step must be the negotiation of a license to transmit with the appropriate authority which will probably include limitation of the transmission output and the VHF equipment manufacturers are next consulted to obtain suitable equipment.

The transmission of information promptly is often of vital importance and the three main methods available are direct wire transmission, audio-frequency transmission and ultrasonic frequency transmission. The first method is suitable for short distances and is the most economical. Miniature components and multi-core telephone-type cables are employed, usually with 50 V dc supply.

In the audio-frequency system a steady signal from an audio-frequency transmitter is received by a band pass filter over the telephone pair of wires catering for about 12 channels. Indication is achieved by interrupting the signal so that 'fail safe' conditions are obtained. A transducer is used to bias the output of a saw-tooth generator to transmit a variable.

An integrating circuit at the receiver displays a reading on a meter proportional to the original signal from the transducer. The cost of terminal equipment is about £60 per channel.

For long distances the ultrasonic frequency transmission is suitable. Transmitters or receivers are connected to a single-core co-axial cable in the same way as the audio-frequency system but up to 300 channels can be accommodated in one cable, the cost per channel being about £75.

Radio communications for North Sea drilling rigs

A communication problem similar but more intense to that experienced in surface mining, particularly in remote situations, was posed by the North Sea rigs drilling for gas.

The North Sea is one of the busiest shipping areas in the world and the normal radio services cannot be expected to meet the special demands of the drilling rigs and in particular exclusive circuits are required rig-to-shore for the exchange of highly confidential information. The British Post Office has therefore provided special radiotelegraph and radiotelephone services. Communications are required between drilling rigs and the shore, between one rig and another and between rigs and their supply vessels and helicopters. Being classified as ships the rigs may use the normal maritime radio services, the most useful of which are the radiotelephone public correspondence channels in the very high frequency, and the medium frequency (or intermediate frequency) bands. The maritime VHF service uses frequency modulation and separate am VHF equipment is required for working with helicopters fitted with conventional airborne VHF equipment. Because of difficulties it was decided to provide exclusive telegraph circuits with a method of modulation requiring the smallest possible bandwidth and a protected telegraph code to offset the effects of interference.

The British Post Office has provided three extra telephone circuits to be shared by all the drilling rigs using frequencies in the high frequency band. Independent side band modulation is used in making it possible for two calls, with different rigs, to be handled simultaneously on the same carrier frequency.

One shore station is at Cullercoats 240 miles from the limit of its service area and the transmitter has an output of 1 kW peak envelope power (pep) and radiates on a frequency of 3750 kc/s with a pilot carrier 26 dB below pep to enable the receivers on the rigs to use carrier automatic gain control (agc). The Cullercoats station receives suppressed carrier signals from the rigs on 3252 kc/s. The transmitter and the receiver are connected to the inland telephone network in the usual way, using radiotelephone terminal equipment to provide two telephone circuits.

The other shore station is the Humber which transmits and receives isb signals on 3264 and 3324 kc/s respectively, the lower side band being used to provide one radiotelephone circuit, the upper side band being used for radiotelegraphy.

Each rig using the service is provided with an isb transmitter of maximum power 500 pep. Since two rigs may be transmitting simultaneously using different side bands, the carrier is fully suppressed.

A simplified block diagram of the equipment at Humber is shown in Fig. 85 and indicates the signals at different points in the system, assuming

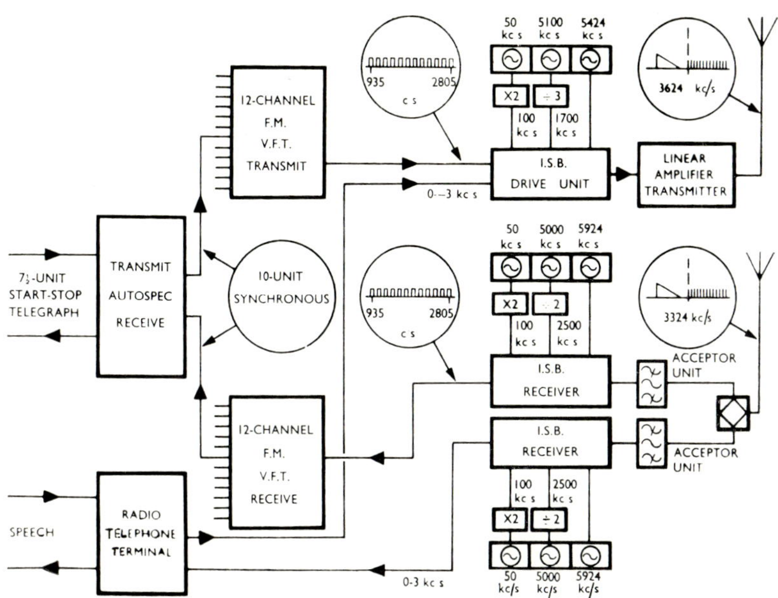

Fig. 85. Simplified block diagram of the installations at the Humber coast station. The radio equipment is duplicated and the vft terminals are only equipped with the individual channels for the drilling rigs which they serve.

that 12 vft circuits and the speech circuit are in use together. In practice some of the vft circuits may at some time be operated through one coast station, the remainder through the other.

A second simplified block diagram of a typical arrangement of equipment for a drilling rig is shown in Fig. 86. In most cases short vertical aerials are likely to be used with a bandpass filter in the feeder to the receiver. Conventional isb transmitters and receivers are suitable, but high-stability oscillators are required to meet the Post Office specification for frequency tolerances. The Post Office specify a maximum transmitter power of 500 W pep and provision for reducing the transmitter power by 24 dB in steps of 3 dB or less.

Closed-circuit television for quarries

In surface mining a reliable means of visually supervising from a remote location the many points where production hold-ups may occur would considerably improve control and supervision. It is necessary on conventional plants to station men at various points where experience has proved trouble is most likely to occur. For example on a plant reducing run-of-quarry material to roadstone aggregate sizes, one man would be

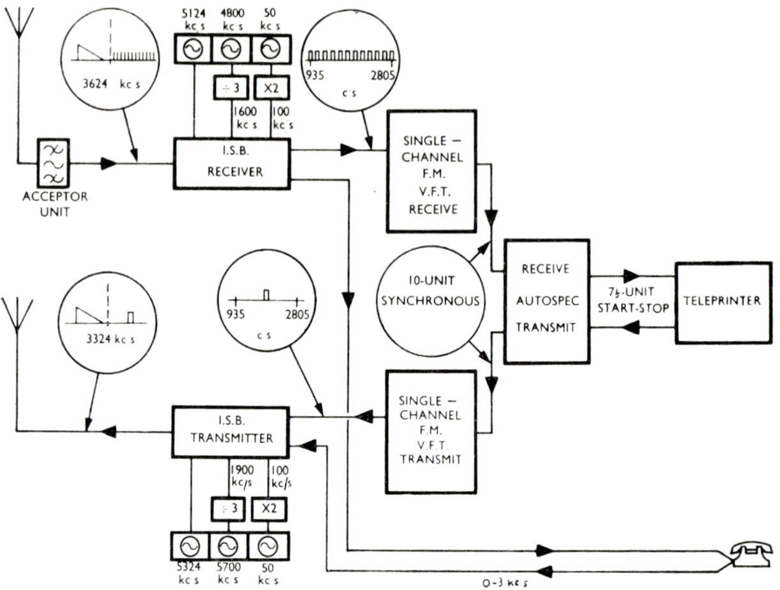

Fig. 86.   *Simplified block diagram of typical equipment for a drilling rig for the special rig–shore telegraph and telephone circuits (Bronsdon and Pilkington).*

located at each of the primary, secondary and tertiary crushers as well as by the main screens. In effect, the plant is a series of isolated units having little or no contact with one another. A sizeable and expensive shut-down could occur should any unit cease to operate.

With a system of remote-controlled television cameras, of which the Pye all-weather TV camera is one, the chances of a stoppage of this nature would be reduced or at least senior management would know of and could help to deal with it expeditiously. The service is applicable in the ready-mixed concrete industry and to bulk handling in general.

Cameras could be arranged to cover any stage in production in quarries from primary crusher feed to product bins or stockpiling storage areas. The cameras would feed back to a control point where a supervisor would select one of several cameras and view the picture of the key point on a monitor screen. Figure 87 is a diagram of a typical crushing plant. A single camera correctly positioned in the primary section would give constant visual supervision of the feeder and the crusher mouth. A similar visual supervision is shown of the secondary section. Additional cameras, as shown, could be located at transfer points in the tertiary section and in the screening section where screen flow, bunker level and low bin level indicator recording dials can be continuously watched. A man would normally be stationed

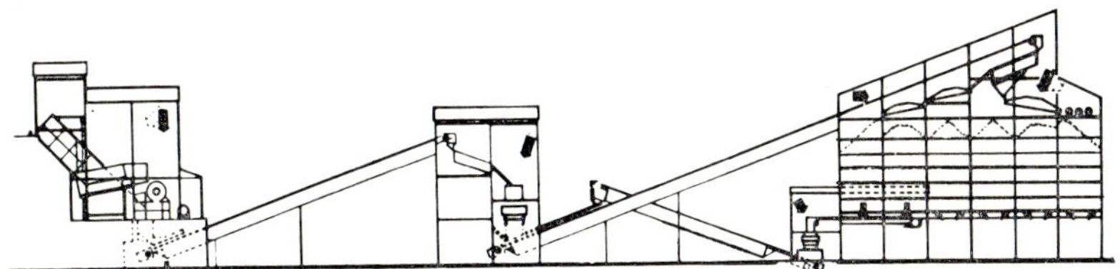

Fig. 87.   Diagram of a typical crushing plant using closed-circuit television.

at each of these points. The basic cost of a camera with a standard lens is £160. A remote control unit for three cameras and a viewing monitor screen would cost from £88, depending on requirements.

REFERENCES

'The Distribution of Electrical Power in Quarries', R. B. Cooke, *Quarry Managers' Journal,* January, 1966, p. 13.

'Electrical Planning for Large Opencast Mines', T. Atkinson, *Opencast Mining, Quarrying and Alluvial Mining*, Institution of Mining and Metallurgy, 1965, p. 505.

'Progress in Plant Control and Automation', L. F. Cohen and A. J. Lyfield, *Quarry Managers' Journal*, March 1966, p. 93.

'Modern Techniques in Remote and Automatic Control', S. H. Jones, *Quarry Managers' Journal*, May, 1967, p. 179.

'Applying Planned Maintenance in Quarries', P. C. M. Bathurst, *Quarry Managers' Journal*, June 1965, p. 244.

'Radio Communications for North Sea Drilling Rigs', E. G. Bronsdon and T. O. Pilkington, *English Electrical Journal*, Vol. 21, No. 4, July-August, 1966, p. 26.

'Computer Applications in the Quarrying Industry', H. B. Warburton, *Quarry Managers' Journal,* July 1967, p. 251.

'Dynamic Positioning of Off-shore Mining Equipment', D. W. Saunders, *Mining Magazine,* October, 1967, Vol. 117, No. 4, p. 236.

'High Voltage Power Distribution for Large Excavators', W. H. Schwedes, *Coal Age,* March 1968, p. 94.

CHAPTER 14

RECLAMATION AFTER SURFACE MINING

Land reclamation after surface mining goes back half a century or more and, in Britain and the Commonwealth, was often a mandatory clause in leases for surface mining undertakings, but frequently an alternative clause was available by which compensation of so much per acre could be paid in place of land restoration.

In recent years, however, the public conscience has been roused and, particularly in the USA, this has resulted in the passing of state laws through adverse criticism in the press, television and the public desire to restore the land to economic usefulness. In 1945, the State of Pennsylvania passed its first strip mine legislation. In 1961, Illinois passed legislation to the same end but 55% of the strip mine operators had been progressive and far-sighted enough to have reclamation schemes already in being. In 1966, Kentucky, as a result of recent legislation in other states, adverse publicity and political pressure, embarked on the most restrictive legislation put into effect but operators have co-operated and reclamation schemes are progressively satisfactory.

In most countries, restoration is now treated as an integral part of surface mining.

RESTORATION OF IRONSTONE WORKINGS IN THE MIDLANDS OF THE UNITED KINGDOM

Leases settled before 1933 dealt with shallow workings, some being full restoring leases making restoration mandatory, and others calling for restoration, but making provision for non-compliance by a compensation penalty clause.

When workings deepened with thickening of the overburden and new leases were negotiated or old leases renewed, there was a trend towards recognizing that full restoration was no longer a viable proposition, and could be covered by a compensation penalty paid to the lessor in lieu of restoration ranging from £20 to £60 per acre in particular cases. As workings became deeper, more use was made of the compensation clause and more land remained unrestored. In 1938, the government called for a report on the position, but World War II intervened. After this the Town and Country Planning Act, 1947, conferred on local planning authorities, usually the County Councils, power to control opencast and underground mining operations. In the Midlands the Minister of Housing and Local Government directed that applications to work iron ore should be referred to him instead of the local planning authority. As an interim measure the Ironstone Areas Special Development Order, 1950, provided for the levelling of all current workings after 1951 and for the removal and replacement of topsoil if the overburden was less than 35 ft. The area worked must be restored to such levels and configuration as will facilitate natural drainage and permit the resumption of normal agricultural operations. In shallow areas the topsoil must be replaced but in deeper areas only within 500 yd of the end of the workings.

In 1951, the Mineral Workings Act reached the Statute Book establishing a fund controlled by the Minister of Housing and Local Government for the restoration of land previously worked for ironstone by opencast. The operator pays $2\frac{1}{4}$d per ton of ironstone produced and the Exchequer $\frac{3}{4}$d and the operator can recoup half from his royalty payments under his surface lease. The Minister determines the rate up to which the operator is required to carry out restoration at his own expense, now fixed at £110 per acre.

Payment from the fund can be obtained for expenditure over £110 per acre, by the operator, for levelling derelict past workings, by a local authority for replacing fences and other fixtures, by owners of restored lands for extra farming costs and for afforestation by owners of the land.

Where the overburden is shallow, restoration immediately follows extraction of the ironstone and the dragline generally strips a 45 ft wide strip, the topsoil being first carefully stripped by the bucket and transferred to previously levelled spoil on the waste side of the face. The subsoil is dealt with in a similar manner, being deposited where the ironstone has immediately previously been excavated. By this means the land is returned almost immediately to agriculture.

Where the depth of overburden is somewhat more than 35 ft, stripping the topsoil is at first done by scraper carrying it away for storage or the dragline places it in windrows clear of the stripping line or loads it into trucks or lorries for storage. Subsequently the stored topsoil is reloaded into dump trucks for respreading on land from under which the ironstone has been worked and the overburden dumped by the dragline and levelled by a bulldozer. Where massive limestone, such as the Lincolnshire limestone, forms part of the overburden it is buried as deeply as possible with the normally underlying clay on top of it. This prevents loss of topsoil which occurs when it is spread on rubbly limestone. Panning or compacting of the clay by the bulldozer is to be avoided or topsoil erosion will occur. Scarifying or deep-rooting should be carried out of the top few inches of the clay before respreading to reduce soil erosion.

With deep overburden when a heavy walking dragline is generally employed and which must have a firm base when operating, the topsoil is first stripped separately by a scraper and either stored or carried round the end of the working and immediately respread on an area of levelled waste. The dragline standing on the limestone takes off the overburden down to its own level and swings it across the quarry to the waste side and then excavates the remaining overburden down to the ironstone. The limestone is always dumped as a base for the material immediately dumped on top of it and bulldozers then complete the levelling and grading in readiness for the topsoil. Five years is generally allowed for deep overburden to settle down after which a drainage scheme is established.

Where the restored land has been topsoiled, very little time elapses before it is again growing crops.

With medium-deep overburden, although topsoil has been respread, the overburden must settle and the period for this depends on the original strata. A drainage scheme is often necessary. Too deep cultivation should not be attempted at first.

Where no topsoil has been respread a 30 lb per acre seeding of timothy, meadow fescue and white clover has been found suitable but the regular application of fertilizer is essential and overgrazing must be avoided.

Afforestation is also suitable for these conditions, alder being the best and sometimes the only crop to thrive. Larch and Corsican pine do well in lighter soils than the clay in which only alder appears to be successful.

RECLAMATION OF OPENCAST COAL SITES IN THE UNITED KINGDOM

Modern earth-moving and tree-planting equipment has made it possible to restore earth on a scale and at a speed impossible ten years ago and at many opencast sites, for example Wentworth in Yorkshire, restoration has resulted in substantial improvements to the agricultural and amenity value of the land from which the coal has been worked. The National Coal Board Opencast Executive is working closely with local authorities in projects for large-scale land restoration. For example it is proposed to reclaim a number of derelict opencast sites in the Seaton Valley while they are working an opencast site near Backworth Lane. The Havelock site is also proposed to be worked to produce some 220,000 tons of coal over two years and simultaneously the derelict sites would be cleaned up from the profits. Several large pit heaps will be dealt with and others completely reclaimed and landscaped. The sites will be prominently in the view of the Tyne Tunnel approach road, the gateway to Northumberland.

The Aberfan spoil heap tragedy has drawn attention to the danger and loss of amenity of spoil heaps generally. Trees are one method of accomplishing two objects, binding the heap together and improving its appearance. Trees are, however, slow growing and there are alternative policies for planting on the tip itself not only trees but grass and other plants. A technique of hydraulic seeding is one by which the tip is sprayed with a mixture of seeds, fertilizer, wood cellulose, fibre, straw and bitumen so that the seeds and a medium in which they can grow are applied together. The process costs from 8d. to 1s. 6d. per yd^2 treated which amounts to £240 per acre at 1s. per yd^2.

At Acorn Bank opencast coal site as work proceeded the topsoil and 3 ft of subsoil were stripped from the area ahead of the other operations and initially placed in dumps on the perimeter of the workings. When sufficient area had been worked, topsoil and subsoil were reclaimed from the dumps and spread on the levelled spoil banks of the de-coaled area, the spoil banks being first carefully graded to planned contours, the subsoil spread in two layers 18 in thick and finally the topsoil was spread in a 12 in layer. Each layer was then rooted and all large stones removed.

At Westfield opencast site, which was originally a bog surrounded by poor agricultural land, the overburden is stripped from the area ahead and the spoil is formed into a hill with a maximum slope of 1 in 8 and blended into the local scenery and as each section is completed it is graded and the subsoil and the topsoil replaced on top of it. After settlement is complete, it is returned to agricultural use.

ASSISTED REVEGETATION

Assisted revegetation can be carried out either by planting trees and shrubs for amenity or timber values, or by the provision of grass lands for pasture.

The assisted method copies precisely the natural process except that the succession is accelerated by deliberate planting which is systematic. Heaps which already show patchy vegetation are in the pioneer stage and will take a higher form of plant life immediately, but bare tips need the establishing of pioneers for stabilizing the surface and forming a layer of nutritive soil. The Forestry Commission some years ago recommended as pioneers for bare tips in Great Britain perennial lupins, broom and black locust or grey alder. Coltsfoot, vetch or other lower forms may be planted first but this may not be necessary if the tip is well weathered.

After 3 to 5 years, the main stand may be planted in the shade of the strong intermediate growth and may consist of poplar, London plane, horse-chestnut, laburnum, lime, sycamore, Corsican pine, Austrian pine, hawthorn, elder and lilac.

In Scotland, on burnt-out tips, European larch, Scots pine, Norway spruce, birch and grey alder have been used successfully and on unburnt tips, Sitha spruce, ash and sycamore may replace the Norway spruce and birch.

For the production of pasture, which takes a relatively long period after fertilization with 8 to 10 cwt of basic slag and 2 cwt of ammonium sulphate per acre, the following seed mixture may be disced or drilled in:

Italian rye grass	4 lb
Perennial rye grass	18 lb
Timothy grass	3 lb
Coltsfoot	3 lb
Meadow fescue	$1\frac{1}{2}$ lb
Dogstail	1 lb
Red clover	2 lb
Late clover	$2\frac{1}{2}$ lb
Alsike clover	2 lb
White clover	1 lb
New Zealand clover	$\frac{1}{2}$ lb per acre

In the USA the reclamation of spoil areas after the coal has been removed and the prevention of stream pollution is considered in the basic planning of a strip mine.

The type of restoration attempted depends on such factors as contours of the area, type of overburden as well as the requirements of state regulations.

Sourness or soil acidity is one of the biggest problems of land reclamation.

Although not planned during the original reclamation of surface mined land, use of lake areas for recreation or sites for houses has received increasing attention. Reclamation of land offers some of the best fishing and hunting in the Midwest of the USA. For most spoil areas, tree planting is usually recommended. Among the tree seedlings, locust, shortleaf Scots pine and white pine have shown the most promise with black locust making the most rapid growth. Other species growing successfully are tulip, poplar, sycamore, cottonwood, sweet gum and river birch. Autumn olive appears to be the most successful shrub, coupling rapid growth with heavy production of fruit for wild life.

A typical example of reclamation in the USA is that at the Robert Bailey Coal Co., an opencast bituminous coal site at Morrisdale, Pennsylvania. It is standard practice to begin backfilling as soon as possible after the removal of the coal seam which has two advantages—the backfilling operations do not get behind or neglected, and performance bond money deposited with the state authority is released more quickly. Under Pennsylvania law, operators of open pits are required to deposit a 500 dollars performance bond for each acre of the royalty to be stripped when the mine is opened in conformity with a stripping plan filed with the state. Replanting follows backfilling. After an acre is backfilled, 400 dollars of the original bond is returned, the remaining 100 dollars is returned when the acre has been replanted to the satisfaction of the state.

Backfilling on the contour and the terrace system are practised depending on the steepness of the original site. Contour backfilling is the more expensive as the land must be restored to its original contour by moving the overburden back to the top of the highwall which must be carried out when the original surface had an inclination of less than 12°.

Terrace backfilling is more usual because much of the stripped land originally exceeded this 12° slope. In this case the pits are not completely filled to the top of the high wall. Overburden is pushed back to give the landscape a gentle rolling appearance but the highwall must be sloped at 45° or less with the terrace graded level with a tolerance of 15°.

Most of the area reclaimed is planted with trees and game food shrubs. Conifers are the commonest trees planted, since Pennsylvania has an oversupply of hardwoods, Douglas fir, white pine, red pine, white and Norwegian spruces being chiefly used.

The company has provided a 1200-acre tract of stripped land near Kylertown as an experimental demonstration area.

The Pennsylvania Mining Association in co-operation with the US Forest Service and Pennsylvania State University employ this area to promote better methods of reclaiming and revegetating mined land and in educating the public with regard to these activities. Various species of plants are tested to measure their adaptability and growth rate on spoil-heap soil and a 180-acre tree farm is also located on the mined area with thirty types of conifers, 13 deciduous varieties and 20 kinds of game-food plants. Two small dams were constructed to provide water for game and to improve the landscape. Feeding stations are provided for food supply for game during the winter months when heavy snowfalls cover the normal food supply. Bucks, does, bears, turkey and grouse have been hunted, American miners being great hunters with their 'bird dogs'.

The largest walking dragline to be designed was constructed by the Bucyrus–Erie Co. for lowering the cost of mining opencast coal at Ohio Power Co.'s Muskingum mine in southern Ohio. The machine, the 4250 W, weighs 30 million pounds, has a 220 yd^3 capacity bucket and a 310 ft long boom. The dragline 'walks' on four shoes each 20 ft by 65 ft and is powered by 170 electric motors developing 48,500 hp. The machine works in a royalty 25 miles by 5 miles. Hand reclamation was commenced in the area in 1947 and 22 million seedling trees have been planted in the area reclaiming some 14,500 acres of opencasted land and 7500 acres of substandard land unsuited for either mining or agriculture. The Ohio Power Co. have opened 30,000 acres of reclaimed land to the public for hunting and fishing in 1800 acres of man-made lakes. Fine public camp sites have been constructed and picnic and roadside information and rest centres are also under construction.

RECLAMATION OF SAND AND GRAVEL PITS

The planning permission conditions with regard to restoration of sites governs the extent to which reclamation is carried out on any particular site. These conditions in the Trent Valley district vary and depend on the date on which approval from the planning authority was received.

Earlier developments were permitted without conditions requiring full restoration, all that was required by the planning authority was that topsoil and subsoil be removed and stacked for respreading; any islands left to be levelled off as directed; growth of weeds to be prevented; farming land to continue until actually required for mining; trees and hedges to be preserved as far as possible; rivers, streams and drainage reinstated successfully; mining to proceed only to within a specified distance of highways, bridle or accommodation roads and the boundaries. Before leaving the site, notice must be given to the planning authority; the whole site to be left clean and tidy, plant, machinery and buildings removed unless it is agreed that certain buildings remain.

In approvals of later date, a further condition was added that land be refilled to its original level and the topsoil respread for ultimate return to agriculture. As the water-table is only 4 ft 6 in below ground level, after mining a lagoon of considerable depth is formed, in some cases 30 to 60 acres. The main problem is the supply of filling material. Four sources are available: (a) local authority household refuse, which is not very successful; (b) colliery refuse, inert material very suitable but not enough is available and transport costs are often prohibitive; (c) Central Electricity Board fly ash, which is the main source and though transport costs are often high no other alternative sites for dumping are available to the Generating Board. Care must be taken to avoid ash pollution of the River Trent, therefore a pipe, 5 to 6 miles in length, has been laid from Stagthorpe Power Station to Hoveringham Quarry and fly ash is pumped into a worked-out area. The fly ash is left to settle and dry enough to take the weight of the machine used to respread the topsoil of which 12 in thickness must be used; (d) sugar beet waste from the British Sugar Corporation's factory adjacent to the Newark quarry. Care has again to be taken to prevent pollution of the River Trent and sugar beet washings have to be 'bunded' to prevent access to them by river water.

Large stretches of water still remain and provide a useful amenity for fishing, yachting and water sports, which are in increasing demand.

On most sites the topsoil is required to be conserved for reclamation in the future and tractors and scrapers are used for taking off this top spit which is then stockpiled. When power station fly ash is to be used as filling, a certain amount of overburden has to be removed and stockpiled for reclamation, to be respread over the filling of fly ash; normally one to two feet of overburden is removed by tractor and scraper. With short hauls greater depths of overburden of 4 to 6 ft are removed by tractors and scrapers and the remaining overburden stripped by dragline cast into the worked out area.

Two methods of depositing fly ash are used. The ash may be brought to the worked-out area in lorries and tipped directly into the void to be levelled by bulldozer when the material rapidly consolidates; no appreciable settlement occurs after levelling and respreading the topsoil.

The second method is to pump the fly ash along a pipeline when it is discharged into the worked-out area. If the water-table at the site is high, water is displaced and this must be conducted away. When the fly ash has displaced the water and filled the void to the original ground level, it dries out and again little further settlement occurs and respreading of the topsoil by a bulldozer follows. Some 5 to 10% of consolidation is allowed for in restoring ground levels but sometimes enough fly ash is not available, the levels are graded out and if excess filling material has to be disposed of the finished area is crowned.

The slimes dams and dumps at a number of South African gold mines are being planted with grass and other vegetation to eliminate the dust nuisance and to prevent pollution of watercourses with mine sand and slimes which are washed into them during the rain storms.

The work of planting is being carried out by the mine authorities themselves at some mines, at others it is being undertaken by teams organized by the Transvaal & Orange Free State Chamber of Mines under the direction of a botanist.

At Rietfontein Consolidated Mines Ltd, in 1960, the sides of the No. 2 slimes dam, with an area of 24 acres, had 6·4 acres planted. The height of the dam is 141 ft, its slope angle is 31° and the slime is very acid with a *p*H of three to four. The area planted, now 20 acres, was covered with 6 in of soil dug out of pits on the mine surface, 12,600 yd^3 being excavated by blasting using light charges. Ash from the native village was mixed with the soil and loaded into 18 ft^3 side-tipping trucks on 18 in gauge rail tracks which were hauled by two single-drum electric winches of 15 hp up the side of the dam to a turn-table on a ledge cut out on one of the steps of the dam, ledges being cut at convenient heights from the bottom toe of the dam to enable the soil to be thrown down the slope. A monthly bonus is paid to 15 African workers and one white mine official to speed up the work.

Kihuya grass is planted and is in abundance. It is a hardy grass. On the slopes trenches 6 in deep and 18 in apart on the horizontal axis are dug in the ash and soil into which the grass roots are firmly embedded.

Water from cooling systems, mine drainage and surplus from the reduction plant is pumped by centrifugal pump to water the planted areas through 1 in diameter hoses daily, the pH of the water being from 7 to 9. The growth of the grass has been satisfactory.

However, the establishment of a permanent cover of vegetation involves not merely growing plants but bringing into being a plant community that will maintain itself indefinitely without further attention or artificial aid. This posed a problem particularly on sand dumps, and expeditions were made to the Kalahari Desert to examine the vegetation and bring back materials for test. A wide range of species can be grown, although not all of them persist, if the pH value of the dump material is raised to 5 or slightly higher and nutrients are supplied. The Kihuya grass appears ideal for the purpose as it grows and spreads rapidly at first and makes a thick, tough cover. However, it dies off unless supplied periodically with additional water and nutrients. Seed of the species finally selected was commercially obtainable but a few species not in commercial use proved so effective that a nursery was established to provide supplies. In tests, a mixture of seeds were sown and the following were used:

Cover grasses—Bromus catharticus, Agrostis tenuis, Dactylus glomerata, Eragrastis curvula, Holcus lanatus, Poa Pratensis, Phalaris tuberosa, Cynodon dactylon, Cynodon plectastachyus, Chloris gayana, Ehrharta calycina, Pennisetum macrourum, Cortaderia selloana.

'Nurse' Plants—Oats, barley, rye.

Legumes—Clover, hairy vetch, lucerne.

Trees—Acacia baileyana, Acacia melanoxylon.

A final trial with the selected species on a dump where the acidity had been corrected with lime and the necessary nutrients had been added, resulted in the production of a satisfactory cover.

A mechanized unit then moved to two sites in the East Rand. Tests were not completely successful, low pH values even after the addition of lime producing failures. Results indicated that even when suitable plant species have been selected and techniques have been developed for preparing the material for their growth, sand dumps still possess problems because the surface moves in the wind and because it is difficult to work on the steep slopes of loose sand. Unless the surface is stabilized temporarily, work on the establishment of vegetation can be lost in a few hours by being buried under several feet of sand or by the surface being blown away. In addition the abrasive action of ground rock blown across the surface will strip plants to a bare stem.

Several methods are available of achieving temporary stability but the most effective and economical is provided by a system of low windbreaks made with cut stems of reeds (Phragmites communis). The area is divided into small paddocks whose size is determined by the slope of the ground, the windbreaks erected and the various procedures leading to seeding and matching can then be carried out, the windbreaks providing excellent protection for the seedlings until they have grown sufficiently to become stable. The ultimate object is to bring into being a self-supporting plant community. What has been achieved so far is the establishment of an initial cover of vegetation composed of species able to grow and persist, at any rate for a time, on the dumps after the pH value has been raised and nutrients supplied which are normally absent. Few, if any, of the plants would survive if conditions reverted to normal in what is only a thin skin on the surface. The initial vegetation has a beneficial effect on the environment since the fast-growing annuals and the legumes included in the seed mixture produce plant debris after a short time and this is continually added to so that after a year or two there is a perceptible improvement in the surface layers of material. As a result, species from the flora of the surrounding countryside begin to appear on the dumps, the first usually being members of the Compositae. Some of the grasses introduced die out and after three or four years the vegetation may consist mainly of a few species which have flourished and spread but there is so far no indication that a cover of vegetation will not be maintained. It seems likely that in the course of years the composition of cover may change until a flora in equilibrium with the environment emerges.

REFERENCES

'Restoration of Ironstone Workings', T. W. Jones, *Opencast Mining, Quarrying and Alluvial Mining*, Institution of Mining and Metallurgy, 1965, p. 584.
'Stabilizing Mine Dumps with Vegetation', A. L. James, *Endeavour,* Vol. XXV, No. 96, September, 1966, p. 154.
'Planting South African Mine Waste Dumps', *Mine and Quarry Engineering,* August, 1963, p. 342.

CHAPTER 15

THE MANAGEMENT OF SURFACE MINES

During the past twenty years, the surface mining industries have passed through major technical and organizational changes. Total mechanization of extractive operations has almost been achieved and through 'take-overs' and amalgamation, capital and production have been concentrated in fewer hands. In addition the industries are expected to increase output by at least 6% per annum up to at least the 1970s. This requires improvement of business and manager performance. Decisions made in business management can affect both the progress of the business and the economy and the society in which the individual business lives.

To work effectively and secure efficient business performance, management needs a well-established pattern of organization with clear lines of accountability and control, all supported by sound management techniques and procedures.

It must be realized that not only are human beings the most valuable asset in a business, they are usually the most expensive and their growth and development can be a critical factor in the survival of a business.

The processes of management have been defined as forecasting, planning, organizing, co-ordinating, controlling and motivating, the first three being defined as the mechanics of management and the last as the dynamics.

In the main, surface mining presents a relatively simple technical situation but these processes still apply whether they concern the top management of a group of projects or the manager of a single unit.

Men need to know the results they are expected to produce and to have a part in establishing their objectives. Consultation on these matters is to show respect for their knowledge and intelligence and forms a strong motive to achieve the sought-after results.

Again, attention cannot be concentrated exclusively on current performance if the services and products produced will be obsolete and fail to meet customer requirements in the future. As a business can only exist as long as it continues to provide goods that the customer requires, there can only be one definition of the purpose of a business, that is to create a customer. Because it is the customer who makes the business what it is, the main functions of any business enterprise are marketing and innovation and both require strong entrepreneurial flair. Marketing does not just mean selling, it should embrace the total business operation and should examine this from the point of view of the customer. To exist in an expanding economy, it must be aware of the changing requirements of its customers, and be ready to meet them.

The concepts of the purpose and functions of business having been promulgated it is easy to recognize that the duty of a business is to survive and in financial terms this means the avoidance of losses. Enough money has to be earned to cover existing and future risks, pay dividends and taxes and provide capital for future expansion. The job of every top manager is to produce an adequate return from the money invested in a business by knowing the markets and, through innovation and entrepreneurial flair, to meet the changing needs of that market. Obviously those managers who fail to produce proper business results must go and new men take over. The company and society at large cannot afford second-rate business management.

Implicit in all this is that business management can no longer be based on hunches and intuition. Things cannot just be allowed to happen, a business like the human beings in it, needs objectives and objectives for success that will define the performance in those key areas which vitally affect the prosperity and even the survival of the business. The main areas in which objectives for performance and results need to be set are:

(a) The marketing position.
(b) Profitability.
(c) Innovation.
(d) Productivity.
(e) Financial and physical resources.
(f) Manager performance and development.
(g) Worker performance and attitude.

Objectives are best considered in two stages, a long-range plan for the next five years which will identify and highlight the milestones that must be reached in all the key areas of performance and will forecast the result of innovation and development in practical and financial terms. In the short term, specific objectives will be defined in an annual profit plan that can be seen as an integral part of the long-range plan and from which annual budgets can be developed which will specify in greater detail the results that are required from each area of the business activity to enable the annual profit plan to be achieved.

In planning the organization of a business, the first problem is to decide what kind of structure the company needs, and the objectives of the organization and the main activities that must be carried out to meet these objectives, hold the key to the organization required. The organization structure should provide the framework for the man at the top to delegate his responsibility right down to the first line of management, the foreman, and establish channels of communication between himself and his subordinates and, most particularly, provide the basis for calling managers to account for the performance of their sphere of influence.

Organization structures have to be tailor-made to meet the needs and circumstances of the individual business. Sales policy, marketing and technical factors and the level at which co-ordination is best established for economic performance all have their influence on the final pattern of organization and the way management should be applied in practice.

Organization in a small mine or quarry can often be informal and rely on good personal relations in the team and a very high standard of morale and team spirit be engendered to give individual attention to customers. This is the strength of the small family business where everyone knows everyone else from the top to the bottom of the ladder, often from boyhood, but when the business grows bigger more formal lines of organization are required. Problems arise from increasing size with the introduction of specialists and limits of their authority and the extent to which their activities can be decentralized along with line management. Most companies now recognize the need to have a logical written pattern of delegation for their managers at all levels in the business.

It has been stated that a business cannot solve its problems merely by buying a technique or system. Every business enterprise is a living organism with its own traditions, its own climate of opinions, its own special make-up. Every situation is different and every system and organization has to be custom built to the individual business. To sustain and improve business and manager performance, therefore, is not a question of simply introducing a series of techniques. It depends finally on the attitude of mind, application and leadership given by the chief executive. He must build up the strengths of his management team and take steps to remedy their weaknesses.

When a clear management structure has been drawn up it is essential to decide the information to be given to each level of management. In the surface mining industries information is required under the broad headings of production, selling, engineering, transport and administration.

In quarrying and surface mining, the production pattern falls into a series of processes and each process may be deemed, from a cost control standpoint, a cost centre. It is fairly easy to measure tonnage through a cost centre or process and the centre forms a convenient allocation area for costs while in the larger units a cost centre is likely to be in the charge of a plant foreman and as such becomes an area of responsibility. By the division of tonnage into cost, the cost per ton of material produced or processed at a cost centre can be quickly obtained and ratios of this type provide a rough yardstick of the performance efficiency of plant, for example the number of hours worked and the resulting production per man-hour.

The sales reports for the upper level of sales management should be in broad outline only concentrating on the main products and particulars of their tonnages, selling values and average ex works prices and these will compare with forecast sales tonnages and values. Lower levels of sales management require more detailed reports, but these, although concerning both major and minor products, will also take account of tonnages, ex works value and average ex works price. These costs are often collected and expressed as a cost per ton sold. This may be arrived at as an overall figure for the cost of sales management, head office sales staff and the salaries and other costs of maintaining sales representatives. They may also be allocated between selling areas.

The high degree of mechanization now reached in the surface mining industries has led to an increased capital investment in plant and machinery. A high level of investment in the extractive industries in the period 1964–70 is visualized in the National Plan and in the private sector of the industry it projects an increase of 7% per annum and this means increasing responsibility upon the engineering staff of the industries. An inventory of plant and machinery is essential and the use of this in planned maintenance has been dealt with on pages 311–313.

If work is undertaken by a central engineering establishment some system of job costing will be necessary so that repairs and maintenance can be costed at the correct cost centres.

The cost of transporting a low-value commodity such as those usually produced by the extractive industries forms a considerable proportion of the total delivered cost of the product and this must be kept as low as possible consistent with efficient and prompt delivery. Performance records of individual vehicles and drivers may improve performance or may form the basis of bonus payments.

Administration costs may be collated in much the same manner as selling costs and expressed as a cost per ton sold. The main interest is in the control of the financial resources and very important is the control of expenditure on fixed assets and also other current assets and liabilities. Reports required concern monthly or quarterly expenditure on capital projects, movements of stocks and work in progress, the level of debtors and the rate of debt collection and bad debts experience, the level of creditors and the rate of payment of creditors and, particularly, the liquid cash resources of the company. From the monthly balance sheet a cash flow statement can be prepared and is a useful method of watching movements in liquid resources and enabling the source as well as expenditure of the funds to be watched and this is even more informative when compared with the overall cash flow budget prepared in accordance with the financial policy of the company. Thus the emphasis in the administration section is on overall financial policy and its control.

The difference between forecasting and budgeting is that forecasting is concerned with probable events, whereas budgeting is related to planned events and is a statement of policy to be carried out. Variances result when actual costs are compared with standard costs. These variances arise where actual circumstances are not exactly as anticipated when the budget was drawn up but because the budget has been drawn up in the same way as the accounting plan, variances can be analysed and a report made to management showing the reasons for variances.

Forecasting and budgeting require a knowledge of anticipated capital expenditure and, to enable management to control this, it is necessary to have a system for approving such expenditure and ensuring that this is in accordance with anticipation as set out in the budget. The practice is increasing of reviewing requests for capital expenditure at various levels of management and in order to ensure adequate control a capital authorization form is used and when it has been approved and work has commenced, progress reports should be submitted to top management at monthly or quarterly intervals.

The large company or group can afford to employ experts to design and operate quite elaborate control systems but these may be beyond the means of the smaller company. However, its professional advisers such as its auditors can advise on the introduction of a system of monthly or quarterly accounts and their comparison with pre-determined budgeted figures and this can often be carried out without adding to administration costs.

Trading conditions in the years ahead are expected to get tougher and the outlook is interesting and challenging. Industry must modernize itself and not only its plant and machinery but also its thinking. Management must make sure that fresh investment will pay off and plan and think ahead.

The quarry management at all levels must be cost conscious and control can only be exercised before not after the event.

The trend towards larger units will force the manager in surface mining industries to become a manager rather than a supervisor. He must have a full knowledge and appreciation of how his work and responsibilities fit into the overall policy of the firm and how his actions and those of management levels beneath him at foreman level affect its prosperity. Top management must realize that the major proportion of cost is the direct responsibility of project managers and foremen and these men must, therefore, really form an integral part of the management team.

WORK STUDY

The basic aims of management, for survival alone, must be not only to produce goods and services but to produce them to acceptable standards in the most efficient way and, therefore, there must be high and improving performance, low costs and rising sales, the maximum yield must be obtained from the resources provided and displayed. To work these lines of attack and to organize and control management must have facts on which to base its judgements. Such are the tasks work study seeks to accomplish and may be defined as: 'The systematic objective and critical examination of all the factors governing the operational efficiency of any specified activity in order to effect improvement'.

Work study is divided into two principal groups:

(1) Method study to improve ways of performing tasks.
(2) Work measurement to establish the time for a task to be performed at a defined level of performance.

These tools or techniques may be used to cover widely different sizes and types of jobs.

There are broad surveying techniques within method study for solving major problems using statistical and mathematical evaluation and there are techniques for solving problems in fine detail if it is economically important to do so. Whether the study covers a whole industry or the movement of a pair of hands, there is no fundamental difference in approach.

Method study, fundamentally, comprises the breakdown of an operation or procedure into component elements and subsequent systematic analysis with the aim of finding better ways of working and it helps to improve efficiency by discarding unnecessary work, avoidable delays and other forms of waste. The techniques are simple, but success in their use calls for self-discipline on the part of the user. Since the activities of work study can impinge on many different functions, and, therefore, people the need for consultation and explanation assumes great importance when it is decided the study should be instituted.

The first few days of instituting a new method are critical and very close and intimate supervision is necessary which should be continued until everybody concerned is thoroughly familiar with the new methods of working.

Work measurement is defined by the British Standards Institution as: 'The application of techniques designed to establish the time for a qualified worker to carry out a specified job at a defined level of performance'. Absolute accuracy cannot be claimed for job times obtained by work measurement techniques but skilled operators can obtain a high degree of consistency in their results and many different types of work can be measured acceptably.

The time taken will depend on two factors: the rate of working and its consistency. Walking unladen at 4 mph along a smooth level surface would attract a standard rating on the British Rating Standards scale of 100, at 2 mph in the same conditions the value would be 50, at 3 mph 75 and at 5 mph 125. The basic time is that for carrying out work at standard rating and an allowance for relaxation is added to the basic time. Thus when a task is completed in the time indicated by the calculated standard minutes or hours of work, standard performance is achieved. Implicit in the definition which the British Standards Institution applies to standard performance is the necessity to include relaxation allowance: 'The rate of output which qualified workers will naturally achieve without over-exertion as an average over the working day or shift provided they know and adhere to the specified method and provided they are motivated to apply themselves to their work'. The definition goes on to say: 'It is recommended that standard performance be denoted by 100 on the BS scale, corresponding to the production of 1 standard hour of work per hour or 60 standard minutes per 60 minutes'.

The scale thus provides a numerical value to be put on performance and enables a level of performance to be specified.

The basic technique of work measurement is time study in which a task, or a portion of a task, is observed while the task is being performed, and two observations are recorded, the observer's concept of the rate of working and the time taken. Basic time can be calculated from these. A worker's ratings may vary not only from cycle to cycle of a job, but often within the job itself. It is good practice to break the job down into 'elements' or distinct parts convenient for observation, measurement and analysis. Each element of a task is studied a sufficient number of times for a realistic and acceptable basic time to be selected.

If time study is not an economic proposition because a job is not repetitive, the technique of analytical approach may be used in which an individual skilled in the type of work to be measured and trained in method study and work measurement can, by using the same analytical approach as in time study, form accurate estimates of the work involved.

Whatever technique of work measurement may be applied, no attempt should be made to start work measurement studies before full explanations have been made to the people concerned and the necessary consultations have been carried out.

The freedom of individuals to work at their own pace has been reduced as industry has changed from individual manual operations to more complex situations with continuous batch processes and a growing interdependence of activities. The output of a worker today is often limited by many circumstances or factors outside his control and there are 'governing times' when the activities of other workers, machines or processes limit what a worker can do. In these circumstances work measurement has to make them known so that potential output and planning calculations can be realistic.

In the past twenty years, the rapid mechanization of surface mining operations has led to fundamental changes in the conduct of operations and, of course, new problems have emerged and the techniques of work study have needed to be applied over many processes to meet the requirements of the new circumstances. Simultaneously, there has been a matching progression from the study of manual operations to the study of more complex situations.

Method study can be applied at any level of activity of an organization and its application to a project as a whole is suitable for the surface mining industries for they are industries with a high level of capital investment and an important aspect of method study is of particular utility since it can be applied with great effect in the design and development stage. It shows clearly how and to what extent different activities are interdependent and enables comprehensive arrangements to be made. The progression of method study from consideration of manual operations on the shop floor to the study of technical problems was a logical one, but method study tended to be associated with manual operations and this was for some time a psychological barrier to having the technique of critical examination accepted in other fields and that this happened was due to the fact that it worked and worked not because it taught anyone to do their jobs, but because it provided them with a basis for logical thinking and helped particularly where the best answer was not the immediately obvious one.

Critical examination can next move down from the wide scene and the overall productive process to the study of particular operations such as: How should the crusher be operated? What means are there for tipping and controlling loads? What should be transported by rubber-tyred vehicles, what type and capacity? What by conveyor belt, what type, number of plies or type, width and speed? Very fine detail may on occasion be required since very small units of time may be important as in the case of a mechanical excavator where movements are repeated so very often.

Accident prevention is a field in surface mining operations in which method study can play an important part. The circumstances that have led or may lead to an accident, can be critically examined with the circumstances charted in the same manner as an ordinary succession of activities can often lead to helpful recommendations—Is the driver of an excavator ever in a potentially dangerous place? What action must be taken in the event of a fall of face?

Two distinct types of problems are presented by work measurement in the surface mining industries, one deriving from the high degree of mechanization and the other from variation of material and the influence of weather and the continuous forward movement of the working faces.

The first type of problem is divisible into four. First is the problem of the driven machine whose productive capacity is known but whose overall effectiveness depends on the skill, intelligence and application of the operator. This applies to excavators, bulldozers, transport and drills. Timing of elements can be carried out as for manual operations but considerable experience is needed to adapt the concept of rating to the recognition of effective handling of machines but satisfactory results can be achieved.

Secondly, the assessment of the work of attendants of units of plant such as crushers or screening plants, where the difficulty is defining the needs and the frequency of these needs and such duties may occur irregularly so that work measured over a shift is generally low. Work measurement needs to be allied to method study in such cases.

Relaxation allowances pose the third problem. Machines have a high capital cost and should have as high a load factor as possible. When driven by a man results depend on the motivation of that man and one suitably motivated will achieve a standard rating and in addition a standard performance over a shift and the difference is the relaxation allowance. The man may need the relaxation allowance but the machine does not. Work measurement can show whether rest breaks are needed and when they should be taken, with a relief driver standing by who may have other alternative duties.

Maintenance poses the fourth problem. Planned maintenance has been dealt with in the preceding chapters and indicates what, how, when and by whom maintenance and inspection needs to be done: The difficulty in dealing with maintenance lies in trying to establish the criteria on which to base any form of payment relating to work done. Measurement is difficult, jobs hard to define and varied in content. The best solution is to try to measure what each maintenance man did in terms of time needed to do the particular job and to base bonus payments on this.

BONUS AND INCENTIVE SCHEMES

In April 1967, the retiring chairman of the Management Consultants Association in his valedictory address in London estimated that the output per man in British industry could be increased on average by at least 25% even with existing capital. This was a result of a random selection of fifty firms which have introduced properly devised incentive schemes. The average increase in productivity of the 7800 employees covered was 41%. This enabled earnings to be increased by 17% and reduced labour costs by 14%. The firms, both large and small, were in a wide variety of industries.

In some industries, notably in gas, electricity and leather, the gains in productivity were markedly higher than the average. Even in the chemical and allied industries, where the rise was least, output per man was still 19% up on the pre-bonus level. Bonus and incentive schemes are introduced in order to improve results, in other words, to pay and it is therefore necessary to have some assessment of the kind of performance at present being obtained without such a scheme in operation. If performance is sufficiently low to make incentives worth while then they should be considered but work measurement and method study should be carried out meticulously before embarking on such incentive schemes. These really put teeth into the application of work study. Within the performance of their own jobs the men who actually do the work are in many ways their own method study experts and if work and pay are tied together it will be found that if the work study has not been properly done, the men doing the job will make a mockery of the proposals, the balance of payment and work will be destroyed and the improvement and achievement sought will not be obtained. The point is well made in a quotation from the book on agreements on productivity negotiated by the Esso Petroleum Company: 'The essential point to grasp is that factory production standards are considered fair not because they are based on work measurement, but because they are judged and accepted as fair by all the parties concerned. It is the mutual agreement between management and workers which matters and which determines what production standards are accepted as right and are adopted. Work measurement is merely a tool for reaching such agreement, not a substitute for it'.

In the broad sense incentives may be divided into three types: non-financial, semi-financial and financial.

Good management provides non-financial incentives such as the pride a man derives from the knowledge that he is working for an organization which is fulfilling a socially useful function, whose aims are clear to him, about whose activities he is well informed and his awareness of the purpose and usefulness of his own contribution.

Unlike non-financial incentives, semi-financial incentives are related to the individual, such as an intelligent promotion policy, the provision of amenities like canteens, changing and drying rooms, holidays with pay and pension schemes.

Financial incentives by definition imply that where they are introduced there will be a departure to some degree from the concept of payment of wages on a time basis. When a situation emerges in which productivity is still below the level which might reasonably be expected, despite the fact that everything else that could be done has been done, financial incentives need to be introduced. On the other hand, not the least important consideration on the subject of financial incentives is that of technological change. If financial incentives are to be a direct reflection of a worker's contribution they must be related always to his contribution. To introduce a larger machine capable of greater output need not call for more effort from an operator, the incentive system must be able to meet the change. If, for example, a larger excavator is introduced, the effort per ton from the operator is reduced. In order to achieve consistency that work and effort rather than output expressed in tons is to be the criterion, the incentive must still depend on effort and not on production.

The improvement of the average rate of working and effectiveness of employees is the aim of financial incentives. To be successful they have to be applied within the atmosphere of co-operation and sense of common purpose and this implies that there must be demonstrably fair dealing as between one man and another. The demands which a job makes in the way of skill, knowledge and physical ability must be acknowledged by the payment structure within which financial incentives are introduced. All people concerned must have made known to them the purpose of the incentives and the way they will operate, and the officers of the trades unions concerned must be kept fully in the picture. The balance of earning opportunity needs to be considered with considerable care in order to achieve demonstrably fair dealing in view of the effect of financial incentives in relation to all employees of an organization. As much opportunity as possible must be given to employees to keep their earnings at a steady level, and this implies selection and training when men are moved from one job to another. The payment of financial incentives implies the payment of money for the achievement of targets, and to be effective the targets must be levels of performance which individuals can recognize easily and which they know they can attain. In addition, the relation between performance and payment must be obvious and capable of being measured, no matter what technological changes take place.

The most satisfactory and convincing basis for financial incentives is reliable quantitative measurement of the work done by an operator and work measurement can provide the degree of reliability which is required. Work done can often be related directly to output and increased production can then be fairly rewarded in proportion to the effective physical effort involved. This overcomes the difficulty of using a yardstick which may have to be changed as equipment changes, it being performance which is rewarded, not output measured in tons, yards or pieces.

A guarantee of a minimum level of payment is usual in most incentive schemes, and, for a scheme to be acceptable, this level in practice is the minimum, a direct financial incentive becomes a reward, additional to the job hourly rate, systematically related to an employee's contribution in effort.

A fixed point in the wages scale must be established for relating an appropriate amount of money to a particular level of effort. For the attainment of standard performance, it has in many cases been found that a bonus payment of a third of the job hourly rate represents an acceptable bargain and one which will achieve its objective, and with this as a starting point, it is then possible to develop a scale of payments appropriate to other levels of performance. A logical system is one in which payment increases in direct proportion to effort and is termed the straight proportion incentive scheme, the scale in practice only operates beyond a performance of 75 because this is the point at which performance equates to the agreed minimum rate.

As far as possible, incentive schemes should recognize the achievement of the particular employee concerned so that earnings are directly related to work done by individuals. Where work passes from hand to hand and employees are interdependent, the existence of such natural teams should be recognized, so that working harmony is most often improved if the task is acknowledged as a team effort, and the reward a team reward. Teams should not be too large or the individual contribution has little effect on the resulting productivity and a degree of incentive is lost. A limit of 7 to 10 employees in a team should be a maximum.

In the application of incentives in the surface mining industries, the use of work measurement is the essential underlying technique and demands that this should be performed in the most accurate and conscientious manner possible, and although a 15% margin of error might be accepted by some managements in planning work it will not be tolerated by men whose pay depends on results. All concerned must be seized of the idea of relating reward to performance, performance expressed in terms of effort, not in terms of tons loaded or feet drilled or miles travelled, since machines and equipment will change, and the effort required will change with them.

Planning by the week and by the year is necessary to minimize fluctuations in earnings from any cause other than the men's own efforts, output rates must be balanced against hours worked and against anticipated sales. If extra reward is to be made for extra effort, it must be possible for extra effort to be made. Waiting time for any reason means that no work is being done and the extra reward cannot be given. If a soft approach is made to this problem and some form of premium payment is made for waiting time, the reward for effort has less significance so that waiting time has to be attacked, and if possible eliminated. One of the peculiar difficulties of surface mining is that of variation in work content, similar jobs requiring different effort on different occasions since conditions in these industries are not static as in factories and different weather and climatic effects are experienced in different periods of the year. Where this variation is encountered, and an average time for a task has been derived from a wide range, it is politic to introduce a different performance to payment relationship, and the most effective and equitable relationship is that known as the 50–100 scheme in which the bonus payment for achievement of a standard performance remains the same but below and above standard the effect on bonus earnings of a change in recorded performance is halved when compared with the straight proportional scheme, so that a performance of 90 on the 50–100 scheme would pay a bonus of 26·7%, and this would require a performance of 95 on the straight proportion scheme. The 'flattening' of the payment graph may appear to reduce the incentive for greater effort, but the bargain which an incentive scheme represents is usually struck at the payment for standard performance and the line on the 50–100 scheme is acknowledged as a 'gearing' device.

The concept of striking a bargain around the standard derived from work measurement brings into focus the question of timing the introduction of incentive schemes. In a situation in which the going rate of wages is already being paid, and wages are raised above that rate, the results from the incentive schemes might be much less favourable than they would be if bonus payments and bargaining took place when the climate was right. The right time to commence incentive schemes introduces a big element of subjective judgment. One ill which is having a very bad effect on work standards and costs is the working of systematic overtime to enhance wages by time and a half or double time payments, often acknowledged and accepted by management. Often work which could have been carried out during the shift is deliberately left. This practice is rife in industry and is the cause of high costs and loss of export and other markets, and therefore it should be definitely prohibited by responsible management. Overtime is for work that cannot be carried out during the normal working week or shift, and to permit it to be used solely as a means of obtaining money and not for present or future output is dishonest.

The introduction of incentive payments to production workers with no provision for similar payments to other men can lead to friction and 'in lieu' payments are not an answer. The first principle of any scheme for maintenance workers is not to have too many men. If there are too many men, work will be manufactured and people will occupy themselves on jobs which are not necessary. Maintenance work should be planned to ensure that materials are always to hand when a job is being done, and where it is decided that work is to be done, and is necessary on a piece of plant, that piece of plant must be available. It is essential to keep two or three days' work at least in front of every man, so that he can see that when he finishes one job there will be another one to go to. Thus, there should be a reserve of non-urgent jobs, to make sure that the men can be kept occupied. The systematic work inherent in schemes of planned maintenance can be a great help in this problem.

OPERATIONAL RESEARCH

F. W. Taylor, Mary Parker Follet and Babbage emphasized the necessity of research in industry generally, the substitution of exact scientific investigation for the hunch, 'the old individual judgement or opinion' and 'the old rule-of-thumb method'. Dudding carried out studies on statistical quality control. Such research became widespread but only obtained its name during World War II when naval and military problems, not previously encountered, arose, and upon whose correct solution the safety of the country depended. Teams of scientists were formed and they assembled all the available data in an orderly manner, analysed them by statistical methods and then presented their conclusions to the executives who made policy decisions. Predictions of this type proved accurate and reliable, and included the hunting of submarines using depth charges, the Early Warning System (radar) and 'saturation bombing'.

The success of this 'operational research' in war attracted the attention of the larger industrial undertakings, including the National Coal Board, which applied the method to a wide range of problems, from small investigations at operating level to large-scale studies of industrial strategy.

By contrast to fundamental scientific research with no particular spur of industrial application, operational research is essentially practical with the definite and limited purpose of finding ways or a way to improve the conduct of certain operations or a line of action, from many possible, as likely to give optimum results.

Operational research has been defined as 'Scientific research into the operation of a going concern, carried out for and with the collaboration of the executives of that concern'. It has been applied to the running of hospitals, colonial administration and to problems in industry. In some cases highly mathematical models are the essence of operational research since many highly complicated problems can be reduced to mathematical forms from which it is possible to deduce the policy to give maximum profit, minimum costs or some other desired result. Where a problem can be reduced to mathematical form, it is easier to be certain that the best solution, or an approximate solution or a plausible solution has been found. Mathematics is a powerful tool where it can be applied, and much of the mathematics is interesting in its own right. In addition there are well-developed conventions for writing up mathematics which is more difficult if a descriptive method is adopted. Many of the most successful studies have been mathematical and have led to fairly standard models.

The Industrial Operations Unit of the Department of Scientific and Industrial Research, through Drs Easterfield and Clements, in 1964 carried out a study of a large quarry firm supplying crushed stone from its own quarries, either natural or coated, to a large number of customers; for use in building, civil engineering and road construction projects with the object of developing methods of devising control systems and improving information flow and analysis within the organization and particularly the most economical size of the lorry fleet.

Early in the investigation, it became clear that delivery dates left so little time for manoeuvre that no improvement in efficiency was likely on the method of allocation of work to the different mixing machines and of the delivery of the stone to different sorts of lorries. It was decided that what was required was a method of recording and processing each after-noon's flow of orders to facilitate the planning of their deliveries and to ascertain whether the orders were within the capacity of the quarry to execute or would need the transfer of lorries from other quarries owned by the firm or the hiring of lorries from local haulage firms. It was assumed that the peak demand for lorries could occur at any time of the day but this was not correct since the demand for lorries rose to a peak around 8 to 9 in the morning and practically never rose any higher and this fact allows for a much simpler method of checking the number of lorries needed.

Much redistribution of orders in the interests of efficient operation is prevented by the tendency of many customer firms to place their orders as late as five or even six in the evening. The attempt to draw a diagram to show what information is flowing and, in particular, where feedback exists, which is useful because it can show up places where a channel ought to exist and where there is a duplication of channels which may lead to confusion and inefficiency. It also may indicate what model operation and feedback are necessary and enable existing channels to be improved.

During the study the information flowing related only to the problems of controlling the scheduling of the mixing plant and the despatch of lorries.

The day before despatch, the quarry office in the light of orders already received and standing information about mixing plant capacity, the size of the lorry fleet and its speed and any policy, explicit or implicit, about conditions when orders may be refused, must now see how conveniently the order will fit into the next day's schedule. On the basis of this infor-mation the order may be accepted or the customer may be asked if he can accept delivery at a different time of the day, or the central sales office may be asked if another quarry can meet the order or, at worst, the order may be refused. It may indicate a necessity to try and borrow lorries from another quarry or hire extra lorries from a haulage firm or whether in certain cases, such as when the extra lorry must be hired for a whole day but could be used for only one load, to refuse the order.

To operate an appropriate control system, the use of two forms was suggested on which to record basic information and plan the schedules for the next day. The basic steps to this end, are: (1) Record details of customers' orders. (2) Prepare the lorry schedule for the despatch of orders. (3) Schedule the production of coated stone from the mixing plant. (4) Record the progress of operating on the day. (5) Control those opera-tions.

After each despatch, the time when the lorry will return must be esti-mated, allowing for expected delays and this is carried out for each order and each load of each order. The total number of loads despatched at any time interval and the customers to which the loads are to go, and the total number of lorries expected to return to the quarry will thus be known.

In addition, the total number of lorries which will be needed to despatch the orders accepted, must also be known. A possible rule-of-thumb method for balancing efficiency against service, might be to accept all orders and to schedule them as well as possible until 4.30 pm which, when graphed, will show either a pattern with one or two noticeable peaks or a fairly level graph. If there are one or two marked peaks, the rule would be that the quarry should manage on as many lorries as the peaks imply, and accept further orders only as far as they will not raise the peaks. The rule should be waived for very large orders, particularly those spread throughout the day. Where the graph is level, extra orders may be accepted for any time and it may be wise to anticipate these by immediately hiring extra lorries, the number deduced from past statistics or the order clerk's judgement.

In planning the mixing schedule, a governing factor is that the mixing plant has a production limit for any half-hour period, so that if the total quantity of coated stone entered under any half-hour period approaches this limit then either part of the stone must be made earlier or no more orders must be accepted that will require production during this time. This will be reflected in a limit to the number of despatches of coated stone that can be made in any time interval in the lorry schedule.

There will be another restriction imposed by the number of lorries that will be expected to return for any size of lorry fleet and will, therefore, be available in any half-hour period. Further, delivery times requested by customers constitute another restriction since most of the coated stone orders have definite delivery times that need to be kept, whereas dry stone orders are much more flexible and because there are many more dry stone loads to be sent out than coated stone loads, the former provide a buffer that assists in ensuring the despatch of coated stone loads. Restrictions on production are difficult and often expensive to alter, so that a schedule which infringes them will need alteration in delivery times, reallocation of existing orders to other quarries, diversion of new orders to other quarries or refusal of orders, the last to be avoided at practically any cost.

Borrowing company lorries from other quarries or hiring extra lorries can overcome restrictions due to the number of lorries available. In the last resort, restrictions imposed by delivery times can be overcome by asking customers to alter delivery times, or reallocating existing orders to other quarries. The finalized mixing schedule is passed to the mixing personnel, the lorry schedule remaining in the despatch office.

The function of a long-term control system will aim at getting the maximum output per unit of the mixing plant and company lorries and drivers. Tonnages per day to be dealt with are fairly easy to calculate, tonnage per hour more difficult since fluctuating although planned tonnage per hour would be calculated from the mixing schedule. In the preliminary review it appeared that the number of lorry loads varied for no apparent reason, and although it might be expected that a high proportion of coated stone in a day's deliveries would make despatching more difficult and lower the index, no such correlation appears to exist. Again, the number of loads per lorry per day at one quarry was consistently higher for company's lorries than for hired lorries, contrary to management expectations.

In almost every case, the planned peak number of lorries was higher than the actual peak, and this was thought to be due to lorries deliberately making good time on the first trip of the day. To test this, the differences between the planned and the actual times of return of lorries on the first trip were analysed, but a wide scatter of results appeared. The distribution of times was roughly normal, with means of about half an hour late and standard deviations of about one hour which indicated that just under half the lorries returned early but there was no general habit of going fast on the first trip. The difference between the actual number of lorries used and the peak of actual demand for lorries was small since there will usually be a number of lorries at the quarry filling up with stone, getting instructions and otherwise engaged.

At each quarry there is preponderance of days on which the excess lies between two and eight as expected, but there are days on which the excess runs much higher, up to 16 lorries in one case and 14 in two. One quarry had both a higher average number of loads per lorry per day and a higher average tonnage per lorry and this may be because the apparently better quarry has a smaller proportion of small orders of under 30 tons which would make the planning of the lorry schedule easier.

One firm has found that when dealing with clean stone, the lorry could go directly under the loading plant and at the other end could simply tip the load off. There was a difference of one hour between coated and clean stone in this respect.

CRITICAL PATH ANALYSIS

This technique of efficient management was inaugurated in the construction industries to decide the most efficient and economical method of planning and carrying out particular operations, but it has also been found of value in a number of other industries including mining and quarrying. In conjunction with a computer, although this is not essential, information becomes available much more quickly than from a manual arithmetical version of the network.

A detailed study of a project or job is broken down into elements of work which logically follow on from one another. These work elements are then represented on a diagram by lines succeeding one another or parallel to each other depending upon whether an element succeeds another or whether two elements can proceed simultaneously. The interdependence of one element with others is indicated by arrows. For example, the walls of a building cannot be built until the foundations are complete, but electric light and power and gas or water could be installed simultaneouly. In Fig. 88 for example, jobs 2, 3 and 4 cannot start until job 1 is finished, and job 5 cannot start until jobs 2 and 4 are completed.

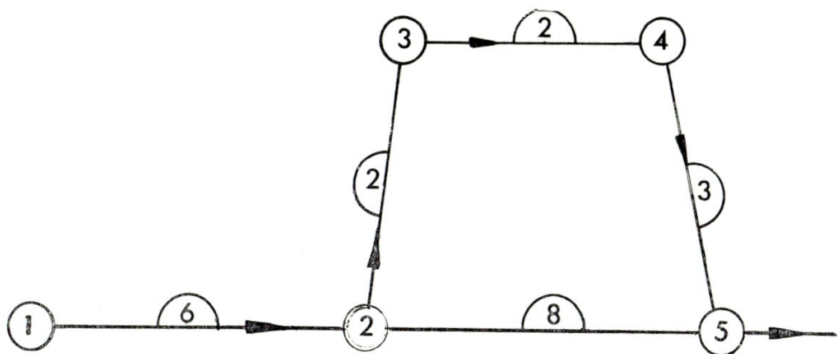

Fig. 88.   *Interdependence of one element with others.*

The duration of each work element is often assessed and indicated in the diagram in the half-circles. By adding the times of the successive work elements, and following the various paths through the network, a total time for the whole job is obtained. The longest path through the network is known as the critical path and is obviously that which the total job will take. Any circumstances such as bad weather which extend a work element in the critical path will extend the time of the completed job so that supervision of these requires to be intense if the target period for completion is to be achieved.

On certain parallel operations, there may be spare time or a float, for example, (2) − (3) − (4) − (5) is seven units, (2) to (5) direct is eight units, so there is a 'float' of one unit in the first succession of jobs. Thus every network has at least one critical path and this is simply the sequence of consecutive jobs whose total time is greater than any other sequence of jobs. These activities have zero 'float' and a strict check must be kept on these. Non-critical jobs may vary within the limits of the 'float' without affecting the duration of the whole project.

COST CONTROL

For optimum results from an industrial undertaking or service cost control is necessary in order to:

(1) Decide what output should be produced to attain maximum profitability, thus involving the determination of marginal costs and revenue.
(2) Consistent with safety and well-being of workers ensure that production is achieved at the lowest possible cost.
(3) Provide information at minimum cost by the development of systems of accounting for the above purposes.
(4) Reveal the expenses making up the production cost of each unit and lead to the tracing of inefficiencies.
(5) Show which activities are profitable and which unprofitable enabling measures to be taken to correct methods of production or decide the cessation of unprofitable lines.
(6) Regulate the activities of a company within a proper system and provide information for comparisons.
(7) Compare the payment of wages and the efficiency of labour.
(8) Reveal the precise cause of an increase or decrease in the profit of a company.
(9) Make available information for a basis of tenders and estimates.

The supply to management of costs and their analysis is the function of the cost accountant in industry and these decisions, affecting the long-term policy of the undertaking, may be founded on the basis of the most reliable and detailed facts available.

Expenses of a business may be allocated under two main headings, 'direct' and 'indirect', the former including wages, materials and other expenses which fluctuate with production or some other factor while the later apply to expenses which cannot be allocated directly to a particular process or operation. The total cost of running an undertaking may be obtained by subdividing and grouping these two main types of expenses which constitute the expenses involved: 'Prime cost' includes the 'direct' costs of labour, raw materials and other fluctuating costs; 'Indirect costs' comprise production overheads, establishment overheads and selling overheads. 'Works cost' is the sum of prime and production overhead costs. 'Production cost' is the sum of works costs and establishment overhead costs. 'Sales cost' is the sum of production cost and selling overhead costs. 'Profit or loss' is the difference between total sales receipts and the sales cost.

These are known as 'elements of cost' and can be applied to an undertaking as a whole, such as the analysis of the activities of a group of quarries, or can be applied to the production or cost centres or departments of a single quarry or works such as transport departments, coated stone plants, ready-mixed concrete plants or any unit which as an entity produces goods or materials or a service within the organization. They can thus be applied to most cost centres which may be defined for purposes of cost control as a location, person, group or item or group of equipment in, or connected with, an organization in relation to which costs may be ascertained and used.

When a cost control system is introduced, an important preliminary is the selection of the number, sites and types of cost centres to be set up. A cost centre may be as wide as a branch quarry operating at a distance from the head office, or as narrow as a single item of equipment such as an excavator working in a quarry, but in all cases, they must be clearly defined so that costs can correctly be allocated to them. A number of cost centres may be grouped together, the total expenses of which will constitute a further and larger cost centre, such as the primary crushing and processing section within a crushing and screening plant being classed as a cost centre and inside this cost centre can be maintained other cost centres pertaining to the primary crusher itself, the screening equipment and the transmission equipment.

When allocating cost centres, a sense of balance should be preserved since the setting up of a large number of centres for a comparatively small production unit will increase the cost of operating the costing system to little purpose. From the costs determined for each cost centre, the unit of cost is calculated, such as cost per ton for a quarry, cost per foot and per ton for drilling.

So that the appropriate cost centres are charged, the cost of labour must be analysed and the portion of their cost allocated to each. Generally, analysis is taken before deductions of income tax, etc., are made from gross wages, usually on a rate per hour with or without bonus, but on account of work done when on a piece-work basis. Normally no problems are experienced in allocating production workers' wages, but care is necessary in dealing with those of maintenance men. Occasions arise when idle time must be allocated and this should be as part of the production overheads.

Materials are charged to cost centres as they are issued from the stores, and in many industries consistute the raw material of the finished product. A large element of cost in the quarrying industry relates to machinery spares and, whereas in other industries these would normally be dealt with under 'other expenses' or 'production overheads', it is more appropriate to cost them as 'materials' and so maintain proper stores control and reconciliation should be made between the amounts charged to cost centres from the stores and the purchases made during the same period after adjusting for opening and closing stocks.

Prime cost is comprised of two main elements, labour and materials, but other expenses are often incurred which must be allocated to particular cost centres and these may relate to the hire of machinery or plant, patent costs, particular designs or tools and the cost centre to which these belong should be easily identifiable.

General indirect expenses are mainly of three types: production, establishment and selling overheads. With the increase in the amount of production plant and capital invested per man directly employed on production and, as productive capacity increases, administrative, selling and distribution expenses form an increasing proportion of total cost.

Production overhead expenses or works expenses are made up of such items as works insurance, rents and rates, power and fuel, depreciation, salaries of works and departmental managers, National Insurance contributions, tools, lubricants, works telephones, works canteen and stores expenses. These will vary from industry to industry but all production overhead expenses must be recouped in production costs. In the case of quarrying the simplest method is to charge on the basis of cost per unit of production at a rate per ton produced, by the appropriate cost centre. An estimate may be made of the overheads for a future financial year by dividing production into total anticipated overheads.

Establishment overhead expenses generally consist of office and administrative expenses which should not be associated with the works or quarries. In the quarrying industry, a rate per ton may be considered appropriate, but where departmental accounts are operated within the costing system of a quarrying firm, such as where a ready-mixed concrete plant or a coated stone plant account is installed, an agreed element of establishment overheads may be charged as a lump sum to these departments in addition to the overheads recouped at a rate per ton charged to the quarry processing cost account.

The selling overheads include all the expenses of sales promotion including advertising, sales managers' and representatives' salaries, showroom expenses and commissions.

Types of cost control

Costing was largely on a historical basis, but increasing attention is now being devoted to future costs in the increasing attention devoted to the application of budgetary control and standard costing together with marginal costing in the control of costs.

There are many kinds of costs, often closely interrelated and, in some respects, representations of the same facts in different ways for particular purposes.

The main types may be classified as follows:

(a) Job or unit costs; (b) contract or terminal costs; (c) single or output costs where only one product or commodity is involved and costs can be expressed per unit of production and this is a type commonly adopted at quarries; (d) multiple costs where a variety of products is manufactured; (e) departmental costs where the cost of the output from each individual department of an organization is required; (f) process costs required where a commodity is fabricated in stages and the cost incurred at each stage is required; (g) operating costs where a service is rendered and the preceding costing methods are inappropriate; (h) standard costs, target or presumed attainable future costs; (i) overhead costs which have long been a bone of contention among accountants sometimes referred to as 'fixed' or 'indirect' in contrast to variable, direct and material costs, but few overhead expenses are absolutely fixed, they generally include general works, administration, selling and distribution expenses; (j) marginal costs, the additional cost of an extra unit of output, are the total cost of the product excluding the fixed overhead expenses.

Budgetary control and standard costs

These have certain elements in common. Both require the establishment, with as much detail and care as possible, of the predetermined target or standard of performance, the measurement of the performance actually achieved and a detailed comparison of the two. Then follows the disclosure of the variances and the reasons for them, together with suggestions for corrective action. It is because of the disadvantages of historical costing that budgetary control and standard costing has emerged.

Budgetary control is a cost control technique by which management sets out to anticipate its volume of trading and consequent production costs for a given period in the future and by comparing the actual results, that is the historical costs, with the budget, variances are brought to light which point to defects in the forecasting and remedial action can be taken as a result. It is a means by which top echelons of management can delegate responsibility and, at the same time exercise central control on the basis of being given information of deviations from the budget.

Although budgets are usually prepared for a year in advance, control is usually exercised at monthly periods. Where returns can be easily prepared at shorter periods, weekly accounts can be used, but since normal commercial practice operates on the basis of monthly accounts—this is a more common subdivision of accounting periods making up the financial year.

Budgetary control is usually applied to an undertaking as a whole or to a department while standard costs are applied to detailed production operations and products. Their application to the extractive industries is probably more difficult than in repetitive industries.

A budget has been defined as 'A financial or quantitive interpretation prior to a defined period of time, of a policy to be pursued for that period to attain a given objective'. Budgetary control may be defined as the establishment of budgets, 'relating the responsibilities of executives to the requirements of a policy and the continuous comparison of actual with budgeted results, either to secure by individual action the objective of that policy or to provide a basis for its revision'. The construction of the budget has been described as a mixture of management policy and decision, of hope and of expectation. A great deal of study and skill is involved in planning of any kind since budgeting is an endeavour to use methodical and scientific methods to organize an undertaking's operations.

Budgetary control seeks as its main purpose the achievement of:

(1) the planning and control of expenditure in line with the expected proceeds from sales so that the maximum profitability of the operations is achieved;
(2) the planning and control of development and, in time, the capital expenditure involved;
(3) the planning and control of other developments and research not involving capital expenditure;
(4) the planning of the resources and finance of the organization so that adequate working capital and capital for development are available when required;
(5) decentralization while control is retained at the centre.

The master budget for the organization is made up from subsidiary budgets of the different activities comprising the whole, depending on their number which will depend on the type of business concerned. The sales budget, production budget, raw materials or purchasing budget, stores, personnel and wages budget, overheads budget and the capital expenditure and cash budget will be included amongst others.

The existing productive capacity, actual and potential financial resources, market and business trends and the availability of raw materials and labour will all determine the level of activity or production it is hoped to attain, and the more accurately and methodically these are studied and elucidated the more realistically and accurately will the target budget be framed.

Corresponding to different levels of production, the scale of overheads should also be determined so that flexibility and efficiency are retained even if the actual capacity employed is altered because of external circumstances or changes in wage rates.

Executives or departments concerned should apply budgets for their level of responsibility and, even if these are amended, should accept the final estimates as practicable. Budgetary control then consists in comparing estimates with the actual results achieved in the budget period. Different levels of management are then circulated with reports concerning their particular levels of authority and their areas of responsibility.

Where standard costing is used in addition to budgetary control, the analysis of variances, or differences with targets, will be carried to greater detail since the object of standard costs is to clarify the financial results of an organization by measuring the variation from targets with the object of maintaining maximum efficiency by executives. They are an intelligent forecast of what costs should be under normal circumstances and their main purpose is to disclose variations from these of costs actually achieved, thus economizing in managerial effort by concentrating attention on such variations, enabling the principle of 'management by exception' to be applied. Cost consciousness should also be stimulated and individual action encouraged to reduce costs and increase production. The fairness of the standards set up must be accepted as fair by both management and operating personnel for success.

Two types of standard costs are in use: normal and expected and ideal. In the first, reasonable allowances are made for errors, waste and spoiled work not up to the standard of quality demanded by the market. The ideal standards represent the best possible results obtainable and are not expected to be attained, variations are always unfavourable but show by how much performance could still be improved. Since they tend to damp enthusiasm and lead to discouragement the normal type is generally adopted.

Standard costing is most usually applied to the expenses of direct labour, direct materials and indirect costs. Cost is a combination of expense and output; if output rises, cost falls, if output falls, cost rises. For any operation, process or product, therefore, a predetermined allowable quantity of labour, materials and fixed and variable services and overhead expenses is agreed, depending on circumstances or conditions and, therefore, varying between one works or quarry and another. Work study may be found useful and, perhaps, essential. Standards should be established at the level of management controlling the operation costed and should be accepted as attainable. At such level of responsibility or cost centre, a standard cost is established but only for expenses under the particular control of that level.

Operating statements are prepared in connection with standard costs and contain *inter alia* the variances from standards which are provided for the level of management concerned. If wage rates or costs of materials alter, costs which, of course, are outside the control of management, a revision variance may be agreed as the amount of alteration of standards which has not been allowed for in the rates used. A general revision of standards is thus avoided.

Marginal costing

Marginal costing is a technique to assist in forecasting costs and profits. It relies on two costing principles. The first is that costs behave broadly in two ways. One kind is fixed for a period and remains unaffected by changes of output, while the other has a direct relation to output. The second is that out of the three factors, costs, prices and volume, the one that changes most and with the greatest frequency is volume. The most variable costs are direct materials and direct labour together with those elements of indirect labour associated with production. Fixed costs are those of supplying machinery with production capacity, supervision, administration, selling and advertising. If it be assumed that sales volumes fluctuate more than costs and that fixed costs remain more or less fixed and variable costs fluctuate with sales, it is necessary to install sufficient productive capacity to meet peak demands. On average this will mean that plant capacity is under-utilized.

The rates of wages in any industry depend upon the competitive pull which other industries are able to exert on its labour, upon the cost of moving to the most attractive alternative industry and upon the strength of the workers' inclination to move—*i.e.* it depends upon the supply price of labour to the industry. The rate of wages depends also on the demand price of employers and for workers of a given productivity, this will be governed by the price, or under imperfect competition, by the marginal revenue of the product of the industry. In the short term, fluctuations in wage rates are closely linked with fluctuations in prices. The most powerful force bringing about an increase or decrease in the price of the materials which are produced in the long run when there is time for a change in the relative attractiveness of an industry to cause workers to move from it or to it, is the supply price of labour; that is the dominant factor, much as, in the long term, it is to cost rather than marginal utility that prices respond.

Fixed costs are fixed and, therefore, do not affect marginal cost. Marginal cost is the addition to total variable costs needed to produce an extra unit. Under perfect competition, marginal costs increase and must increase as the output of the firm expands. It is increasing marginal costs that limit the expansion of any one firm and keep it small relative to the industry as a whole and the reason for increasing marginal costs is diminishing returns. In the short term the capacity of each firm, comprised of land, buildings, plant and machinery and other fixed capital is finite and cannot be altered. As more workers, or some other variable factor, are employed in conjunction with this total capacity, their marginal productivity will fall. A firm will try to maximize its profits and will expand its output so long as the extra output adds more to its receipts or revenue. It will expand up to the point at which its marginal revenue equals its marginal costs; that is marginal cost will equal price. One of the main tasks of marginal costing is the calculation of what costs are fixed and for what length of time.

REFERENCES

'Manager and Company Performance', G. E. W. Peart, *Quarry Managers' Journal*, February, 1967, p. 47.

'Management Information and Control', H. B. Morris, *Quarry Managers' Journal*, March, 1967, p. 87.

'The Application of Operational Research to Quarrying', T. E. Easterfield and D. W. G. Clements, *Quarry Managers' Journal*, February 1965, p. 53.

'Work Study and the Design of Incentive Schemes', S. Bremelow, *Quarry Managers' Journal,* July 1965, p. 273.

'Company Financial and Cost Accounts with Special Reference to the Quarrying Industry', T. F. Walton, *Quarry Managers' Journal,* June 1965, p. 225.

Coal Mining Economics, Organisation and Management, John Sinclair, Pitman, London, Second Edition, 1967.

'Planning and Financing Capital Projects', Symposium of Production Engineering Research Association of Great Britain, *Mining Magazine,* April 1967, Vol. 116, No. 4, p. 254.

'Computer Applications in the Quarrying Industry', H. B. Warburton, *Quarry Managers' Journal,* July 1967, p. 251.

ELECTROMAGNETIC PROSPECTING

Electromagnetic methods may be classified in two general groups. The first includes methods in which the source of the electromagnetic field remains stationary while the receivers are moved to explore the area. The second includes procedures in which the energizing and receiving systems are moved together.

Essentially a fixed-source method consists of the measurement of electromagnetic fields about the source. The mutual coupling between the source and the earth is constant, but the mutual coupling between the receiver and the earth (unless the earth is homogeneous) and also between the source and the receiver changes at each station. The results are usually normalized by relating the field data to the calculated free space or primary field.

The Turam or two-frame is probably the most common fixed-source method. The energizing source is an insulated cable grounded at both ends or formed into a large rectangular loop. Measurements are taken along a traverse at 5-metre to 50-metre intervals using two small receiving coils, the lagging coil being placed at the position previously occupied by the leading coil. The complex ratio, *i.e.* in-phase and out-of-phase ratios, of the voltages induced in the two coils is measured. Operating frequency range is about 100 to 800 cps.

The Turam method and its modifications have a greater working depth than the other electromagnetic procedures used in ore prospecting. Under favourable conditions conductors have been located at depths of 200 metres. A modified Turam method with one of the electrodes grounded in the upper end of a plunging ore-body was used to follow the extension of this body to a depth of 200 metres beneath a layer of conducting schists.

The compensator method employs large energizing layouts similar to those used with the Turam method. The voltage induced in a single receiving coil is measured relative to a reference signal carried by a cable from the energizing source. Usually both the vertical and a horizontal component of the field are measured at each station at two or more frequencies in the range from 10 cps to 300 cps. Interpretative procedures make it possible to estimate depth to conducting horizons. The compensator method is used chiefly to map horizons in sedimentary rocks, although it is sometimes employed in searching for flat-lying metallic-ore deposits. It is used in Scandinavia in searching for coal and similar economic resources of sedimentary origin.

The cross-ring method utilizes a small portable energizing loop oriented with its plane roughly parallel to the ground. Two identical receiving coils are fastened together in a perpendicular arrangement, one oriented so that it is in the plane of the energizing coil. The complex ratio of the voltages induced in the two receiving coils is measured and the data are plotted directly in profile form without preliminary calculations. Short profiles (up to 100 metres long) can be measured by moving the receiving coils and holding the energizing coil in a fixed position. Usual frequency is 3500 cps.

Depth range of the cross-ring equipment now in use is limited to some tens of metres.

Bore-hole electromagnetic methods have not been used widely in Scandinavia, but limited applications have been successful and additional work is planned by various organizations. A combination surface-hole method uses a single coil in the bore-hole as a receiver and a large low-frequency energizing loop or grounded cable placed on the surface of the ground. The complex ratio of the voltage induced in the receiving coil to the current in the energizing source is measured.

In a moving-source method changes in the mutual coupling between source and receiver are measured. The coupling between both the source and receiver and the earth (unless the earth is homogeneous) changes at each station, but the free-space coupling between source and receiver remains constant. Usually results are normalized in the instrument by relating the readings to the free-space coupling.

The most common moving-source method is the Slingram or loop frame, which utilizes a small portable energizing coil and one receiving coil. A reference voltage is brought from the energizing coil to the receiving coil by a cable. The ratio of the mutual impedance between the two coils in the presence of the earth to their mutual impedance in free space is measured by finding the complex ratio between the voltage induced in the receiving coil and the reference voltage. The coil spacing, held constant for each series of measurements, may range from 20 metres to 100 metres. Usually the coils are oriented so that they are co-planar and horizontal or (on steep slopes) parallel to the ground; however, in techniques used less frequently, the coils are held in perpendicular positions or in vertical, co-axial, or co-planar positions. Operating frequency ranges from 500 cps to 3600 cps. The complex mutual impedance ratios are presented in the form of individual profiles or contour maps without preliminary calculations.

The maximum working depth for the Slingram method, using two horizontal coils, is equal roughly to the coil spacing. The greatest depth at which a poorly conducting vertical ore-body will cause a significant anomaly may be less than one-half the coil spacing, but a large, horizontal, highly conductive body at a depth of twice the coil spacing may give a significant anomaly. The working depth for using a perpendicular coil arrangement is about the same or a little greater than with two horizontal coils, but with vertical coils it may be only half as great.

Cross-ring equipment is sometimes used in a moving-source technique simply by maintaining a fixed distance between energizing and receiving loops. The coil spacing used may range from 20 metres to 80 metres.

An airborne system having some similarity to the cross-ring method uses two perpendicular energizing coils whose planes are parallel to the axis of the aircraft. Two receiving coils are oriented in similar directions to the transmitting coils and are towed in a bird, which is positioned as closely as possible along the axis of the aircraft. A relatively large source-to-receiver distance of 200 metres or more is sometimes obtained by towing the bird in a second aircraft. The energizing coils are excited by two voltages of the same frequency, but 90° out-of-phase with respect to each other, thereby creating a rotating or circularly polarized free space or primary field. The complex ratio of the voltages induced in the receiving coils is measured. In the absence of conducting material this ratio is independent of the distance between aircraft and bird, axial rotation of the bird, and to a considerable extent other misorientations, of the bird.

A moving-source method is used in conductivity logging of bore-holes. Frequencies in the order of 20,000 cps are used in a sonde having a co-axial Slingram arrangement. The coil spacing is some tens of centimetres; hence the radius of investigation is quite small.

Before discussing certain aspects of various specific methods it is desirable to point out some of the relative merits of fixed-source methods. With a fixed-source method two identical disturbing bodies do not, in general, give rise to equal anomalies because the free-space coupling between source and receiver (and hence the depth of penetration of the normal fields, and other factors) is unequal at successive stations. The shapes of the anomalies obtained with fixed-source methods, however, are relatively simple; the anomalies tend to be similar in shape to magnetic anomalies.

When a moving-source technique is used, the free-space coupling between source and receiver is constant; therefore, two identical disturbing bodies give rise to equal anomalies. In general, however, anomalies obtained with a moving-source method are somewhat complicated because the coupling between the source and the conductors varies. An anomaly obtained with a moving source usually has more maxima and minima than one obtained with a fixed-source method. The additional maxima and minima in moving-source anomalies are sometimes referred to as edge effects, because they occur near the edges of conductors.

Usually, several conducting bodies in proximity are most easily located and defined by fixed-source methods because of their relative freedom from edge effects. When moving-source techniques are employed the occurrence of more than one shallow conducting body within a distance equal to or less than the source-to-receiver distance results in data that are very difficult to interpret correctly.

When edge effects are not too troublesome moving-source methods are often superior to fixed-source methods, since the importance of one anomaly compared to another is more easily evaluated. Edge effects are not too troublesome: (1) if the horizontal extent of the conducting units is somewhat larger than the source-to-receiver distance; (2) if depth of burial and separation between units is great; or (3) if the boundary between units of different conductivities is not abrupt.

With a moving-source method, if vegetation and terrain do not interfere with lines of sight or connecting cables, the source and receiver may be placed broad-side rather than in-line to the traverse. Edge effects are thereby reduced and, if the traverse is perpendicular to the strike, individual conductors among a group of several parallel linear conductors may be resolved more easily.

In areas that are very complex geologically, especially when graphitic schists and slates are involved, the horizontal range of fixed-source equipment may be limited greatly by high attenuation of the normal field and the results may be complicated by the mutual coupling between adjacent conductors. Measurements with a fixed-source method become very laborious and in interpreting the results, it may be impossible to resolve all the separate conductors as distinct units. In such areas, a contour map prepared from moving-source data may give the interpreter the most useful information. This is especially true when the purpose of the survey is to map geological trends rather than to locate individual ore deposits.

Among the fixed-source methods, the radio-reference-signal system requires the fewest men to make the measurements. It offers an advantage over the Turam method in obtaining the fields directly instead of by calculation. These measured values are often more accurate than those obtained by calculation, especially in regions where the field is near zero and the ratios may approach zero or infinity.

However, the ratio curve is more sensitive to small features than the field curve. By considering the ratios, therefore, it is possible to detect weaker conductors with the Turam method than with the radio-reference-signal system. Also since the magnitude of a field anomaly depends on the position of the disturbing body with respect to the cable and to other bodies, a field anomaly cannot be evaluated on the basis of its magnitude alone. Ratio anomalies, however, depend on the shape and not the magnitude of the field anomaly. Thus, it may be advantageous to have the ratios as well as the fields.

The depth range and speed of operation of the cross-ring method are inferior to those of the Turam and radio-reference signal, but the cross-ring is sometimes the most useful for working out details of a group of shallow conductors in proximity to each other.

When a moving source is used cross-ring equipment has an advantage over Slingram because it does not require a cable connexion from transmitting to receiving coil. Also the maintenance of a constant distance between the two coils is not critical unless response from the overburden is great. Results are not as easily interpreted as those obtained by the Slingram method because, in general, anomalies occur simultaneously in both the horizontal and vertical fields and thus the data are not related to any fixed reference.

Slingram methods, either ground or airborne, are probably less expensive per traverse kilometre than their counterparts in other systems. Slingram is especially adaptable to reconnaissance work because no large energizing layouts are necessary. When a sizeable area is to be surveyed in detail by ground work the radio-reference signal may be nearly as inexpensive as the Slingram method, since the expense of laying out cables is partially offset by the fact that each set of measuring apparatus can be operated by one man.

For detailed work over areas of particular interest, it may be advisable to use the method that yields the most information regardless of cost. In such cases, one of the fixed-source methods will usually be chosen because of its greater sensitivity, depth of penetration, and resolving power. The variety of problems in prospecting is so great that any one of the systems discussed may be the best in a particular instance.

A reference signal from the cable is used to modulate the radio transmitter placed near the energizing layout. The voltage demodulated in a radio receiver located in the measuring apparatus is sent to a phase splitter and then applied as the reference voltage to each of the two diode-transformer phase discriminators. The voltage induced in the receiving coil is amplified by a calibrated amplifier and fed to the phase discriminators. Direct current meters connected to the outputs of the phase discriminators then give the in-phase and out-of-phase components of the magnetic field at the receiver relative to the current in the energizing cable.

One of the airborne electromagnetic methods using ground energization utilizes essentially this same system, but with the output of the phase discriminators fed to two continuously recording meters.

A simple airborne Slingram arrangement also uses this measuring system, except that the reference voltage is sometimes transmitted by a wire connexion. Also, the normal or primary signal from the receiver is compensated for by a reference voltage, so that in the absence of conductors there is no signal voltage applied to the phase discriminators.

APPENDIX II

PERFORMANCE OF MEDIUM AND LARGE DRAGLINES

In the Mercure mine in Italy, the dragline used is a Rapier W 1800 with a 265 ft boom and a 43 yd^3 bucket. In order to obtain optimum results in open pit mining it was considered essential to carry out a study of overburden handling from the technical and related financial points of view.

The main advantage of the dragline is the possibility of digging and excavating overburden and depositing it in one operation in its final resting place on the spoil bank within the maximum dumping radius, which is the horizontal projection of the boom length.

The overburden is removed and the ore-body or other desired mineral is uncovered in a series of narrow strips or cuts of width determined by the thickness of the overburden, its nature, angle of rest and other factors. The cut is bounded on one side by the high wall or working face of the excavation and the other by the spoil bank and at the bottom by the top of the ore-body or other mineral deposit being worked which is then excavated and the exploited cut filled with overburden or spoil from the next cut. No further equipment is required for spreading the spoil, or for other purposes in this simple system. The *in situ* volume of material removed in a given time will decrease with increase of the 'swell' or expansion of the excavated material.

With a dragline with a boom of a given length at a given inclination, there is a maximum height of overburden which can be excavated and beyond which the dumping radius of the dragline is no longer sufficient to enable the whole of the spoil to be dropped on to the spoil bank. In this case, with one dragline only in use, the 'bridging' method may be used in which a 'bridge' of excavated spoil is used to enable the dragline to traverse and reach the geometrical summit of the spoil heap (Fig. 89). It will be noted that part of the material in the 'bridge' must be rehandled subsequently by the dragline. The percentage of spoil rehandled in terms of the total spoil removed measured *in situ* before 'swell' or expansion increases, as the thickness of overburden increases, and decreases for a given thickness with increase of strip or cut width.

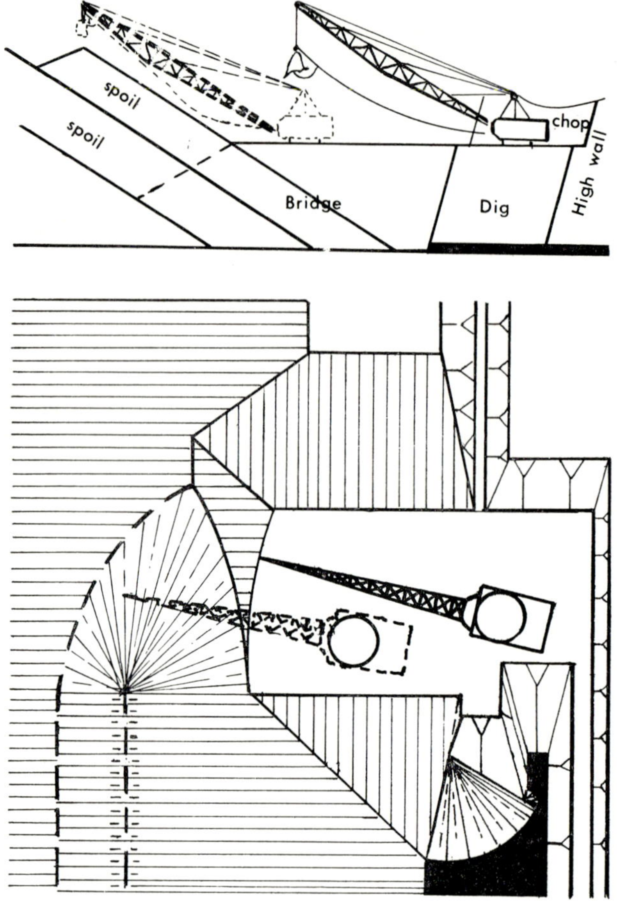

Fig. 89.   Limitations on overburden removal with constant boom inclination and length.

In Fig. 90,

B = dragline boom length

γ = inclination of boom to the horizontal

m = distance of centre-line of dragline from high wall

h = thickness of overburden

l = strip or 'cut' width.

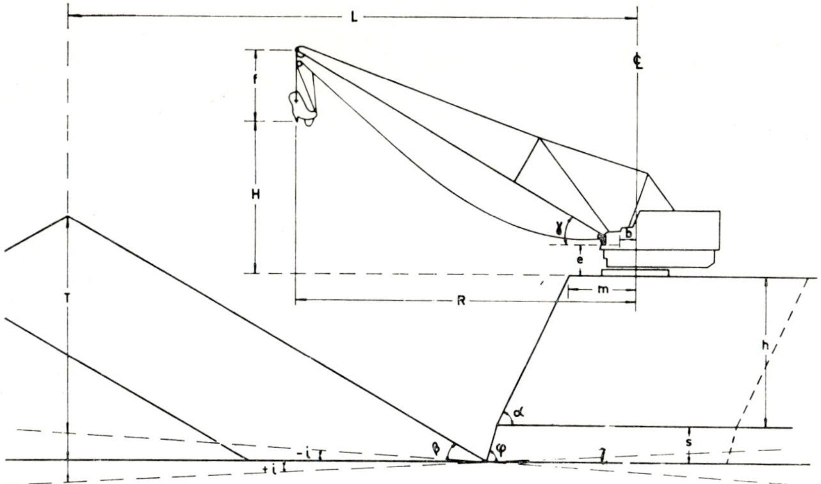

Fig. 90.   Parameters in the geometrical study of dragline operation.

c = expansion or swell factor generally 115 to 120% of the *in situ* material

s = thickness of ore body or other mineral deposit

α = slope angle of high wall

β = slope angle of spoil

φ = slope angle of the ore body or other mineral deposit

i = inclination of ore body or other mineral deposit across the strip or cut, + if inclination is towards the spoil, − if away from the spoil

R = dragline dumping radius

H = dragline dumping height

T = maximum spoil height

L = distance of centre line of dragline to top of spoil

Rh = percentage of spoil rehandled as percentage of overburden handled *in situ* or 'bank' measure

b = distance between the boom footpin and the machine centre line

e = the height of the boom footpin above the dragline bearing surface

f = the length of the bucket and its chains plus a clearance.

Then the dumping radius $R = B \cos \gamma + b$

and the dumping height $H = B \sin \gamma + e - f$.

The maximum height T of the spoil capable of containing the excavated spoil for a unit length of 'cut' is determined from Fig. 91 by the area of parallelogram MNN'M' (the material removed) times the swelling factor

$$c = \text{A BB'C (the spoil) } c. \text{ Hence } T = Ch + \frac{l}{4} \tan \beta \left(1 - \frac{\tan^2 i}{\tan^2 \beta}\right).$$

From Fig. 91 area ABB'C = area MM'N'N $\times$ c but MM'N'N $\times$ c = $cl'h'$ and ABB'C = $T'l'$ − area AB'C. Also $l' = l'/\cos i$, $h' = h \cos i$,

$$T' = T \cos i, \text{ and area A'B'C} = \frac{l'y \sin (\beta + i)}{2} \text{ where } y = \frac{l' \sin (\beta - i)}{\sin 2\beta}$$

whence

$$T'l' = ch'l' + \frac{l'^2 \sin (\beta - i) \sin (\beta + i)}{4 \sin \beta \cos \beta}$$

and

$$T = ch + \frac{l}{4} \frac{\sin^2 (\beta \cos^2 i - \cos^2 \beta \sin^2 i)}{\cos^2 i}$$

$$= ch + \frac{l}{4} \tan \beta \left(1 - \frac{\tan^2 i}{\tan^2 \beta}\right)$$

so that T is a parabolic function of $\tan i$ and a linear function of h and l.

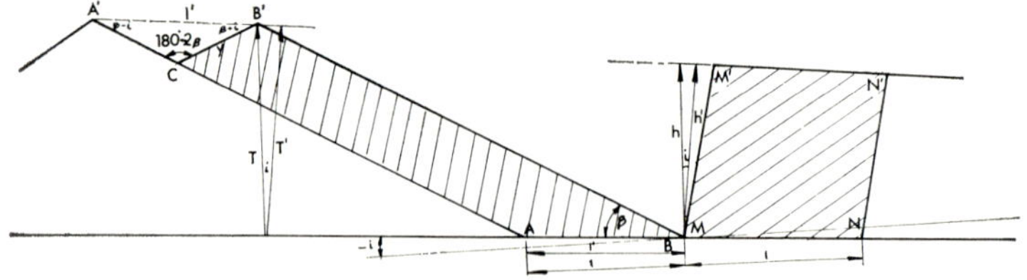

Fig. 91.   Geometrical representation of maximum spoil height.

The distance L between the centre line and the apex of the spoil varies with the thickness of the overburden, h. When L becomes greater than the dumping radius R rehandling becomes necessary. It is, therefore, important to determine how L varies with the thickness of the overburden, h. From Fig. 92 it can be shown that

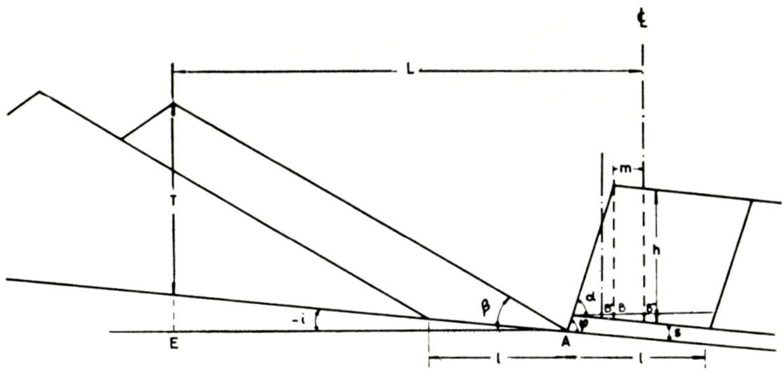

Fig. 92.   *Geometrical determination of distance from unit centre line to spoil apex.*

$$L = EA + AB + m$$

$$= h \, \frac{c(\tan \alpha - \tan i) + (\tan \beta + \tan i)}{(\tan \beta + \tan i)(\tan \alpha - \tan i)} +$$

$$+ \frac{l}{4} \, \frac{(\tan \beta - \cot \beta \tan^2 i)}{\tan \beta + \tan i} + \frac{s}{\tan \varphi} + m$$

$$= \frac{T \cot \beta}{1 + \cot \beta \tan i} + \frac{h \cot \alpha}{1 - \cot \alpha \tan i} + \frac{S}{\tan \varphi} + m$$

$$= \frac{Ch \cot \beta + \dfrac{l}{4}\left(1 - \dfrac{\tan^2 i}{\tan^2 \beta}\right)}{1 + \cot \beta + \tan i} + \frac{h \cot \alpha}{1 - \cot \alpha \tan i} +$$

$$+ \frac{S}{\tan \varphi} + m$$

$$\text{whence } L = \frac{c(\tan \alpha - \tan i) + (\tan \beta + \tan i)}{(\tan B + \tan i)(\tan \alpha - \tan i)} +$$

$$+ \frac{\dfrac{l}{4}(\tan \beta - \cot \beta \tan^2 i)}{\tan \beta + \tan i} + \frac{S}{\tan \varphi} + m$$

For a constant value of L, the height h decreases as i goes from positive to negative.

From this the overburden thickness at which rehandling becomes necessary and the choice of the best direction for strip mining may be determined. When planning the layout of the stripping operations, it is generally best to orientate the workings so that the base of the deposit is level and dips towards the spoil bank and away from the high wall.

In determining the relations between the thickness of the deposit, S, and the maximum height of the spoil bank, T, and L, the distance between this and the centre line of the dragline, S has no effect on T, while its effect on L is $S/(\tan \varphi)$. If φ is 90°, that is if the slope angle of the deposit is vertical, L is independent of S.

Since L is the distance between the apex of the spoil bank and the centre line of the dragline, it is also the dumping radius needed so that the overburden block of thickness h, can be dumped directly. Similarly, by

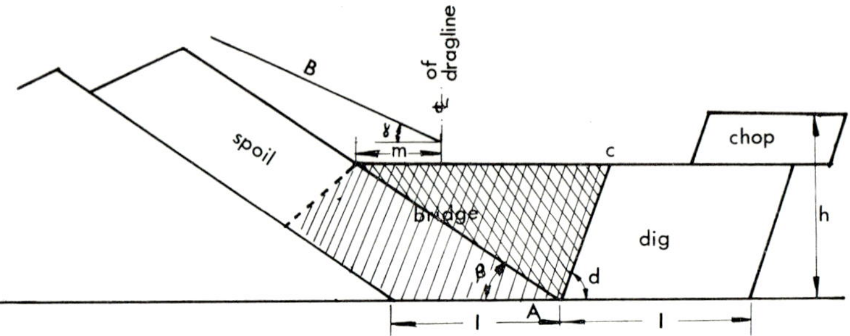

Fig. 93.   *Construction of overburden bridge.*

substituting R for L and h_r for h in order to determine the maximum overburden thickness, h_r, which can be discharged directly without rehandling by a dragline with a dumping radius R. When the overburden thickness, h, is greater than h_r, the length L becomes greater than the dumping radius R, and the dragline must move towards the spoil heap in order to dump the whole volume of the expanded overburden of thickness h, and strip width l. When a single dragline only is to be employed, a bridge may be built up as mentioned previously and as shown in Fig. 93 at the working level, upon which the dragline may work and move. A portion of the bridging material will later be rehandled and dumped by the same dragline.

When the thickness of the overburden to be removed, h, is less than h_r, it can be excavated without rehandling, but if h is greater than h_r, some rehandling will be necessary. The rehandling percentage required increases as the slope angle β increases, and as the thickness of the useful deposit increases.

REFERENCE

'Performance of Medium and Large Draglines', M. Sappa and C. Santostefano, *Mining Magazine,* September 1966, p. 164.

AGGREGATES IN CONCRETE

Concrete is being used for an ever-increasing variety of purposes, and in many fields is being pushed to the limit. In this situation, more and more technical difficulties are likely to be met, and the users' difficulties would be reduced if the producer would manufacture a uniform product. In the past, some variation in grading and particle shape was not of great importance, but now, in the field of high-strength concrete for example, close uniformity is absolutely essential if the designers' requirements are to be met.

Structures are being designed for an average strength of 13,000 to 14,000 lb/in^2 and not only the type but the source of the aggregate would need to be specified, and this might have to be brought a distance of as much as 100 miles. The concrete produced is inevitably of very low workability and requires very strong compactive effort, the material fully compacted into position, and this implies slow rates of construction and very strong formwork.

It is necessary, therefore, that producers of aggregates should have up-to-date information of the requirements of architects, engineers and contractors so that data may be available on which to base future production plans.

With the increase in the strength and other quality requirements of concrete, there is a corresponding increase required in the quality of the aggregates. Aggregates for concrete are specified to comply with British Standard 882 'Coarse and fine aggregates from natural sources', 1965. The methods of sampling, describing, classifying and testing aggregates are provided, but without any guide to acceptance figures for the various test methods which it specifies and it is purely descriptive of test methods, in British Standard 812, 1967, 'Methods for sampling and testing of mineral aggregates, sands and fillers'.

BS 882 divides fine aggregates into four grading zones in which the fine aggregates in each zone are suitable for most, but not all, uses. The mix proportions must be chosen according to the grading characteristics of the aggregates used to make concrete of high strength and durability. As the fine aggregate becomes finer the ratio of the fine to the coarse aggregates is reduced: also as the maximum size of the aggregate increases so the workability of the concrete decreases. As the grading of the fine aggregate approaches the coarse outer limit of zone 1 or the fine outer limit of zone 4, the exact design of a concrete mix becomes increasingly important and the suitability of a given fine aggregate grading in some cases depends on the grading and shape of the coarse aggregates.

Some architects and engineers are wary of using a fine aggregate within zone 4, and, to a lesser extent, zone 1. Coarse sands result in harsh concrete which is more difficult to work into place and to finish to a smooth, fine surface. On the other hand, the inclusion of an excess of zone 4 material can result in a surface skin which is mechanically weak and lacks durability.

Zone 4 material is stated in BS 882 not to be suitable for concrete in reinforced concrete structures unless tests have been made to ascertain its suitability for the proposed mix.

British Standard 882 requires that no harmful materials such as iron pyrite, coal, mica, shale or other laminated materials or flaky or elongated particles shall be contained in aggregates in such a form or in sufficient quantity to affect adversely the strength or durability of the concrete including the resistance to frost and corrosion of the reinforcement. The specification does not prohibit the presence of any of these substances at all, but it is essential that aggregates should be free from certain very active chemical impurities such as those which inhibit or retard the correct hydration of cement.

The British Standard limits the maximum amounts of clay and fine silt and uses a test with known quality sand when dealing with organic matter.

Cleanliness, particle shape, texture and grading are all becoming increasingly important for much concrete work as the uses of concrete widen. All-in aggregate is less used because of variations in grading which are present or which arise due to segregation during handling. Mechanical loading from stockpiles tends to show up deficiencies due to segregation because the machine is non-selective whereas manual loading is. There will be an increasing call for separate sizes of aggregates, and this request is made with good reason, the reduced margins of error now written into specifications and the increased demands being made by the user upon concrete as a material require this greater insistence on uniformity. Variations in particle shape are noticeable from sources of gravel where oversize material is crushed and fed back into the main stream of aggregate being processed.

Several problems arise since aggregates are won from the sea round the coasts of the British Isles at an increasing rate, one concerns sea shells which are generally of hard material which can be used to produce good quality concrete, but often only by using a rather higher proportion of cement than normally, because more cement paste is required to cover the particles and obtain adequate workability in the concrete.

Marine-dredged aggregate has been wetted by sea water containing $3\frac{1}{2}\%$ of salt, mainly sodium chloride with some magnesium sulphate. After draining typical salt contents show 0·05 to 0·10% of chloride; coarse aggregates usually have lower chloride content. Such chloride content would be likely to affect the concrete by altering the rate of stiffening and hardening, by increasing the drying shrinking and by increasing the risk of corrosion to reinforcement and efflorescence. These would have little or no effect in most structures but could be of importance as in pretensioned prestressed concrete. Some sea-borne aggregates are washed with fresh water to reduce the salt content.

Dust in crushed rock aggregates is generally deprecated but where a test was taken of material through a No 100 sieve, comparative concrete mixes were made with various additions of dust and the results showed that the presence of dust did not affect the relation between strength and water–cement ratio but when the dust exceeded a particular proportion, the workability was adversely affected.

Since the variety of architectural finishes used is now great, aggregate suppliers can expect architects and contractors to request that the whole of the requirements for a particular project shall be supplied from the same source, and to guarantee that the supplies will be of the same quality and colour as an approved sample.

Strength requirements of 5000 lb/in² in concrete are now commonplace, and, as already mentioned, 'high strength' is generally regarded as 8000 to 10,000 lb/in² and designers and research organizations are thinking in terms of high strengths up to 14,000 lb/in². Average strengths between 9000 and 12,000 lb/in² require the use of a good quality crushed rock coarse aggregate with a natural sand fine aggregate. Strengths over 12,000 lb in² require a specially selected crushed rock aggregate and trials might be necessary using several sources, and the rate of gain of strength of the particular source of cement might also have to be considered.

On large jobs such as a dam, the local rock is normally used suitably crushed as aggregate. If results of trial mixes indicate the aggregate to be of poor particle shape or grading, the aggregate production technique should be examined and the crushing ratios reduced or impact crushers considered.

In connection with roads and runways of concrete, the subject of skidding resistance is of great importance and it would appear that the texture of the concrete surface is more important than the properties of the aggregate although the latter will assume more importance as the surface wears. Specifications require a deeply brushed finish to concrete with a texture depth of about 0·020 in. Skidding tests on well-worn surfaces have shown no difference between gravel and crushed rock concrete in skidding resistance but the properties of the fine aggregate seem to be of considerable importance and higher skid resistances are obtained when a fair proportion of particles between $\frac{1}{8}$ in and $\frac{3}{8}$ in are present at the surface zone, and sand should be avoided on high-speed roads.

Unlike bituminous surfacings, the polished stone coefficient is of little value in predicting the suitability of a particular aggregate in a normal dense concrete surface. Another factor in the choice of aggregate is the ease with which the hardened concrete can be sawn if this method is used for joint forming and, in this respect, limestone gives the easiest cutting and flint gravel the most difficult with basalts, granites and quartzites in between.

Most insurance companies complain of the alarming increase in fires in buildings and this indicates the advantages of structural materials with a high degree of fire resistance and this includes concrete which also has a low coefficient of thermal conductivity. The temperature falls quickly away from the surface exposed to high temperature and adjoining materials are less likely to be ignited.

For granolithic concrete floor toppings, the basic property required is high abrasion resistance without risk of the floor becoming slippery and this can be obtained by using a 'granite' coarse aggregate consisting of a wide range of crushed igneous rocks. Gravel coarse aggregate can be used but at the risk of producing a slippery floor.

A low drying shrinkage is required and shrinkage increases with increased water content.

As concrete may be called upon to withstand chemical liquids and the lower the water/cement ratio, the greater the density of the cement paste used, and on this depends the rate of attack so that values of this ratio are sometimes specified when chemical attack is to be resisted. For example, in resisting potential sulphate attack in addition to the use of a suitable sulphate-resisting Portland cement, a maximum water/cement ratio of 0·55 is recommended and this leads indirectly to a preference for aggregates of good particle shape and grading. Concrete is an alkaline material and has a good resistance to alkalis with pH values between 7 and 13. Acid liquors with a pH value less than 7 attack all types of Portland cement and concrete used in such conditions must be given a protective coating. A limestone aggregate which is itself attacked and neutralizes some of the acid, may be preferable.

The maximum size of aggregate for a particular job depends on many factors such as reinforcement, spacing of reinforcement and cost of aggregate, but the usual maximum sizes are $1\frac{1}{2}$ in, $\frac{3}{4}$ in, and $\frac{3}{8}$ in. A further need at times, is to reduce the heat of hydration of the concrete which entails using a large aggregate size.

The relation between grading and mix properties for concretes of similar workability is a function of the aggregate specific surface, the relation applying to over-sanded rather than under-sanded mixes.

REFERENCES

'Aggregates in Concrete', D. R. Sharp and B. W. Shacklock, *Quarry Managers' Journal,* May 1966, p. 181.
'Methods for Sampling and Testing of Mineral Aggregates, Sands and Fillers', *BS* 812, 1967. BSI Sales Office, 113 Pentonville Road, London W.1.